Springer Series in Advanced Manufacturing

Series Editor

Professor D. T. Pham
Intelligent Systems Laboratory
WDA Centre of Enterprise in Manufacturing Engineering
University of Wales Cardiff
PO Box 688
Newport Road
Cardiff
CF2 3ET
UK

Other titles in this series

Assembly Line Design
B. Rekiek and A. Delchambre

Advances in Design
H.A. ElMaraghy and W.H. ElMaraghy (Eds.)

Effective Resource Management in Manufacturing Systems: Optimization Algorithms in Production Planning
M. Caramia and P. Dell'Olmo

Condition Monitoring and Control for Intelligent Manufacturing
L. Wang and R.X. Gao (Eds.)

Optimal Production Planning for PCB Assembly
W. Ho and P. Ji

Trends in Supply Chain Design and Management
H. Jung, F.F. Chen and B. Jeong (Eds.)

Process Planning and Scheduling for Distributed Manufacturing
L. Wang and W. Shen (Eds.)

Collaborative Product Design and Manufacturing Methodologies and Applications
W.D. Li, S.K. Ong, A.Y.C. Nee and C. McMahon (Eds.)

R. Venkata Rao

Decision Making in the Manufacturing Environment

Using Graph Theory and Fuzzy Multiple Attribute Decision Making Methods

Springer

R. Venkata Rao, PhD
Department of Mechanical Engineering
Sardar Vallabhbhai National Institute of Technology, Surat
Ichchanath
Surat 395 007
Gujarat State
India

British Library Cataloguing in Publication Data
Rao, R. Venkata
Decision making in the manufacturing environment : using graph theory and fuzzy multiple attribute decision making methods. - (Springer series in advanced manufacturing)
1. Production management - Decision making 2. Graph theory
3. Fuzzy decision making 4. Multiple criteria decision making
I. Title
658.5'036

Springer Series in Advanced Manufacturing ISSN 1860-5168
ISBN 978-1-84996-653-5 e-ISBN 978-1-84628-819-7

Printed on acid-free paper

9 8 7 6 5 4 3 2 1

Springer Science+Business Media
springer.com

Dedicated to my parents,
Sujatha Rao (wife), and Jaya Lakshmi (daughter)

Foreword

Manufacturing technology plays a vital role for the development of a country's industrial growth and largely dictates the trend of the economy. Liberalization and globalization with reformed industrial trade policies have made manufacturing a key element to address global competition.

In the last 20 years, strategic thinking has overtaken single-minded cost reduction and cost minimization in manufacturing. Consequently, the pursuit of cost, quality, flexibility, dependability and timeliness has replaced the single-minded cost reduction in manufacturing firms, which was the norm in manufacturing until the 1970s. Now, manufacturers find competitive advantage through better design, improved customer satisfaction, quick response, faster new-product introduction, and other goals overshadowed in the past by the sole pursuit of cost reduction. The new engineering challenges require systematic and integrated planning and optimization approaches in the manufacturing environment. In this context, the aim of a manufacturing system is to achieve overall performance, utilizing resources in development, design, production, delivery and support of products.

Decision making in the manufacturing environment is a strategic topic, especially in connection with the complexity of driving forces and factors influencing manufacturing systems dynamics. The decision-making exercise can be implemented in the manufacturing environment at different stages, if appropriate procedures are made available to the designers, manufacturing engineers, production planners, and managers. These aspects are considered in the present book using graph theory and fuzzy MADM methods.

Professor R. Venkata Rao has become known as one of the leading experts in the field of decision making related to manufacturing environment. I congratulate him on his achievement, and believe that the book is highly appropriate for use by academicians, designers and practitioners, manufacturing engineers, production planners, marketing managers, applied researchers in industry, academic institutes, R&D organizations, and all decision makers in the manufacturing environment.

Surat, Gujarat, India
4th December 2006

(Prof. P. D. POREY)
Director, S. V. National Institute of Technology

Preface

The purpose of this book is to demonstrate how the graph theory and matrix approach as well as fuzzy multiple attribute decision-making methods can be effectively used for decision making in various situations of the manufacturing environment. The book is divided into two parts. In Part 1, an introduction to the decision-making situations in the manufacturing environment, graph theory and matrix approach as a decision-making method, classical MADM methods, and a logical approach to solve fuzzy MADM problems are presented. In Part 2, the applications of these methods to various real manufacturing situations are presented. The book documents the latest research works, and a significant number of these are original studies of mine published in various national and international journals and conference proceedings. As can be seen from the topics covered, the book deals with most situations in the manufacturing environment (*e.g.*, manufacturing processes such as machining, welding, casting, forming and modern machining methods; advanced manufacturing technologies such as CAD/CAM, robotics, FMS, CIMS, and rapid prototyping; environmentally conscious design and manufacturing, environmental impact assessment; vendor selection, *etc.*). Both graph theory and fuzzy MADM approaches have been successfully applied to various manufacturing situations, and the results are presented. A thorough literature survey on each topic, real case studies, and computer codes have also been included. Thus, the book is expected to become essential reading for the industry and academia, as it makes decision making easier, logical, systematic, efficient, and effective.

I am grateful to Anthony Doyle and Simon Rees of Springer-Verlag, London, for their support and help in producing this book. I wish to thank various researchers and the publishers of international journals for giving me the permission to reproduce certain portions of their published research works. I gratefully acknowledge Prof. P. D. Porey who has written a nice foreword. My special thanks go to my colleagues at SVNIT, Surat.

While every attempt has been made to ensure that no errors (printing or otherwise) enter the book, the possibility of these creeping is always there. I would be grateful to the readers if these errors are pointed out. Suggestions for further improvement of the book will be thankfully acknowledged.

(R. Venkata Rao)

Contents

Part 1

Introduction to Decision Making

1

Introduction to Decision Making in the Manufacturing Environment

1.1 Introduction

Manufacturing is the backbone of any industrialized nation. Its importance is emphasized by the fact that, as an economic activity, it comprises approximately 20 to 30% of the value of all goods and services produced. A country's level of manufacturing activity is directly related to its economic health. In general, the higher the level of manufacturing activity in a country, the higher the standard of living of its people.

Manufacturing can be defined as the application of mechanical, physical, and chemical processes to modify the geometry, properties and/or appearance of a given starting material in the making of new, finished parts or products. This effort includes all intermediate processes required for the production and integration of a product's components. The ability to produce this conversion efficiently determines the success of the company. The type of manufacturing performed by a company depends on the kinds of products it makes. Manufacturing is an important commercial activity carried out by companies that sell products to customers. In the modern sense, manufacturing involves interrelated activities that include product design and documentation, material selection, process planning, production, quality assurance, management, and marketing of products. These activities should be integrated to yield viable and competitive products.

Manufacturing technologies have continually gone through gradual but revolutionary changes. These advancements in manufacturing technologies have brought about a metamorphism in the world industrial scene. They include CNC, CAD/CAM, FMS, robotics, rapid prototyping, environmentally sustainable technologies, *etc.*, which have become an integral part of manufacturing. Parallel to this are rapid strides in the development of new products, and the emergence of an open economy leading to global competition. Manufacturing industries are compelled to move away from traditional setups to more responsive and dynamic ones. Many new concepts have emerged from these changes, sustained by strategies aimed at meeting the challenges arising from global markets. Product attributes like quality, reliability, cost, life-cycle prediction, and the organizational ability to meet market pressures like delivery and service, have come into focus.

A long array of emerging technologies have opened up the potential for a variety of new products. Fast-changing technologies on the product front cautioned the need for an equally fast response from the manufacturing industries. The old, traditional model of 'unfocused, short-term views and non-holistic vision' is becoming replaced by the enlightened approach of 'focused, holistic and strategic vision'. To meet the challenges, manufacturing industries have to select appropriate manufacturing strategies, product designs, manufacturing processes, work piece and tool materials, machinery and equipment, *etc.* The selection decisions are complex, as decision making is more challenging today. Necessary conditions for achieving efficient decision making consist in understanding the current and upcoming events and factors influencing the whole manufacturing environment, in exploring the nature of decision-making processes and the reach of different typologies of methods and techniques, and finally in structuring appropriately the decision-making approach based on a wide range of issues related to manufacturing systems design, planning, and management.

Decision makers in the manufacturing sector frequently face the problem of assessing a wide range of alternative options, and selecting one based on a set of conflicting criteria. Some of the important decision-making situations in the manufacturing environment are listed below:

- Material selection for a given engineering application
- Evaluation of alternative product designs
- Machinability evaluation of work materials
- Cutting fluid selection for a given machining application
- Evaluation and selection of modern machining methods
- Evaluation and selection of flexible manufacturing systems
- Machine group selection in a flexible manufacturing cell
- Failure cause analysis of machine tools
- Robot selection for a given industrial application
- Selection of automated inspection systems
- Selection of material handling equipment
- Selection of a rapid prototyping process in rapid product development
- Selection of software for design and manufacturing applications
- Selection of the most appropriate welding process for a given job
- Mouldability analysis of parts
- Evaluation of metal stamping layouts
- Selection of forging conditions for a given component
- Evaluation of environmentally conscious manufacturing programs
- Environmental impact assessment of manufacturing processes
- Evaluation of aggregate risk in green manufacturing
- Selection of best product end-of-life scenario
- Integrated project evaluation and selection
- Facility location selection

- Operational performance evaluation of competing companies
- Vendor selection in a supply chain environment

It must be noted that in choosing the right alternative, there is not always a single definite criterion of selection, and decision makers have to take into account a large number of criteria including technological, economic, ethical, political, legal, and social factors. There is a need for simple, systematic, and logical methods or mathematical tools to guide decision makers in considering a number of selection criteria and their interrelations. The objective of any selection procedure is to identify appropriate selection criteria, and obtain the most appropriate combination of criteria in conjunction with the real requirement. Thus, efforts need to be extended to identify those criteria that influence an alternative selection for a given problem, using simple and logical methods, to eliminate unsuitable alternatives, and to select the most appropriate alternative to strengthen existing selection procedures. This book presents such simple, systematic and logical methods.

1.2 Decision-making Methods Used

The methods included in this book for decision making in the manufacturing environment are:

(i) Graph theory and matrix approach
(ii) Fuzzy multiple attribute decision-making methods.

Graph theory is a logical and systematic approach. The advanced theory of graphs, and its applications are very well documented. Graph/digraph model representations have proved to be useful for modeling and analyzing various kinds of systems and problems in numerous fields of science and technology. If the graph/digraph is complex, it becomes difficult to analyze it visually. This can be done by computer through the use of the matrix method. An equivalent matrix of the graph/digraph model can be defined. Graph theory and the matrix approach help in identifying attributes, and offer a better visual appraisal of the attributes and their interrelations. This approach is capable of handling the inherent errors, and can deal with any number of qualitative and quantitative attributes simultaneously. The method has axiomatic foundation, involves less computation, provides great emphasis on decision-making methodology, and offers a more objective, simple and consistent decision-making approach. In addition, identification and comparison of alternatives in terms of their similarity/ dissimilarity can be carried out. The application of graph theory and the matrix approach as a decision-making tool in manufacturing situations is relatively new, and this approach has not been used by previous researchers.

In addition to graph theory and the matrix approach, some other important methods, known as multiple attribute decision-making (MADM) methods, are also used in this book for decision making in the manufacturing environment. These methods fall under the category of multiple criteria decision making (MCDM), *i.e.*, decision making in the presence of multiple, generally conflicting criteria. Depending on the domain of alternatives, MCDM problems are usually subdivided into continuous and discrete types. MCDM problems have two classifications: multiple objective decision making (MODM), and multiple attribute decision making (MADM). MODM methods have decision variable values that are determined in a continuous or integer domain with either an infinitive or a large number of alternative choices, the best of which should satisfy the decision maker's constraints and preference priorities. MADM methods, on the other hand, are generally discrete, with a limited number of pre-specified alternatives. These methods require both intra- and inter-attribute comparisons, and involve explicit tradeoffs that are appropriate for the problem considered.

Each decision matrix in MADM methods has four main parts, namely: (a) alternatives, (b) attributes, (c) weight or relative importance of each attribute (*i.e.*, weight), and (d) measures of performance of alternatives with respect to the attributes. Of the many MADM methods, five methods are commonly used: the weighted sum method (WSM), weighted product method (WPM), four modes of the analytic hierarchy process (AHP), Revised AHP, and technique for order preference by similarity to ideal solution (TOPSIS). A compromise ranking method (VIKOR) is also included in this book as an MADM method. However, one of the most crucial problems in many decision-making methods is the precise evaluation of pertinent data. Often, the data are imprecise and fuzzy. It is desirable to develop decision-making methods to deal with this aspect. Classical MADM methods can not effectively handle problems with such imprecise information. To resolve this difficulty, fuzzy MADM methods are used.

The purpose of this book is to demonstrate how graph theory and the matrix approach as well as fuzzy multiple attribute decision-making methods can be effectively used for decision making in various situations of the manufacturing environment. Some of the situations have been mentioned above. Further, the book presents the concept of group decision making, the process of making a judgment based upon the opinion of different individuals. Such decision-making is a key component to the functioning of an organization, because organizational performance involves more than just one individual's action. Moving from a single decision maker to a multiple decision maker setting introduces a great deal of complexity into the analysis. However, this book suggests simple and efficient methods to make the analysis less complex.

The book documents the latest research works related to each of the manufacturing situations listed. Further, it presents the real case studies under most of the topics, as well as results of application of the proposed methods and the comparisons. The methods described in this book will be very useful to the decision makers in the manufacturing sector, as these methods make decision making easier, logical, systematic, efficient, and effective.

The next chapter describes the graph theory and matrix approach as a decision-making method in the manufacturing environment.

2

Graph Theory and Matrix Approach as a Decision-making Method

2.1 Introduction

A graph G = (V, E) consists of a set of objects V = $\{v_1, v_2, \ldots\}$ called vertices or nodes, and another set E = $\{e_1, e_2, \ldots\}$, of which the elements are called edges, such that each edge e_k is identified with a pair of vertices. The vertices v_i and v_j associated with edge e_k are called the end vertices of e_k. The most common representation of a graph is by means of a diagram, in which the vertices are represented by small points or circles, and each edge as a line segment joining its end vertices.

The application of graph theory was known centuries ago, when the long-standing problem of the Konigsberg bridge was solved by Leonhard Euler in 1736 by means of a graph. Since then, graph theory has proved its mettle in various fields of science and technology such as physics, chemistry, mathematics, communication science, computer technology, electrical engineering, sociology, economics, operations research, linguistics, internet, *etc.* Graph theory has served an important purpose in the modeling of systems, network analysis, functional representation, conceptual modeling, diagnosis, *etc.* Graph theory is not only effective in dealing with the structure (physical or abstract) of the system, explicitly or implicitly, but also useful in handling problems of structural relationship. The theory is intimately related to many branches of mathematics including group theory, matrix theory, numerical analysis, probability, topology, and combinatorics. The advanced theory of graphs and their applications are well documented (Harary, 1985; Wilson and Watkins, 1990; Chen, 1997; Deo, 2000; Jense and Gutin, 2000; Liu and Lai, 2001; Tutte, 2001; Pemmaraju and Skiena, 2003; Gross and Yellen, 2005; Biswal, 2005).

This chapter presents the details of graph theory and the matrix approach as a decision-making method in the manufacturing environment. To demonstrate the approach, an example of machinability evaluation of work materials for a given machining operation is considered. Machinability is a measure of ease with which a work material can satisfactorily be machined. The machinability aspect is of considerable importance for the manufacturing engineer to know in advance, so that the processing can be planned in an efficient manner. The study can also be a basis for cutting tool and cutting fluid performance evaluation, and machining parameter optimization. In the process of product design, material selection is important for realizing the design objective, and for reducing the production cost. The machinability of engineering materials, owing to the marked influence on the production cost, needs to be taken into account in the product design, although it will not always be a criterion considered top priority in the process of material selection. If there is a finite number of work materials from among which the best material is to be chosen, and if each work material satisfies the required design and functionality of the product, then the main criterion to choose the work material is its operational performance during machining, *i.e.*, machinability.

Machinability evaluation is based on the evaluation of certain economic and technical objectives (*e.g.*, higher production rate, low operational cost, good product quality, *etc.*), which are the consequences of the machining operation on a given work material. Machining process output variables (*e.g.*, cutting tool life, cutting tool wear, cutting forces, power consumption, processed surface finish, processed dimensional accuracy, *etc.*) are nothing but the behavioral properties of the work materials during machining operations in terms of economic and technical consequences and are directly related to machining operations, and hence to machinability. Thus, the machining process output variables are the pertinent and most commonly accepted measures of machinability, and are also called pertinent machinability attributes.

2.2 Machinability Attributes Digraph

A directed graph (or a digraph) is nothing but a graph with directed edges. A machinability attributes digraph models the machinability attributes and their interrelationship for a given machining operation. This digraph consists of nodes and edges. A node $\{V_i\}$ represents presence or measure of an i-th machinability attribute. The number of nodes considered is equal to the number of machinability attributes considered for a given machining operation. The directed edge represents the relative importance among the attributes. If node 'i' has a relative importance over another anode 'j' in the machinability evaluation of work materials for the given machining operation, then a directed edge or arrow is drawn from node i to node j (*i.e.*, e_{ij}). If j has relative importance over i, then the directed edge or arrow is drawn from node j to node i (*i.e.*, e_{ji}).

To demonstrate a machinability attributes digraph, an example of machinability evaluation of work materials in cylindrical grinding operation is considered. Grinding is a machining process of material removal in the form of small chips by the mechanical action of abrasive particles bonded together in a grinding wheel. In this operation, wheel wear is most important, so as to reduce the cost of production. The wheel wear is measured in terms of a ratio known as 'grinding ratio', which is defined as the ratio of amount of work material removed to the amount of wheel wear. Higher values of grinding ratio are desired for economic reasons. Two components of the cutting force, namely, normal force and tangential force, significantly affect the grinding process. Higher values of normal force increase the roughness of the processed surfaces, and the geometric and dimensional inaccuracy of the processed parts. Tangential force affects the rating of the motors driving the wheel and the work piece, and higher values of tangential force mean increased power consumption. The grinding process imparts high-grade surface finish and good dimensional accuracy to the job. However, the temperature encountered in the grinding process is very high, and adversely affects the process. So, every care is to be taken to reduce the grinding temperature. All these variables described are the machining process output variables and are the pertinent machinability attributes and these attributes refer to the performance of work material during machining operations in terms of technical and economic consequences, and can be used for objective comparison. A work material is said to possess good machinability in cylindrical grinding operation if it offers higher grinding ratio, and lower values of normal force, tangential force, surface roughness, dimensional inaccuracy, and grinding temperature.

Based on the above discussion, the machinability attributes considered for the cylindrical grinding operation are: grinding ratio (GR), normal force (NF), tangential force (TF), surface finish (SF), dimensional accuracy of the produced job (DA), and grinding temperature (GT). A machinability attributes digraph for the cylindrical grinding operation is shown in Figure 2.1. As six machinability attributes are considered here, there are six nodes in the machinability attributes digraph with nodes 1, 2, 3, 4, 5, and 6 representing the machinability attributes GR, NF, TF, SF, DA, and GT, respectively. The attribute GR is more important than the other machinability attributes in cylindrical grinding. Every effort should be made to increase the grinding ratio, as it greatly affects the cost of production. So, directed edges are drawn for the attribute GR (*i.e.*, node 1) to the other attributes (*i.e.*, nodes 2, 3, 4, 5, and 6). NF is more important than the attributes TF, SF, DA, and GT in cylindrical grinding operation, as it affects the surface roughness, and the geometric and dimensional accuracy of the processed parts. So, directed edges are drawn from node 2, representing NF, to the nodes 3, 4, 5, and 6. SF is more important than TF, so a directed edge is drawn from node 4 to node 3. DA is more important than TF, so a directed edge is drawn from node 5 to node 3. GT is more important than TF, SF, and DA in cylindrical grinding operation, so directed edges are drawn from node 6 to the nodes 3, 4, and 5 representing TF, SF, and DA, respectively.

A machinability attributes digraph gives a graphical representation of the attributes and their relative importance for quick visual appraisal. As the number of nodes and their interrelations increase, the digraph becomes more complex. In such a case, the visual analysis of the digraph is expected to be difficult and complex. To overcome this constraint, the digraph is represented in a matrix form.

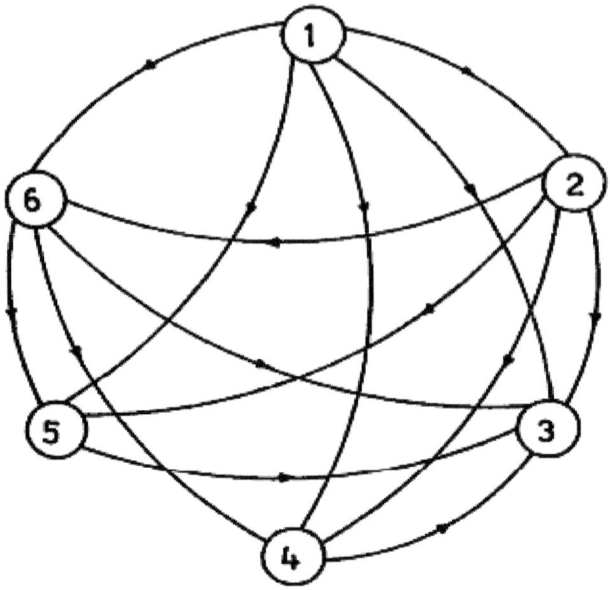

Figure 2.1. Machinability attributes digraph for the cylindrical grinding operation (attributes: 1. grinding ratio, 2. normal force, 3. tangential force, 4. surface finish, 5. dimensional accuracy, and 6. grinding temperature)

2.3 Matrix Representation of the Digraph

Matrix representation of the machinability attributes digraph gives one-to-one representation. A matrix called the machinability attributes relative importance matrix is defined. This is represented by a binary matrix (a_{ij}), where a_{ij} represents the relative importance between attributes i and j such that,

a_{ij} = 1, if the i-th machinability attribute is more important than the j-th machinability attribute for a given machining operation

= 0, otherwise.

It is noted that a_{ii} = 0 for all i, as an attribute can not have relative importance over itself. The machinability attributes relative importance matrix (RIM) for the machinability attributes digraph shown in Figure 2.1 is written as:

$$B = \begin{array}{c} \text{Attributes} \\ \text{GR} \\ \text{NF} \\ \text{TF} \\ \text{SF} \\ \text{DA} \\ \text{GT} \end{array} \begin{array}{c} \begin{array}{cccccc} \text{GR} & \text{NF} & \text{TF} & \text{SF} & \text{DA} & \text{GT} \end{array} \\ \begin{bmatrix} 0 & 1 & 1 & 1 & 1 & 1 \\ 0 & 0 & 1 & 1 & 1 & 1 \\ 0 & 0 & 0 & 0 & 0 & 0 \\ 0 & 0 & 1 & 0 & 0 & 0 \\ 0 & 0 & 1 & 0 & 0 & 0 \\ 0 & 0 & 1 & 1 & 1 & 0 \end{bmatrix} \end{array} \quad (2.1)$$

The machinability attributes relative importance matrix (RIM) is analogous to the adjacency matrix in graph theory. It is noted from the RIM that all diagonal elements have value 0 and all off-diagonal elements have value either 0 or 1. This means that in this matrix only relative importance among the machinability attributes is considered, and the measures of the machinability attributes is not considered. To incorporate this, another matrix, called 'characteristic machinability attributes presence and relative importance matrix (CPRIM)', is defined and this, for the machinability attributes digraph of Figure 2.1, is written as C given by:

$$
\begin{array}{ll}
 & \begin{array}{cccccc} \text{GR} & \text{NF} & \text{TF} & \text{SF} & \text{DA} & \text{GT} \end{array} \\
C = [AI-B] = \begin{array}{c} \text{GR} \\ \text{NF} \\ \text{TF} \\ \text{SF} \\ \text{DA} \\ \text{GT} \end{array} & \begin{bmatrix} A & -1 & -1 & -1 & -1 & -1 \\ 0 & A & -1 & -1 & -1 & -1 \\ 0 & 0 & A & 0 & 0 & 0 \\ 0 & 0 & -1 & A & 0 & 0 \\ 0 & 0 & -1 & 0 & A & 0 \\ 0 & 0 & -1 & -1 & -1 & A \end{bmatrix}
\end{array} \quad (2.2)
$$

where I is an identity matrix, and A is a variable representing the measure of the machinability attribute. Matrix C is analogous to the characteristic matrix in graph theory. Referring to the matrix in Equation 2.2, it is noted that the value of all diagonal elements is identical, *i.e.*, the presence or measure of each machinability attribute is taken to be the same. In practice, this is not true. Also, the relative importance of one machinability attribute over the other machinability attribute, *i.e.*, a_{ij}, may take any value other than the extreme value 0 or 1. Thus, there is a need for considering a general attribute value representing attribute presence or measure as well as relative importance value to develop a matrix equation leading to a broad-based machinability evaluation. To consider these aspects, another matrix, D, called 'variable characteristic machinability attributes presence and relative importance matrix (VCPRIM)', is developed.

$$
\begin{array}{ll}
 & \begin{array}{cccccc} \text{GR} & \text{NF} & \text{TF} & \text{SF} & \text{DA} & \text{GT} \end{array} \\
D = [E-F] = \begin{array}{c} \text{GR} \\ \text{NF} \\ \text{TF} \\ \text{SF} \\ \text{DA} \\ \text{GT} \end{array} & \begin{bmatrix} A_1 & -a_{12} & -a_{13} & -a_{14} & -a_{15} & -a_{16} \\ 0 & A_2 & -a_{23} & -a_{24} & -a_{25} & -a_{26} \\ 0 & 0 & A_3 & 0 & 0 & 0 \\ 0 & 0 & -a_{43} & A_4 & 0 & 0 \\ 0 & 0 & -a_{53} & 0 & A_5 & 0 \\ 0 & 0 & -a_{63} & -a_{64} & -a_{65} & A_6 \end{bmatrix}
\end{array} \quad (2.3)
$$

where E is a diagonal matrix with diagonal element A_i representing a variable of presence or measure of the i-th machinablity attribute. If a machinability attribute is excellent, then it is assigned a maximum value. If a machinability attribute is not very significant, then it is assigned a minimum value. In general, most of the machinability attributes are assigned intermediate values of the interval scale, as attributes may be moderately present. These judgments are to be made based on an appropriate test of the machinability attribute. In the absence of this test, a subjective value based on experience is assigned. F is a matrix of which the off-diagonal elements are represented as a_{ij}, instead of 1, wherever the i-th machinability attribute has more relative importance than the j-th machinability attribute.

It may be noted that the matrix VCPRIM considers the presence or measures of the machinability attributes, and their relative importance for the given machining operation. The characteristic multinomial of the matrix VCPRIM is nothing but the determinant of the matrix VCPRIM, and may be written as:

$$\det(D) = A_1 A_2 A_3 A_4 A_5 A_6 \tag{2.4}$$

Equation 2.4 contains only one term, *i.e.*, $A_1 A_2 A_3 A_4 A_5 A_6$, which is a set of six machinability attributes measures. It is evident that the relative importance among the machinability attributes is not represented by this characteristic multinomial. It is therefore necessary to look into the aspect of relative importance representation in the machinability attributes digraph and its matrix to identify the reasons. If the i-th machinablity attribute is more important than the j-th machinability attribute, then a directed edge is drawn from i to j to represent this relative importance. Similarly, if the j-th machinability attribute is more important than the i-th machinability attribute, then a directed edge is drawn from j to i to represent their relative importance. But if the i-th machinability attribute is less important than the j-th machinability attribute, then no directed edge is drawn from i to j, and *vice versa.* In that case, a_{ij} (or a_{ji}) becomes 0 in the matrix representation of the digraph. This 0 causes many terms of the characteristic multinomial to become 0 (as there are no relative importance loops in the corresponding machinability attributes digraph), thus leading to the loss of a fair amount of information useful during the machinability evaluation. Hence, the relative importance between i, j and j, i is distributed on a scale 0 to L and is defined as:

$$a_{ji} = L - a_{ij} \tag{2.5}$$

It means that a scale is adapted from 0 to L on which the relative importance values are compared. If a_{ij} represents the relative importance of the i-th machinability attribute over the j-th machinability attribute, then the relative importance of the j-th machinability attribute over the i-th machinability attribute is evaluated using Equation 2.5. The modified machinability attributes digraph showing the presence or measures of the machinabilty attributes, and all the possible relative importance among these is shown in Figure 2.2.

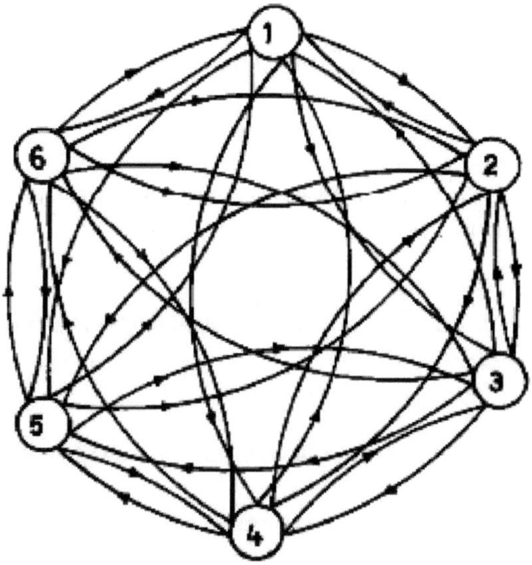

Figure 2.2. Modified machinability attributes digraph for the cylindrical grinding operation (attributes: 1. grinding ratio, 2. normal force, 3. tangential force, 4. surface finish, 5. dimensional accuracy, and 6. grinding temperature)

The modified VCPRIM for this digraph for the cylindrical grinding operation is represented as:

$$G = \begin{array}{c} \text{Attributes} \\ \text{GR} \\ \text{NF} \\ \text{TF} \\ \text{SF} \\ \text{DA} \\ \text{GT} \end{array} \begin{array}{c} \begin{array}{cccccc} \text{GR} & \text{NF} & \text{TF} & \text{SF} & \text{DA} & \text{GT} \end{array} \\ \begin{bmatrix} A_1 & -a_{12} & -a_{13} & -a_{14} & -a_{15} & -a_{16} \\ -a_{21} & A_2 & -a_{23} & -a_{24} & -a_{25} & -a_{26} \\ -a_{31} & -a_{32} & A_3 & -a_{34} & -a_{35} & -a_{36} \\ -a_{41} & -a_{42} & -a_{43} & A_4 & -a_{45} & -a_{46} \\ -a_{51} & -a_{52} & -a_{53} & -a_{54} & A_5 & -a_{56} \\ -a_{61} & -a_{62} & -a_{63} & -a_{64} & -a_{65} & A_6 \end{bmatrix} \end{array} \quad (2.6)$$

where A_i is the measure of the i-th machinability attribute represented by node v_i, and a_{ij} the relative importance of the i-th machinability attribute over the j-th, represented by the edge e_{ij}. The characteristic multinomial of this matrix G is defined as 'variable characteristic machinability function (VCF)', and is written as Equation 2.6.

$$\det(G) = \prod_{i=1}^{6} A_i - \sum_{i=1}^{5} \sum_{j=i+1}^{6} \sum_{k=1}^{3} \sum_{l=k+1}^{4} \sum_{m=l+1}^{5} \sum_{\substack{n=m+1 \\ k,l,m,n \neq \text{pus}}}^{6} (a_{ij}a_{ji})A_k A_l A_m A_n$$

$$- \sum_{i=1}^{4} \sum_{j=i+1}^{5} \sum_{k=j+1}^{6} \sum_{l=1}^{4} \sum_{m=l+1}^{5} \sum_{\substack{n=m+1 \\ k,l,m,n \neq \text{pus}}}^{6} (a_{ij}a_{jk}a_{ki} + a_{ik}a_{kj}a_{ji})A_l A_m A_n$$

$$+\Big[\sum_{i=1}^{3}\sum_{j=i+1}^{6}\sum_{k=i+1}^{5}\sum_{l=i+2}^{6}\sum_{m=1}^{5}\sum_{n=m+1}^{6}(a_{ij}a_{ji})(a_{kl}a_{lk})A_mA_n \qquad k,l,m,n \neq \text{pus}$$

$$-\sum_{i=1}^{3}\sum_{j=i+1}^{5}\sum_{k=i+1}^{6}\sum_{l=j+1}^{6}\sum_{m=1}^{5}\sum_{n=m+1}^{6}(a_{ij}a_{jk}a_{kl}a_{li}+a_{il}a_{lk}a_{kj}a_{ji})A_mA_n\Big] \qquad k,l,m,n \neq \text{pus}$$

$$+\Big[\sum_{i=1}^{4}\sum_{j=i+1}^{5}\sum_{k=j+1}^{6}\sum_{l=1}^{5}\sum_{m=l+1}^{6}\sum_{n=1}^{6}(a_{ij}a_{jk}a_{ki}+a_{ik}a_{kj}a_{ji})(a_{lm}a_{ml})A_n \qquad k,l,m,n \neq \text{pus}$$

$$-\sum_{i=1}^{2}\sum_{j=i+1}^{5}\sum_{k=i+1}^{6}\sum_{l=i+1}^{6}\sum_{m=j+1}^{6}\sum_{n=1}^{6}(a_{ij}a_{jk}a_{kl}a_{lm}a_{mi}+a_{im}a_{ml}a_{lk}a_{kj}a_{ji})A_n\Big] \qquad k,l,m,n \neq \text{pus}$$

$$+\Big[\sum_{i=1}^{3}\sum_{j=i+1}^{5}\sum_{k=i+1}^{6}\sum_{l=j+1}^{6}\sum_{m=1}^{5}\sum_{n=m+1}^{6}(a_{ij}a_{jk}a_{kl}a_{li}+a_{il}a_{lk}a_{kj}a_{ji})(a_{mn}a_{nm}) \qquad k,l,m,n \neq \text{pus}$$

$$+\sum_{i=1}^{1}\sum_{j=i+1}^{5}\sum_{k=j+1}^{6}\sum_{l=1}^{4}\sum_{m=l+1}^{5}\sum_{n=m+1}^{6}(a_{ij}a_{jk}a_{ki}+a_{ik}a_{kj}a_{ji})(a_{lm}a_{mn}a_{nl}+a_{ln}a_{nm}a_{ml}) \qquad k,l,m,n \neq \text{pus}$$

$$-\sum_{i=1}^{1}\sum_{j=i+1}^{6}\sum_{k=i+1}^{3}\sum_{l=i+2}^{6}\sum_{m=k+1}^{5}\sum_{n=k+2}^{6}(a_{ij}a_{ji})(a_{kl}a_{lk})(a_{mn}a_{nm}) \qquad k,l,m,n \neq \text{pus}$$

$$-\sum_{i=1}^{1}\sum_{j=i+1}^{5}\sum_{k=i+1}^{6}\sum_{l=i+1}^{6}\sum_{m=i+1}^{6}\sum_{n=j+1}^{6}(a_{ij}a_{jk}a_{kl}a_{lm}a_{mn}a_{ni}+a_{in}a_{nm}a_{ml}a_{lk}a_{kj}a_{ji})\Big] \qquad k,l,m,n \neq \text{pus} \tag{2.7}$$

‘pus’ stands for ‘previously used subscripts’, *i.e.*, in Equation 2.7, k, l, m and n take those subscripts that are other than previously used subscripts. The multinomial Equation 2.7 in symbolic form is a complete expression for the considered cylindrical grinding operation, as it considers measures of the attributes and all possible relative importance among the attributes. Mathematically, each term is a product of six different matrix elements. If this function is interpreted from a combinatorial point of view, it is found that different terms are the sets of distinct diagonal elements (A_i) and loops of off-diagonal elements of different sizes (*i.e.*, $a_{ij}a_{ji}$, $a_{ij}a_{jk}a_{ki}$, *etc.*).

The variable characteristic machinability function (VCF) contains terms arranged in (6 + 1) groupings and these groupings represent the measures of attributes and the relative importance loops. The first grouping represents the measures of the machinability attributes. The second grouping is absent, as there is no self-loop in the digraph. The third grouping contains 2-attribute relative importance loops and measures of four attributes. Each term of the fourth grouping represents a set of a 3-attribute relative importance loop, or its pair, and measures of three attributes. The fifth grouping contains two sub-groupings. Each term of the first sub-grouping is a set of two 2-attribute relative importance loops and the measures of two attributes. Each term of the second sub-grouping is a set of a 4-attribute relative importance loop, or its pair, and the measures of two attributes. The sixth grouping contains two sub-groupings. Each term of the first sub-grouping is a set of a 3-attribute relative importance loop, or its pair, and a 2-attribute relative importance loop and the measure of one attribute. Each term of the second sub-grouping is a set of 5-attribute relative importance loop, or its pair, and the measure of one attribute. The seventh grouping contains four sub-groupings. Each term of the first sub-grouping is a set of a 4-attribute relative importance loop, or its pair, and a 2-attribute relative importance loop. Each term of the second sub-grouping is a set of a 3-attribute relative importance loop, or its pair, and another 3-attribute relative importance loop, or its pair. Each term of the third sub-grouping is a set of three 2-attribute relative importance loops. Each term of the fourth sub-grouping is a set of a 6-attribute relative importance loop, or its pair. After identifying these combinatorial terms, and by associating a proper physical meaning with these, a new mathematical meaning of the multinomial is obtained.

The variable characteristic machinability function is the characteristic of the work material, and a powerful tool for machinability evaluation. However, a close look at the multinomial reveals that its various characteristic coefficients carry both positive and negative signs. The variable characteristic machinability function may not be able to provide the total objective value, when the numerical values for A_i and a_{ij} are substituted in the multinomial, because some of the information is lost by subtraction and addition operations in the determinant function. Considering these factors, the 'variable permanent machinability function (VPF)' is defined. This function is derived from a new matrix called the 'machinability permanent matrix'. The machinability permanent matrix, H, for the machinability attributes digraph (Figure 2.2) is written as Equation 2.8.

$$
\begin{array}{ll}
 & \begin{array}{cccccc} \text{GR} & \text{NF} & \text{TF} & \text{SF} & \text{DA} & \text{GT} \end{array} \\
H= \begin{array}{c} \text{GR} \\ \text{NF} \\ \text{TF} \\ \text{SF} \\ \text{DA} \\ \text{GT} \end{array} &
\left[\begin{array}{cccccc}
A_1 & a_{12} & a_{13} & a_{14} & a_{15} & a_{16} \\
a_{21} & A_2 & a_{23} & a_{24} & a_{25} & a_{26} \\
a_{31} & a_{32} & A_3 & a_{34} & a_{35} & a_{36} \\
a_{41} & a_{42} & a_{43} & A_4 & a_{45} & a_{46} \\
a_{51} & a_{52} & a_{53} & a_{54} & A_5 & a_{56} \\
a_{61} & a_{62} & a_{63} & a_{64} & a_{65} & A_6
\end{array} \right]
\end{array}
\qquad (2.8)
$$

(Attributes listed along rows and columns.)

The permanent of H may be called the 'variable permanent machinability function (VPF)'.

$$
\text{per}(H) = \prod_{i=1}^{6} A_i + \sum_{i=1}^{5} \sum_{j=i+1}^{6} \sum_{k=1}^{3} \sum_{l=k+1}^{4} \sum_{m=l+1}^{5} \sum_{n=m+1}^{6} (a_{ij}a_{ji}) A_k A_l A_m A_n \quad k,l,m,n \neq \text{pus}
$$

$$
\begin{aligned}
&+\sum_{i=1}^{4}\sum_{j=i+1}^{5}\sum_{k=j+1}^{6}\sum_{l=1}^{4}\sum_{m=l+1}^{5}\sum_{\substack{n=m+1\\ k,l,m,n\neq pus}}^{6}(a_{ij}a_{jk}a_{ki}+a_{ik}a_{kj}a_{ji})A_lA_mA_n\\
&+\Big[\sum_{i=1}^{3}\sum_{j=i+1}^{6}\sum_{k=i+1}^{5}\sum_{l=i+2}^{6}\sum_{m=1}^{5}\sum_{\substack{n=m+1\\ k,l,m,n\neq pus}}^{6}(a_{ij}a_{ji})(a_{kl}a_{lk})A_mA_n\\
&+\sum_{i=1}^{3}\sum_{j=i+1}^{5}\sum_{k=i+1}^{6}\sum_{l=j+1}^{6}\sum_{m=1}^{5}\sum_{\substack{n=m+1\\ k,l,m,n\neq pus}}^{6}(a_{ij}a_{jk}a_{kl}a_{li}+a_{il}a_{lk}a_{kj}a_{ji})A_mA_n\Big]\\
&+\Big[\sum_{i=1}^{4}\sum_{j=i+1}^{5}\sum_{k=j+1}^{6}\sum_{l=1}^{5}\sum_{m=l+1}^{6}\sum_{\substack{n=1\\ k,l,m,n\neq pus}}^{6}(a_{ij}a_{jk}a_{ki}+a_{ik}a_{kj}a_{ji})(a_{lm}a_{ml})A_n\\
&+\sum_{i=1}^{2}\sum_{j=i+1}^{5}\sum_{k=i+1}^{6}\sum_{l=i+1}^{6}\sum_{m=j+1}^{6}\sum_{\substack{n=1\\ k,l,m,n\neq pus}}^{6}(a_{ij}a_{jk}a_{kl}a_{lm}a_{mi}+a_{im}a_{ml}a_{lk}a_{kj}a_{ji})A_n\Big]\\
&+\Big[\sum_{i=1}^{3}\sum_{j=i+1}^{5}\sum_{k=i+1}^{6}\sum_{l=j+1}^{6}\sum_{m=1}^{5}\sum_{\substack{n=m+1\\ k,l,m,n\neq pus}}^{6}(a_{ij}a_{jk}a_{kl}a_{li}+a_{il}a_{lk}a_{kj}a_{ji})(a_{mn}a_{nm})\\
&+\sum_{i=1}^{1}\sum_{j=i+1}^{5}\sum_{k=j+1}^{6}\sum_{l=1}^{4}\sum_{m=l+1}^{5}\sum_{\substack{n=m+1\\ k,l,m,n\neq pus}}^{6}(a_{ij}a_{jk}a_{ki}+a_{ik}a_{kj}a_{ji})(a_{lm}a_{mn}a_{nl}+a_{ln}a_{nm}a_{ml})\\
&+\sum_{i=1}^{1}\sum_{j=i+1}^{6}\sum_{k=i+1}^{3}\sum_{l=i+2}^{6}\sum_{m=k+1}^{5}\sum_{\substack{n=k+2\\ k,l,m,n\neq pus}}^{6}(a_{ij}a_{ji})(a_{kl}a_{lk})(a_{mn}a_{nm})\\
&+\sum_{i=1}^{1}\sum_{j=i+1}^{5}\sum_{k=i+1}^{6}\sum_{l=i+1}^{6}\sum_{m=i+1}^{6}\sum_{\substack{n=j+1\\ k,l,m,n\neq pus}}^{6}(a_{ij}a_{jk}a_{kl}a_{lm}a_{mn}a_{ni}+a_{in}a_{nm}a_{ml}a_{lk}a_{kj}a_{ji})\Big]
\end{aligned}
\tag{2.9}
$$

It may be noted that the only difference between the VPF, *i.e.*, per (H), and the determinant polynomial det (G), *i.e.*, VCF, is that the former does not carry negative signs with its terms, while both positive and negative signs appear in the latter. Comparing Equations 2.8 and 2.9, it is noted that each term of the grouping/sub-grouping is the same in both cases, the only difference being in the signs of the coefficients. Both the functions are basically the same, and have the same physical meaning, except for the difference in signs. It may be mentioned that the permanent is a standard matrix function, and is used in combinatorial mathematics (Marcus and Minc, 1965; Jurkat and Ryser, 1966; Nijenhuis and Wilf, 1975).

Use of the permanent concept in machinability evaluation will help in representing machinability attributes of work materials as obtained from combinatorial consideration. Application of the permanent concept will lead to a better appreciation of machinability attributes of the work materials. Moreover, using this, no negative sign will appear in the equation, and hence no information will be lost.

The adjacency matrix, incidence matrix, characteristic matrix, *etc.*, could also be used for machinability evaluation, but these matrices have their own drawbacks. The adjacency matrix makes no provision for parallel-directed edges in both directions (*i.e.*, relative importance in both directions), and the elements of the matrix are either 0 or 1. On expanding the adjacency matrix, only some numbers can be obtained that do not reveal much physical information associated with the machinability attributes and their relative importance. The incidence matrix contains the elements either 0 or 1, and it requires more computer storage than needed for an adjacency matrix, as the number of edges is usually greater than the number of nodes. Moreover, as the incidence matrix is a non-square matrix, its further use for machinability evaluation is not possible. The characteristic matrix is not an invariant of the system, as a new matrix can be obtained by changing the labeling, but one matrix can be obtained from the other by proper permutations of rows and columns. The characteristic multinomial or characteristic function, which is nothing but the determinant of the characteristic matrix, contains both positive and negative signs, and may not be able to provide the total objective value when the numerical values for A_i and a_{ij} are substituted in the multinomial, because some of the information is lost by subtraction and addition operations in the determinant function, as explained above. Due to these reasons, researchers have used the permanent function of a matrix, which does not contain any negative terms, and thus provides the complete information without any loss (Gandhi *et al.*, 1991; Gandhi and Agrawal, 1992, 1994; Venkatasamy and Agrawal, 1996, 1997; Rao and Gandhi, 2001, 2002a, 2002b; Rao, 2004, 2006a, 2006b, 2006c, 2006d; Grover *et al.*, 2004; Rao and Padmanabhan, 2006).

In general, if there is M number of machinability attributes, and the relative importance exists among all the machinability attributes, then the machinability attributes matrix, J, for the considered machinability attributes digraph is written as Equation 2.10.

$$
\begin{array}{cc}
 & \text{Attributes} \\
 & \begin{array}{c} 1 \\ 2 \\ 3 \\ - \\ - \\ M \end{array}
\end{array}
\begin{array}{c}
\begin{array}{cccccc} 1 & 2 & 3 & - & - & M \end{array} \\
\left[\begin{array}{cccccc}
A_1 & a_{12} & a_{13} & - & - & a_{1M} \\
a_{21} & A_2 & a_{23} & - & - & a_{2M} \\
a_{31} & a_{32} & A_3 & - & - & a_{3M} \\
- & - & - & - & - & - \\
- & - & - & - & - & - \\
a_{M1} & a_{M2} & a_{M3} & - & - & A_M
\end{array}\right]
\end{array}
$$

$$J = \quad (2.10)$$

The VPF for this matrix J contains factorial M (M!) number of terms. In sigma form, it is written as Equation 2.11.

$$
\text{per}(J) = \prod_{i=1}^{M} A_i + \sum_{i=1}^{M-1} \sum_{j=i+1}^{M} \cdots\cdots\cdots \sum_{\substack{M=t+1\\ \ldots,\, M \neq \text{pus}}}^{M} (a_{ij}a_{ji})A_kA_lA_mA_nA_o \ldots\ldots A_tA_M
$$

$$
+\sum_{i=1}^{M-2} \sum_{j=i+1}^{M-1} \sum_{k=j+1}^{M} \cdots\cdots\cdots_{l=1} \sum_{\substack{M=t+1\\ k,\, \ldots,\, M \neq \text{pus}}}^{M} (a_{ij}a_{jk}a_{ki} + a_{ik}a_{kj}a_{ji})A_lA_mA_nA_o \ldots\ldots A_tA_M
$$

$$
+\Big[\sum_{i=1}^{M-3} \sum_{j=i+1}^{M} \sum_{k=i+1}^{M-1} \sum_{l=i+2}^{M} \cdots\cdots\cdots \sum_{\substack{M=t+1\\ k,l,\, \ldots,\, M \neq \text{pus}}}^{M} (a_{ij}a_{ji})\,(a_{kl}a_{lk})A_mA_nA_o \ldots\ldots A_tA_M
$$

$$
+\sum_{i=1}^{M-3} \sum_{j=i+1}^{M-1} \sum_{k=i+1}^{M} \sum_{l=j+1}^{M} \cdots\cdots\cdots \sum_{\substack{M=t+1\\ k,l,\, \ldots,\, M \neq \text{pus}}}^{M} (a_{ij}a_{jk}a_{kl}a_{li} + a_{il}a_{lk}a_{kj}a_{ji})A_mA_nA_o \ldots\ldots A_tA_M\Big]
$$

$$
+\Big[\sum_{i=1}^{M-2} \sum_{j=i+1}^{M-1} \sum_{k=j+1}^{M} \sum_{l=1}^{M-1} \sum_{m=l+1}^{M} \cdots\cdots\cdots \sum_{\substack{M=t+1\\ k,l,m,\, \ldots,\, M \neq \text{pus}}}^{M} (a_{ij}a_{jk}a_{ki} + a_{ik}a_{kj}a_{ji})(a_{lm}a_{ml})A_nA_o \ldots\ldots A_tA_M
$$

$$
+\sum_{i=1}^{M-4} \sum_{j=i+1}^{M-1} \sum_{k=i+1}^{M} \sum_{l=i+1}^{M} \sum_{m=j+1}^{M} \cdots\cdots\cdots \sum_{\substack{M=t+1\\ k,l,m,\, \ldots,\, M \neq \text{pus}}}^{M} (a_{ij}a_{jk}a_{kl}a_{lm}a_{mi} + a_{im}a_{ml}a_{lk}a_{kj}a_{ji})A_nA_o \ldots\ldots A_tA_M\Big]
$$

$$
+\Big[\Big(\sum_{i=1}^{M-3} \sum_{j=i+1}^{M-1} \sum_{k=i+1}^{M} \sum_{l=j+1}^{M} \sum_{m=1}^{M-1} \sum_{n=m+1}^{M} \cdots\cdots\cdots \sum_{\substack{M=t+1\\ k,l,m,n,\, \ldots,\, M \neq \text{pus}}}^{M} (a_{ij}a_{jk}a_{kl}a_{li} + a_{il}a_{lk}a_{kj}a_{ji})(a_{mn}a_{nm})A_o \ldots\ldots A_tA_M
$$

$$
+\sum_{i=1}^{M-5} \sum_{j=i+1}^{M-1} \sum_{k=j+1}^{M} \sum_{l=1}^{M-2} \sum_{m=l+1}^{M-1} \sum_{n=m+1}^{M} \cdots\cdots \sum_{\substack{M=t+1\\ k,l,m,n,\, \ldots,\, M \neq \text{pus}}}^{M} (a_{ij}a_{jk}a_{ki} + a_{ik}a_{kj}a_{ji})(a_{lm}a_{mn}a_{nl} + a_{ln}a_{nm}a_{ml})A_o \ldots A_tA_M
$$

$$
+\sum_{i=1}^{M-5} \sum_{j=i+1}^{M} \sum_{k=i+1}^{M-3} \sum_{l=i+2}^{M} \sum_{m=k+1}^{M-1} \sum_{n=k+2}^{M} \cdots\cdots\cdots \sum_{\substack{M=t+1\\ k,l,m,n,\, \ldots,\, M \neq \text{pus}}}^{M} (a_{ij}a_{ji})\,(a_{kl}a_{lk})\,(a_{mn}a_{nm})\,A_o \ldots\ldots A_tA_M
$$

$$
+\sum_{i=1}^{M-5} \sum_{j=i+1}^{M-1} \sum_{k=i+1}^{M} \sum_{l=i+1}^{M} \sum_{m=i+1}^{M} \sum_{n=j+1}^{M} \cdots \sum_{\substack{M=t+1\\ k,l,m,n,\, \ldots,\, M \neq \text{pus}}}^{M} (a_{ij}a_{jk}a_{kl}a_{lm}a_{mn}a_{ni} + a_{in}a_{nm}a_{ml}a_{lk}a_{kj}a_{ji})A_o \ldots\ldots A_tA_M\Big)\Big]
$$

$$
+ \text{----------} \tag{2.11}
$$

'pus' stands for 'previously used subscripts', *i.e.*, in the Equation 2.11, k, l, m, n, … , M take those subscripts that are other than previously used subscripts. The VPF contains terms arranged in (M + 1) groups, and these groups represent the measures of attributes and the relative importance loops. The first group represents the measures of M attributes. The second group is absent as there is no self-loop in the digraph. The third group contains 2-attribute relative importance loops and measures of (M-2) attributes. Each term of the fourth group represents a set of a 3-attribute relative importance loop, or its pair, and measures of (M-3) attributes. The fifth group contains two sub-groups. The terms of the first sub-group is a set of two 2-attribute relative importance loops and the measures of (M-4) attributes. Each term of second sub-group is a set of a 4-attribute relative importance loop, or its pair, and the measures of (M-4) attributes. The sixth group contains two sub-groups. The terms of the first sub-group is a set of a 3-attribute relative importance loop, or its pair, and 2-attribute relative importance loop and the measures of (M-5) attributes. Each term of the second sub-group is a set of a 5-attribute relative importance loop, or its pair, and the measures of (M-5) attributes. Similarly other terms of the equation are defined. Thus, the VPF fully characterizes the considered machinability evaluation problem, as it contains all possible structural components of the attributes and their relative importance. It may be mentioned that this equation is nothing but the determinant of an M * M matrix but considering all the terms as positive.

The computer program written in C^{++} language to calculate the permanent function of a square matrix of M * M size is given in Appendix A.

2.4 Machinability Index

The machinability index is a measure of the ease with which a work material can satisfactorily be machined in a given machining operation. The machinability function defined above, *i.e.*, Equation 2.11, contains measures of attributes and their relative importance, and is hence appropriate, and can be used for evaluation of the machinability index. As the machinability function contains only positive terms, higher values of A_i and a_{ij} will result in increased value of the machinability index. To calculate this index, the required information is the values of A_i and a_{ij}.

The value of A_i should preferably be obtained from a standard or specific test. If such objective value is not available, then a ranked value judgment on a scale, *e.g.*, 0 to 1, is adapted. Table 2.1 represents the machinability attribute on a subjective scale. It holds for a given machining operation, some of the A_i will be subjective, and the others objective. Moreover, these objective values will have different units. It is therefore desirable to convert, or normalize, the objective values of A_i on the same scale as the subjective values, *i.e.*, 0 to 1. If A_i has range A_{il} and A_{iu}, the value 0 is assigned to the lowest range value A_{il} and 1 is assigned to the highest range value A_{iu}. The other, intermediate value A_{ii} of the machinability attribute is assigned a value in between 0 and 1, as per the following:

$$A_i = (A_{ii} - A_{il}) / (A_{iu} - A_{il}) \tag{2.12}$$

Equation 2.12 is applicable for general beneficial attributes only. A beneficial attribute (*e.g.*, grinding ratio) is one of which higher attribute value is more desirable for the given machining operation. A non-beneficial attribute (*e.g.*, normal force) is one of which the lower attribute value is desirable. Therefore, in the case of non-beneficial machinability attributes, the attribute value 0, on scale 0 to 1, is assigned to the highest range value A_{iu}, and the value 1 is assigned to the lower range value A_{il}. The other intermediate value A_{ii} of the machinability attribute is assigned a value in between 0 and 1, as per the following:

$$A_i = (A_{iu} - A_{ii}) / (A_{iu} - A_{il}) \tag{2.13}$$

Alternatively, the normalized value A_i can be calculated by A_{ii} / A_{iu} in the case of the beneficial attribute, and by A_{il} / A_{ii} in the case of the non-beneficial attribute. This alternative method is better than the method described by Equations 2.12 and 2.13 as it does not contain '0' as the normalized attribute value, and hence no information will be lost subsequently in machinability index calculation.

The relative importance between two attributes (*i.e.*, a_{ij}) for a given machining operation is also assigned value on the scale 0 to 1, and is arranged into six classes. The relative importance implies that an attribute 'i' is compared with another attribute 'j' in terms of relative importance for the given machining operation. The relative importance between i, j and j, i is distributed on the scale 0 to 1, and is defined similarly to Equation 2.5 in which L is taken as 1. If a_{ij} represents the relative importance of the i-th attribute over the j-th attribute, then the relative importance of the j-th attribute over the i-th attribute is evaluated using Equation 2.5. For example, if the i-th attribute is slightly more important than the j-th attribute, then $a_{ij} = 6$ and $a_{ji} = 4$.

Table 2.2 aids in assigning a_{ij} values based on the above. The relative importance is expressed in six classes, which lead to minimization of subjectivity while deciding the relative importance between machinability attributes.

Table 2.1. Value of attribute

Subjective measure of attribute	Assigned value
Exceptionally low	0.0
Extremely low	0.1
Very low	0.2
Low	0.3
Below average	0.4
Average	0.5
Above average	0.6
High	0.7
Very high	0.8
Extremely high	0.9
Exceptionally high	1.0

Table 2.2. Relative importance of attributes

Class description	Relative importance a_{ij}	$a_{ji} = 1 - a_{ij}$
Two attributes are equally important	0.5	0.5
One attribute is slightly more important over the other	0.6	0.4
One attribute is strongly more important over the other	0.7	0.3
One attribute is very strongly important over the other	0.8	0.2
One attribute is extremely important over the other	0.9	0.1
One attribute is exceptionally more important over the other	1.0	0.0

It may be mentioned that one may choose any scale, *e.g.*, 0 to 1, 0 to 5, 1 to 5, 0 to 10, 1 to 10, 1 to 11, 0 to 50, 0 to 100, 1 to 100, 1 to 110, 0 to 1000, 1 to 1000, or any other scale for A_i and a_{ij}. But the final ranking will not change, as these are relative values. It is, however, desirable to choose a lower scale for A_i and a_{ij} to obtain a manageable value of machinability index. It may be further mentioned that the scales adapted for A_i and a_{ij} can be independent of each other. Whenever the machinability index is calculated for a work material, only the diagonal elements will change, *i.e.*, (A_i), and the off-diagonal elements (a_{ij}) remain the same.

The machinability index for each material is evaluated using Equation 2.11, and substituting the value of A_i and a_{ij}. The work materials are arranged in the descending or ascending order of the machinability index to rank these for a given machining operation. These are called the machinability ranking values of the work materials for the given machining operation. The work material, for which the value of machinability index is highest, is the best choice for the machining operation considered. However, the final decision depends on factors such as cost, availability, environmental constraints, economical constraints, political constraints, *etc.* Compromise, however, should be made to select the work material having the highest value of machinability index.

The next section describes the identification and comparison of work materials.

2.5 Identification and Comparison of Work Materials

2.5.1 Identification of Work Materials

The variable permanent machinability function, *i.e.*, Equation 2.11, is useful for the identification and comparison of work materials for a given machining operation. The number of terms in each grouping of the machinability function for all the work materials for a given machining operation will be the same. However, their values will be different. This aspect is used for the purpose. Let T_{ij} represent the total value of terms of the j-th sub-grouping of i-th grouping of the machinability function. In case there is no sub-grouping, then $T_{ij} = T_i$, *i.e.*, total value of terms of the i-th grouping. The identification set for a work material for the given machining operation is:

$$/ T_1 / T_2 / T_3 / T_4 / T_{51} + T_{52} / T_{61} + T_{62} / \ldots\ldots\ldots \quad (2.14)$$

Two work materials can be compared using Equation 2.14.

2.5.2 Comparison of Work Materials

In general, two work materials are never identical from the performance (*i.e.*, machinability) point of view. If two work materials are similar, then they must be similar in performance, and *vice versa*. Comparison of two work materials is also carried out by evaluating the coefficient of similarity/dissimilarity based on the numerical value of the terms of the machinability function in its grouping/sub-grouping. The coefficient of similarity/dissimilarity lies in the range 0 – 1. If two work materials are of similar performance, then the coefficient of similarity is 1 and coefficient of dissimilarity is 0. In the same manner, if two work materials are of dissimilar performance, then the coefficient of dissimilarity is 1 and coefficient of similarity is 0. Based on performance dissimilarity, the coefficient of dissimilarity for two work materials is proposed as Equation 2.15.

$$C_d = (1/Q) \left(\sum_{i=1}^{M-1} \sum_{j=i+1}^{M} \psi_{ij} \right) \quad (2.15)$$

where, Q = maximum of $\sum_{i=1}^{M-1} \sum_{j=i+1}^{M} T_{ij}$ and $\sum_{i=1}^{M-1} \sum_{j=i+1}^{M} T'_{ij}$

T_{ij} and T'_{ij} denote the values of the terms for the machinability function of the two work materials under comparison, and $\psi_{ij} = |T_{ij} - T'_{ij}|$. It may be noted that the absolute difference between the values of the terms for the machinability function of the two work materials is considered for proposing C_d. The coefficient of similarity is proposed as:

$$C_s = 1 - C_d \quad (2.16)$$

Equations 2.15 and 2.16 are useful for comparing two work materials, based upon their performance in a given machining operation. The coefficients of similarity and dissimilarity, and the identification sets are also useful for work materials documentation, and for easy storage and retrieval of the work materials data for various machining operations.

Thus, graph theory and the matrix approach can be used as a decision-making method for choosing an appropriate alternative work material from amongst the given alternatives, based on machinability. The proposed method offers a general procedure that can be used for any type of decision-making problem involving any number of selection attributes and alternatives. The next section describes the general methodology of graph theory and matrix approach as a decision-making method.

2.6 Methodology of GTMA as a Decision-making Method

The main steps are given below:

Step 1: Identify the pertinent attributes and the alternatives involved in the decision-making problem under consideration. Obtain the values of the attributes (A_i) and their relative importance (a_{ij}). An objective or subjective value, or its range, may be assigned to each identified attribute as a limiting value or threshold value for its acceptance for the considered decision-making problem. An alternative with each of its selection attributes, meeting the acceptance value, may be short-listed. After short-listing the alternatives, the main task in choosing the alternative is to see how it serves the considered attributes.

Step 2:

1. Develop the attributes digraph considering the identified pertinent attributes and their relative importance. The number of nodes shall be equal to the number of attributes considered in Step 1 above. The edges and their directions will be decided upon based on the interrelations among the attributes (a_{ij}). Refer to Section 2.2 for details.
2. Develop the attributes matrix for the attributes digraph. This will be the M*M matrix with diagonal elements as A_i and off-diagonal elements as a_{ij}. Refer to Section 2.3 for details.
3. Obtain the permanent function for the attributes matrix, on the lines of Equation 2.11.
4. Substitute the values of A_i and a_{ij}, obtained in step 1, in Equation 2.11 above to evaluate the index for the short-listed alternatives.
5. Arrange the alternatives in the descending order of the index. The alternative having the highest value of index is the best choice for the decision-making problem under consideration.
6. Obtain the identification set for each alternative, using Equation 2.14. Refer to Section 2.5 for details.
7. Evaluate the coefficients of dissimilarity and similarity using Equations 2.15 and 2.16. List also the values of the coefficients for all possible combinations.
8. Document the results for future analysis/reference.

Step 3: Take a final decision, keeping practical considerations in mind. All possible constraints likely to be experienced by the user are looked into during this stage. These include constraints such as: availability or assured supply, management constraints, political constraints, economic constraints, environmental constraints, *etc.* However, compromise may be made in favor of an alternative with a higher index.

From the above, it is clear that the graph theory and matrix approach as a decision-making method is relatively new, and offers a generic, simple, easy, and convenient decision-making method that involves less computation. The method lays emphasis on decision-making methodology, gives much attention to the issues of identifying the attributes, and to associating the alternatives with the attributes, *etc.* The method enables a more critical analysis and any number of objective and subjective attributes can be considered. In the permanent procedure, even a small variation in attributes leads to a significant difference in the selection index, and hence it is easy to rank the alternatives in the descending order, with clear-cut difference in the selection index. Further, the proposed procedure not only provides the analysis of alternatives, but also enables the visualization of various attributes present and their interrelations, using graphical representation. The measures of the attributes and their relative importance are used together to rank the alternatives, and hence provides a better evaluation of the alternatives. The permanent concept fully characterizes the considered selection problem, as it contains all possible structural components of the attributes and their relative importance.

The decision-making capability of graph theory and the matrix approach can be utilized for making decisions in the manufacturing environment, and Chapters 5-30 of this book present those details.

The next chapter gives an introduction to the multiple attribute decision-making methods.

References

Biswal PC (2005) Discrete mathematics and graph theory. Prentice Hall India, New Delhi

Chen WK (1997) Graph theory and its engineering applications. Advanced Series in Electrical and Computer Engineering, University of Illinois, Chicago

Deo N (2000) Graph theory with applications to engineering and computer science. Prentice Hall India, New Delhi

Gandhi OP, Agrawal VP (1992) FMEA - A digraph and matrix approach. Reliability Engineering and System Safety 35:147–158

Gandhi OP, Agrawal VP (1994) A digraph approach to system wear evaluation and analysis. Journal of Tribology 116:268–274

Gandhi OP, Agrawal VP, Shishodia KS (1991) Reliability analysis and evaluation of systems. Reliability Engineering and System Safety 32:283–305

Gross J, Yellen J (2005) Graph theory and its applications. CRC Press, Florida

Grover S, Agrawal VP, Khan IA (2004) A digraph approach to TQM evaluation of an industry. International Journal of Production Research 42:4031–4053

Harary F (1985) Graphs and applications. Wiley, New York

Jense JB, Gutin G (2000) Digraph theory, algorithms, and applications. Springer, London

Jurkat WB, Ryser HJ (1966) Matrix factorisation of determinants and permanents. Journal of Algebra 3:1–11

Liu LB, Lai LHJ (2001) Matrices in combinatorics and graph theory. Kluwer Academic Publishers, Dordrecht

Marcus M, Minc H (1965) Permanents. American Mathematics Monthly 72:571–591

Nijenhuis A, Wilf HS (1975) Combinatorial algorithms. Academic Press, New York

Pemmaraju S, Skiena S (2003) Computational discrete mathematics: combinatorics and graph theory with mathematica. Cambridge University Press, UK

Rao RV (2004) Digraph and matrix methods for evaluating environmentally conscious manufacturing programs. International Journal of Environmentally Conscious Design and Manufacturing 12:23–33

Rao RV (2006a) A decision making framework model for evaluating flexible manufacturing systems using digraph and matrix methods. International Journal of Advanced Manufacturing Technology 30:1101–1110

Rao RV (2006b) A material selection model using graph theory and matrix approach. Materials Science and Engineering A 431:248–255

Rao RV (2006c) Machine group selection in a flexible manufacturing cell using digraph and matrix methods. International Journal of Industrial and Systems Engineering 1:502–518

Rao RV (2006d) Plant location selection using fuzzy digraph and matrix methods. International Journal of Industrial Engineering 13:357–362

Rao RV, Gandhi OP (2001) Digraph and matrix method for selection, identification and comparison of metal cutting fluids. Proc. IME, Journal of Engineering Tribology 212:307–318

Rao RV, Gandhi OP (2002a) Digraph and matrix methods for machinability evaluation of work materials. International Journal of Machine Tools and Manufacture 42:321–330

Rao RV, Gandhi OP (2002b) Failure cause analysis of machine tools using digraph and matrix methods. International Journal of Machine Tools & Manufacture 42:521–528

Rao RV, Padmanabhan KK (2006) Selection, identification and comparison of industrial robots using digraph and matrix methods Robotics and Computer Integrated Manufacturing 22:373–383

Tutte WT (2001) Graph theory. Cambridge University Press, UK

Venkatasamy R, Agrawal VP (1996) Selection of automobile vehicle by evaluation through graph theoretical methodology. International Journal of Vehicle Design 17:449–470

Venkatasamy R, Agrawal VP (1997) A digraph approach to quality evaluation of an automotive vehicle. Quality Engineering 9:405–417

Wilson RJ, Watkins JJ (1990) Graphs, an introductory approach. Wiley, New York

3

Introduction to Multiple Attribute Decision-making (MADM) Methods

3.1 Introduction

Multiple criterion decision making (MCDM) refers to making decisions in the presence of multiple, usually conflicting criteria. The problems of MCDM can be broadly classified into two categories: multiple attribute decision making (MADM) and multiple objective decision making (MODM), depending on whether the problem is a selection problem or a design problem. MODM methods have decision variable values that are determined in a continuous or integer domain, with either an infinitive or a large number of choices, the best of which should satisfy the decision maker's constraints and preference priorities. MADM methods, on the other hand, are generally discrete, with a limited number of predetermined alternatives. MADM is an approach employed to solve problems involving selection from among a finite number of alternatives. An MADM method specifies how attribute information is to be processed in order to arrive at a choice. MADM methods require both inter- and intra-attribute comparisons, and involve appropriate explicit tradeoffs.

Each decision table (also called decision matrix) in MADM methods has four main parts, namely: (a) alternatives, (b) attributes, (c) weight or relative importance of each attribute, and (d) measures of performance of alternatives with respect to the attributes. The decision table is shown in Table 3.1. The decision table shows alternatives, A_i (for i = 1, 2, ….. , N), attributes, B_j (for j = 1, 2, ….. , M), weights of attributes, w_j (for j=1, 2, ….., M) and the measures of performance of alternatives, m_{ij} (for i= 1, 2, ….., N; j=1, 2, ….., M). Given the decision table information and a decision-making method, the task of the decision maker is to find the best alternative and/or to rank the entire set of alternatives. It may be added here that all the elements in the decision table must be normalized to the same units, so that all possible attributes in the decision problem can be considered.

Table 3.1. Decision table in MADM methods

Alternatives	Attributes					
	B_1 (w_1)	B_2 (w_2)	B_3 (w_3)	- (-)	- (-)	B_M (w_M)
A_1	m_{11}	m_{12}	m_{13}	-	-	m_{1M}
A_2	m_{21}	m_{22}	m_{23}	-	-	m_{2M}
A_3	m_{31}	m_{32}	m_{33}	-	-	m_{3M}
-	-	-	-	-	-	-
-	-	-	-	-	-	-
A_N	m_{N1}	m_{N2}	m_{N3}	-	-	m_{NM}

Of the many MADM methods reported in the literature (Saaty, 1980, 2000; Hwang and Yoon, 1981, Chen and Hwang, 1992; Yoon and Hwang 1995; Olson, 1996; Triantaphyllou and Sanchez, 1997; Zanakis *et al.*, 1998; Gal *et al.*, 1999; Triantaphyllou, 2000; Figueira *et al.*, 2004), few important methods that have a higher potential to solve decision-making problems in the manufacturing environment are presented in this chapter.

3.2 Multiple Attribute Decision-making Methods

3.2.1 Simple Additive Weighting (SAW) Method

This is also called the weighted sum method (Fishburn, 1967) and is the simplest, and still the widest used MADM method. Here, each attribute is given a weight, and the sum of all weights must be 1. Each alternative is assessed with regard to every attribute. The overall or composite performance score of an alternative is given by Equation 3.1.

$$P_i = \sum_{j=1}^{M} w_j m_{ij} \tag{3.1}$$

Previously, it was argued that SAW should be used only when the decision attributes can be expressed in identical units of measure (*e.g.*, only dollars, only pounds, only seconds, *etc.)*. However, if all the elements of the decision table are normalized, then SAW can be used for any type and any number of attributes. In that case, Equation 3.1 will take the following form:

$$P_i = \sum_{j=1}^{M} w_j (m_{ij})_{normal} \tag{3.2}$$

where $(m_{ij})_{normal}$ represents the normalized value of m_{ij}, and P_i is the overall or composite score of the alternative A_i. The alternative with the highest value of P_i is considered as the best alternative.

The attributes can be beneficial or non-beneficial. When objective values of the attribute are available, normalized values are calculated by $(m_{ij})_K/(m_{ij})_L$, where $(m_{ij})_K$ is the measure of the attribute for the K-th alternative, and $(m_{ij})_L$ is the measure of the attribute for the L-th alternative that has the highest measure of the attribute out of all alternatives considered. This ratio is valid for beneficial attributes only. A beneficial attribute (*e.g.*, profit) means its higher measures are more desirable for the given decision-making problem. By contrast, non-beneficial attribute (*e.g.*, cost) is that for which the lower measures are desirable, and the normalized values are calculated by $(m_{ij})_L/(m_{ij})_K$.

If the restriction that the sum of all weights is to be equal to 1 is relaxed, then Equation 3.3 can be used and this method is called simple multiple attribute rating technique (SMART).

$$P_i = [\sum_{j=1}^{M} w_j (m_{ij})_{normal}] / \sum_{j=1}^{M} w_j \tag{3.3}$$

Edwards *et al.* (1982) proposed a simple method to assess weights for each attribute to reflect its relative importance to the decision. For a start, the attributes are ranked in order of importance and 10 points are assigned to the least important attribute. Then, the next-least important attribute is chosen, more points are assigned to it, and so on, to reflect their relative importance. The final weights are obtained by normalizing the sum of the points to one.

3.2.2 Weighted Product Method (WPM)

This method is similar to SAW. The main difference is that, instead of addition in the model, there is multiplication (Miller and Starr, 1969). The overall or composite performance score of an alternative is given by Equation 3.4.

$$P_i = \prod_{j=1}^{M} [(m_{ij})_{normal}]^{w_j} \tag{3.4}$$

The normalized values are calculated as explained under the SAW method. Each normalized value of an alternative with respect to an attribute, *i.e.*, $(m_{ij})_{normal}$, is raised to the power of the relative weight of the corresponding attribute. The alternative with the highest P_i value is considered the best alternative.

3.2.3 Analytic Hierarchy Process (AHP) Method

One of the most popular analytical techniques for complex decision-making problems is the analytic hierarchy process (AHP). Saaty (1980, 2000) developed AHP, which decomposes a decision-making problem into a system of hierarchies of objectives, attributes (or criteria), and alternatives.

An AHP hierarchy can have as many levels as needed to fully characterize a particular decision situation. A number of functional characteristics make AHP a useful methodology. These include the ability to handle decision situations involving subjective judgements, multiple decision makers, and the ability to provide measures of consistency of preference (Triantaphyllou, 2000). Designed to reflect the way people actually think, AHP continues to be the most highly regarded and widely used decision-making method. AHP can efficiently deal with tangible (*i.e.*, objective) as well as non-tangible (*i.e.*, subjective) attributes, especially where the subjective judgements of different individuals constitute an important part of the decision process. The main procedure of AHP using the radical root method (also called the geometric mean method) is as follows:

Step 1: Determine the objective and the evaluation attributes. Develop a hierarchical structure with a goal or objective at the top level, the attributes at the second level and the alternatives at the third level.

Step 2: Determine the relative importance of different attributes with respect to the goal or objective.

- Construct a pair-wise comparison matrix using a scale of relative importance. The judgements are entered using the fundamental scale of the analytic hierarchy process (Saaty 1980, 2000). An attribute compared with itself is always assigned the value 1, so the main diagonal entries of the pair-wise comparison matrix are all 1. The numbers 3, 5, 7, and 9 correspond to the verbal judgements 'moderate importance', 'strong importance', 'very strong importance', and 'absolute importance' (with 2, 4, 6, and 8 for compromise between these values). Assuming M attributes, the pair-wise comparison of attribute i with attribute j yields a square matrix $B_{M \times M}$ where a_{ij} denotes the comparative importance of attribute i with respect to attribute j. In the matrix, $b_{ij} = 1$ when $i = j$ and $b_{ji} = 1/b_{ij}$.

$$\begin{array}{c c} \text{Attributes} & \begin{array}{ccccccc} B_1 & B_2 & B_3 & - & - & B_M \end{array} \\ B_{MxM} = \begin{array}{c} B_1 \\ B_2 \\ B_3 \\ - \\ - \\ B_M \end{array} & \left[\begin{array}{cccccc} 1 & b_{12} & b_{13} & - & - & b_{1M} \\ b_{21} & 1 & b_{23} & - & - & b_{2M} \\ b_{31} & b_{32} & 1 & - & - & b_{3M} \\ - & - & - & - & - & - \\ - & - & - & - & - & - \\ b_{M1} & b_{M2} & b_{M3} & - & - & 1 \end{array} \right] \end{array} \quad (3.5)$$

- Find the relative normalized weight (w_j) of each attribute by (i) calculating the geometric mean of the i-th row, and (ii) normalizing the geometric means of rows in the comparison matrix. This can be represented as:

$$GM_j = \left[\prod_{j=1}^{M} b_{ij} \right]^{1/M} \quad (3.6)$$

and

$$w_j = GM_j / \sum_{j=1}^{M} GM_j \quad (3.7)$$

  The geometric mean method of AHP is commonly used to determine the relative normalized weights of the attributes, because of its simplicity, easy determination of the maximum Eigen value, and reduction in inconsistency of judgements.

- Calculate matrices A3 and A4 such that A3 = A1 * A2 and A4 = A3 / A2, where $A2 = [w_1, w_2, \ldots, w_j]^T$.
- Determine the maximum Eigen value λ_{max} that is the average of matrix A4.

- Calculate the consistency index CI = (λ_{max} - M) / (M - 1). The smaller the value of CI, the smaller is the deviation from the consistency.
- Obtain the random index (RI) for the number of attributes used in decision making. Refer to Table 3.2 for details.
- Calculate the consistency ratio CR = CI/RI. Usually, a CR of 0.1 or less is considered as acceptable, and it reflects an informed judgement attributable to the knowledge of the analyst regarding the problem under study.

Step 3: The next step is to compare the alternatives pair-wise with respect to how much better (*i.e.,* more dominant) they are in satisfying each of the attributes, *i.e.*, to ascertain how well each alternative serves each attribute. If there is N number of alternatives, then there will be M number of N x N matrices of judgements, since there are M attributes. Construct pair-wise comparison matrices using a scale of relative importance. The judgements are entered using the fundamental scale of the AHP method (Saaty, 1980, 2000). The steps are the same as those suggested under main step 2.

Table 3.2. Random index (RI) values

Attributes	3	4	5	6	7	8	9	10
RI	0.52	0.89	1.11	1.25	1.35	1.4	1.45	1.49

In the AHP model, both the relative and absolute modes of comparison can be performed. The relative mode can be used when decision makers have prior knowledge of the attributes for different alternatives to be used, or when objective data of the attributes for different alternatives to be evaluated are not available. The absolute mode is used when data of the attributes for different alternatives to be evaluated are readily available. In the absolute mode, CI is always equal to 0, and complete consistency in judgements exists, since the exact values are used in the comparison matrices.

Step 4: The next step is to obtain the overall or composite performance scores for the alternatives by multiplying the relative normalized weight (w_j) of each attribute (obtained in step 2) with its corresponding normalized weight value for each alternative (obtained in step 3), and summing over the attributes for each alternative. This step is similar to the SAW method.

Kwiesielewicz and Uden (2004) stated that even if the pair-wise comparison matrix B_{MxM} is of acceptable consistency, the matrix may still be contradictory. If a matrix is contradictory, then it is difficult to derive weights that satisfy *all* the judgements expressed in B_{MxM}. Hence, it is imperative to remove any such contradictory matrix from the decision-making process. For example, if b_{ij} = 1 and b_{ik} = 1, then b_{jk} must be equal to 1. If any judgement is made such that b_{jk} > 1, then contradiction is present in the matrix, and needs to be removed. Kwiesielewicz and Uden (2004) formulated an algorithm to check for the presence of any contradiction in B_{MxM}.

It may be added here that the AHP method can also be used for assigning the values of relative importance (a_{ij}) to the attributes in graph theory and the matrix approach (GTMA). Refer to Sections 2.3 and 2.4.

3.2.4 Revised Analytic Hierarchy Process (RAHP) Method

The revised AHP (RAHP) method was suggested by Belton and Gear (1983). They observed that sometimes it is possible for AHP to yield unjustifiable ranking reversals. The problem is that if a new alternative, identical to a non-optimal alternative, is introduced, then the ranking of the existing alternatives changes. Belton and Gear (1983) argued that the reason for this ranking inconsistency was that the relative performance measures of all alternatives in terms of each attribute (obtained in step 3 of Section 3.2.3) summed to one. Instead of having the relative performance values sum up to one, dividing each relative performance value by the maximum value in the corresponding vector of relative values was suggested. This avoids the rank reversals when a new non-optimal alternative is introduced. This method is also called 'ideal mode AHP'. Saaty, the author of the original AHP, had accepted this revised version.

3.2.5 Multiplicative Analytic Hierarchy Process (MAHP) Method

Barzilai and Lootsma (1994) and Lootsma (1999) proposed a multiplicative version of the AHP. In this MAHP method, the normalized weight value for each alternative (obtained in step 3 of Section 2.2.3) is raised to the power of the relative normalized weight (w_j) of each attribute (obtained in step 2 of Section 3.2.3), with multiplication over all the attributes for each alternative. This step is similar to WPM.

3.2.6 Technique for Order Preference by Similarity to Ideal Solution (TOPSIS) Method

The TOPSIS method was developed by Hwang and Yoon (1981). This method is based on the concept that the chosen alternative should have the shortest Euclidean distance from the ideal solution, and the farthest from the negative ideal solution. The ideal solution is a hypothetical solution for which all attribute values correspond to the maximum attribute values in the database comprising the satisfying solutions; the negative ideal solution is the hypothetical solution for which all attribute values correspond to the minimum attribute values in the database. TOPSIS thus gives a solution that is not only closest to the hypothetically best, that is also the farthest from the hypothetically worst. The main procedure of the TOPSIS method for the selection of the best alternative from among those available is described below:

Step 1: The first step is to determine the objective, and to identify the pertinent evaluation attributes.

Step 2: This step represents a matrix based on all the information available on attributes. This matrix is nothing but the decision table shown in Table 3.1. Each row of this matrix is allocated to one alternative, and each column to one attribute.

Therefore, an element m_{ij} of the decision table 'D' gives the value of the j-th attribute in original real values, that is, non-normalized form and units, for the i-th alternative.

In the case of a subjective attribute (*i.e.*, objective value is not available), a ranked value judgement on a scale is adopted. Table 2.1, as explained in Chapter 2, may be used for this purpose. Once a subjective attribute is represented on a scale, then the normalized values of the attribute assigned for different alternatives are calculated in the same manner as that for objective attributes.

Step 3: Obtain the normalized decision matrix, R_{ij}. This can be represented as

$$R_{ij} = m_{ij} / [\sum_{j=1}^{M} m^2_{ij}]^{1/2} \tag{3.8}$$

Step 4: Decide on the relative importance (*i.e.,* weights) of different attributes with respect to the objective. A set of weights w_j (for j=1, 2, ….. , M) such that $\sum w_j = 1$ may be decided upon.

Step 5: Obtain the weighted normalized matrix V_{ij}. This is done by the multiplication of each element of the column of the matrix R_{ij} with its associated weight w_j. Hence, the elements of the weighted normalized matrix V_{ij} are expressed as:

$$V_{ij} = w_j R_{ij} \tag{3.9}$$

Step 6: Obtain the ideal (best) and negative ideal (worst) solutions in this step. The ideal (best) and negative ideal (worst) solutions can be expressed as:

$$\begin{aligned} V^+ &= \{(\overset{\max}{\underset{i}{\sum}} V_{ij} / j \in J), (\overset{\min}{\underset{i}{\sum}} V_{ij} / j \in J') / i = 1,2, \ldots, N\}, \\ &= \{V_1^+, V_2^+, V_3^+, \ldots\ldots, V_M^+\} \end{aligned} \tag{3.10}$$

$$\begin{aligned} V^- &= \{(\overset{\min}{\underset{i}{\sum}} V_{ij} / j \in J), (\overset{\max}{\underset{i}{\sum}} V_{ij} / j \in J') / i = 1,2, \ldots, N\}, \\ &= \{V_1^-, V_2^-, V_3^-, \ldots\ldots, V_M^-\} \end{aligned} \tag{3.11}$$

where J = (j = 1, 2, …, M) /j is associated with beneficial attributes, and
J' = (j = 1, 2, …, M) /j is associated with non-beneficial attributes.

V_j^+ indicates the ideal (best) value of the considered attribute among the values of the attribute for different alternatives. In the case of beneficial attributes (*i.e.*, those of which higher values are desirable for the given application), V_j^+ indicates the higher value of the attribute. In the case of non-beneficial attributes (*i.e.*, those of which lower values are desired for the given application), V_j^+ indicates the lower value of the attribute.

V_j^- indicates the negative ideal (worst) value of the considered attribute among the values of the attribute for different alternatives. In the case of beneficial attributes (*i.e.*, those of which higher values are desirable for the given application), V_j^- indicates the lower value of the attribute. In the case of non-beneficial attributes (*i.e.*, those of which lower values are desired for the given application), V_j^- indicates the higher value of the attribute.

Step 7: Obtain the separation measures. The separation of each alternative from the ideal one is given by the Euclidean distance in the following equations.

$$S_i^+ = \{\sum_{J=1}^{M} (V_{ij} - V_j^+)^2\}^{0.5}, \qquad i = 1, 2, \ldots, N \tag{3.12}$$

$$S_i^- = \{\sum_{J=1}^{M} (V_{ij} - V_j^-)^2\}^{0.5}, \qquad i = 1, 2, \ldots, N \tag{3.13}$$

Step 8: The relative closeness of a particular alternative to the ideal solution, P_i, can be expressed in this step as follows.

$$P_i = S_i^- / (S_i^+ + S_i^-) \tag{3.14}$$

Step 9: A set of alternatives is generated in the descending order in this step, according to the value of P_i indicating the most preferred and least preferred feasible solutions. P_i may also be called the overall or composite performance score of alternative A_i.

It may be added here that in step 4 of the TOPSIS method, even though the weights of different attributes with respect to the objective, w_j (for j=1, 2, ….. , M), are decided by the decision maker rather arbitrarily, only few systematic methods can be used. The systematic methods of deciding the weights of attributes are explained below.

3.2.6.1 Entropy Method

Shannon and Weaver (1947) proposed the entropy concept and this concept has been highlighted by Zeleny (1982) for deciding the objective weights of attributes. Entropy is a measure of uncertainty in the information formulated using probability theory. It indicates that a broad distribution represents more uncertainty than does a sharply peaked one. To determine weights by the entropy measure, the normalized decision matrix R_{ij}, given by Equation 3.8, is considered. The amount of decision information contained in Equation 3.8 and associated with each attribute can be measured by the entropy value e_j as:

$$e_j = -k \sum_{i=1}^{N} R_{ij} \ln R_{ij} \tag{3.15}$$

where k = 1/ln N is a constant that guarantees $0 \leq e_j \leq 1$. The degree of divergence (d_j) of the average information contained by each attribute can be calculated as:

$$d_j = 1 - e_j \tag{3.16}$$

The more divergent the performance ratings R_{ij} (for i = 1, 2, ….., N) for the attribute B_j, the higher its corresponding d_j, and the more important the attribute B_j for the decision-making problem under consideration (Zeleny, 1982).

The objective weight for each attribute B_j (for j = 1, 2, ….., M) is thus given by:

$$w_j = d_j / \sum_{k=1}^{M} d_k \tag{3.17}$$

3.2.6.2 Standard Deviation Method

The standard deviation (SD) method calculates objective weights of the attributes by Equation 3.18.

$$w_j = \sigma_j / \sum_{k=1}^{M} \sigma_k \tag{3.18}$$

where σ_j is the standard deviation of the normalized vector $R_j = (R_{1j}, R_{2j}, R_{3j}, \ldots.., R_{Nj})$ in Equation 3.8.

Both the entropy method and standard deviation method calculate the objective weights of the attributes without giving any consideration to the preferences of the decision maker.

3.2.6.3 AHP Method

Step 2 of the AHP method, described in Section 3.2.3, can be used for deciding the weights of attributes. In this case, the weights obtained are not objective but subjective, giving consideration to the preferences of the decision maker.

3.2.7 Modified TOPSIS Method

In the TOPSIS method, the normalized decision matrix R_{ij} is weighted by multiplying each column of the matrix by its associated attribute weight. The overall performance of an alternative is then determined by its Euclidean distance to V_j^+ and V_j^-. However, this distance is interrelated with the attribute weights, and should be incorporated in the distance measurement. This is because all alternatives are compared with V_j^+ and V_j^-, rather than directly among themselves. Deng *et al.* (2000) presented the weighted Euclidean distances, rather than creating a weighted decision matrix. In this process, the positive ideal solution (R^+) and the negative ideal solution (R^-), which are not dependent on the weighted decision matrix, are defined as:

$$R^+ = \{(\sum \max_i R_{ij} / j \in J), (\sum \min_i R_{ij} / j \in J') / i = 1,2, \ldots, N\},$$
$$= \{R_1^+, R_2^+, R_3^+, \ldots\ldots, R_M^+\} \tag{3.19}$$

$$R^- = \{(\sum \min_i R_{ij} / j \in J), (\sum \max_i R_{ij} / j \in J') / i = 1,2, \ldots, N\},$$
$$= \{R_1^-, R_2^-, R_3^-, \ldots\ldots, R_M^-\} \tag{3.20}$$

where J = (j = 1, 2, …, M) /j is associated with beneficial attributes, and
J' = (j = 1, 2, …, M) /j is associated with non-beneficial attributes.

The weighted Euclidean distances are calculated as

$$D_i^+ = \{\sum_{J=1}^{M} w_j(R_{ij} - R_j^+)^2\}^{0.5}, \qquad i = 1, 2, \ldots, N \tag{3.21}$$

$$D_i^- = \{\sum_{J=1}^{M} w_j (R_{ij} - R_j^-)^2\}^{0.5}, \qquad i = 1, 2, \ldots, N \tag{3.22}$$

The relative closeness of a particular alternative to the ideal solution, $P_{i\text{-}mod}$, can be expressed in this step as follows.

$$P_{i\text{-}mod} = D_i^- / (D_i^+ + D_i^-) \tag{3.23}$$

A set of alternatives is made in the descending order, according to the value of $P_{i\text{-}mod}$ indicating the most preferred and least preferred feasible solutions.

It may be mentioned here that instead of using vector normalization in the TOPSIS (or modified TOPSIS) method, linear normalization may be used (Lai *et al.*, 1994). In that case, normalization is carried out as per Equation 3.24.

$$R_{ij} = m_{ij} / [(m_{ij})_{max} - (m_{ij})_{min}] \tag{3.24}$$

where $(m_{ij})_{max}$ is the best value and $(m_{ij})_{min}$ the worst value of an attribute corresponding to the considered alternatives.

3.2.8 Compromise Ranking Method (VIKOR)

The foundation for compromise solution was established by Yu (1973) and Zeleny (1982) and later advocated by Opricovic and Tzeng (2002, 2003, 2004, 2007) and Tzeng *et al.* (2002a, 2002b, 2005). The compromise solution is a feasible solution that is the closest to the ideal solution, and a compromise means an agreement established by mutual concession. The compromise solution method, also known as the VIKOR (VIšekriterijumsko KOmpromisno Rangiranje) method, was introduced as one applicable technique to implement within MADM. The multiple attribute merit for compromise ranking was developed from the L_p-metric used in the compromise programing method (Zeleny, 1982).

$$L_{p,i} = \{\sum_{j=1}^{M} (w_j [(m_{ij})_{max} - (m_{ij})] / [(m_{ij})_{max} - (m_{ij})_{min}])^p\}^{1/p} \tag{3.25}$$

$1 \leq p \leq \infty$; $i = 1, 2, \ldots, N$

Within the VIKOR method $L_{1,i}$ (as E_i in Equation 3.26) and $L_{\infty,i}$ (as F_i in Equation 3.27) are used to formulate the ranking measure. The main procedure of the VIKOR method is described below:

Step 1: The first step is to determine the objective, and to identify the pertinent evaluation attributes. Also determine the best, *i.e.*, $(m_{ij})_{max}$, and the worst, *i.e.*, $(m_{ij})_{min}$, values of all attributes.

Step 2: Calculate the values of E_i and F_i:

$$E_i = \sum_{j=1}^{M} w_j \, [(m_{ij})_{max} - (m_{ij})] / [(m_{ij})_{max} - (m_{ij})_{min}] \quad (3.26)$$

$$F_i = Max^m \text{ of } \{w_j \, [(m_{ij})_{max} - (m_{ij})] / [(m_{ij})_{max} - (m_{ij})_{min}] \mid j = 1, 2, \ldots, M\} \quad (3.27)$$

Step 3: Calculate the values of P_i:

$$P_i = v \, ((E_i - E_{i-min}) / (E_{i-max} - E_{i-min})) + (1 - v) \, ((F_i - F_{i-min}) / (F_{i-max} - F_{i-min})) \quad (3.28)$$

where E_{i-max} is the maximum value of E_i, and E_{i-min} the minimum value of E_i; F_{i-max} is the maximum value of F_i, and F_{i-min} is the minimum value of F_i. v is introduced as weight of the strategy of 'the majority of attributes'. Usually, the value of v is taken as 0.5. However, v can take any value from 0 to 1.

Step 4: Arrange the alternatives in the ascending order, according to the values of P_i. Similarly, arrange the alternatives according to the values of E_i and F_i separately. Thus, three ranking lists can be obtained. The compromise ranking list for a given v is obtained by ranking with P_i measures. The best alternative, ranked by P_i, is the one with the minimum value of P_i.

Step 5: For given attribute weights, propose a compromise solution, alternative A_k, which is the best ranked by the measure P, if the following two conditions are satisfied (Tzeng *et al.*, 2005):

- Condition 1: 'Acceptable advantage' $P(A_k) - P(A_l) \geq (1/(N-1))$. A_l is the second-best alternative in the ranking by P.
- Condition 2: 'Acceptable stability in decision making' aternative A_k must also be the best ranked by E and/or F. This compromise solution is stable within a decision-making process, which could be: 'voting by majority rule' (when $v > 0.5$ is needed) or 'by consensus' (when $v \approx 0.5$) or 'with veto' (when $v > 0.5$).

If one of the conditions is not satisfied, then a set of compromise solutions is proposed, which consists of:

- Alternatives A_k and A_l if only condition 2 is not satisfied
- Alternatives A_k, A_l, ….. , A_p if condition 1 is not satisfied; A_p is determined by the relation $P(A_p) - P(A_l) \approx (1/(N-1))$.

VIKOR is a helpful tool in MADM, particularly in a situation where the decision maker is not able, or does not know how to express preference at the beginning of system design. The obtained compromise solution could be accepted by the decision makers because it provides a maximum 'group utility' (represented by E_{i-min}) of the 'majority', and a minimum of individual regret (represented by F_{i-min}) of the 'opponent' (Opricovic and Tzeng, 2002, 2003, 2004, 2007). The compromise solutions could be the basis for negotiations, involving the decision makers' preference by attribute weights.

3.3 Sensitivity Analysis

In sensitivity analysis, the ranking reversal of the alternatives is checked by changing the weights of relative importance of the attributes. The decision maker can check the ranking reversals by changing the weights (of relative importance) of the attributes by a percentage. However, it is obvious that if the assigned weights are changed, then the chances for rank reversals of the alternatives increase. Once the decision maker is clear about the relative importance of the attributes and assigns accordingly, then there is no need to check the ranking reversals simply by changing the weights. Hence, this is not developed further in this book. If the decision maker wishes to conduct sensitivity analysis, then he or she can do so.

3.4 Group Decision Making (GDM)

Group decision making is the process of making a judgement based upon the opinion of different individuals. Such decision making is a key component to the functioning of an organization, because organizational performance involves more than only one individual's action. Moving from a single decision maker to a multiple decision-maker setting introduces a great deal of complexity into the analysis. Various methods of group decision making are used on a wide set of attributes ranging from the strictly technical, to the psychophysical and social, and finally to the logical and scientifically valid. The group decision-making concept can be applied to the MADM techniques described in Section 3.2. There are different ways in which GDM can be carried out (Yu, 1973; Chen and Hwang, 1992; Dyer and Forman, 1992; Csáki *et al.*, 1995; Forman and Penewati, 1998; Chen, 2000; Lai *et al.*, 2002; Jaganathan *et al.*, 2006). Two ways were described by Forman and Peniwati (1998) and Jaganathan *et al.* (2006) for achieving group consensus in AHP. The two ways are:

1. Aggregation of individual judgements (AIJ), and
2. Aggregation of individual priorities (AIP).

In the first case, it is assumed that several individuals act as one individual and their judgements, *i.e.*, the opinions expressed regarding the relative importance (or weights) of the attributes, are aggregated using the weighted geometric mean to form a single composite attribute weight representing the total view of the group. In the second case, the group members act individually, and their final priorities are aggregated using the weighted arithmetic mean or weighted geometric mean. If there are n decision makers (g(k), k = 1, 2, ... , n), then mathematically

$$b_{ij}\ (AIJ) = \prod_{k=1}^{n} (b_{ij\ g(k)})^{l_{g(k)}} \tag{3.29}$$

$$P_i\ (AIP) = \prod_{k=1}^{n} (P_{i\ g(k)})^{l_{g(k)}} \quad \text{or} \quad \prod_{k=1}^{n} l_{g(k)}\ P_{i\ g(k)} \tag{3.30}$$

where $l_{g(k)}$ is the importance of the decision maker in the group, and $\sum l_{g(k)} = 1$. P_i is the performance score of alternative A_i.

The same approaches can be extended to other MADM methods, where group consensus is required. Csáki *et al.* (1995) presented a group decision support system. In this system, the method of calculating the group utility (group composite performance score) of alternative A_i (for i = 1, 2, ..., N) is as follows.

For each attribute B_j (for j = 1, 2, ….., M), the individual weights of importance of the attributes are aggregated into the group weights w_j (for j = 1, 2, ….. , M):

$$w_j = [\sum_{k=1}^{n} l_{g(k)} w_j / \sum_{k=1}^{n} l_{g(k)} \quad j = 1, 2, ….., M \tag{3.31}$$

The group qualification Q_{ij} of alternative A_i against attribute B_j is:

$$Q_{ij} = [\sum_{k=1}^{n} l_{g(k)} m_{ij} / \sum_{k=1}^{n} l_{g(k)} \quad j = 1, 2, ….., M; i = 1, 2, ….., N \tag{3.32}$$

$\sum l_{g(k)}$ need not be equal to 1 in Equations 3.31 and 3.32.

The group utility P_i of alternative A_i is determined as the weighted algebraic mean of the aggregated qualification values with the aggregated weights:

$$P_i = [\sum_{j=1}^{M} w_j Q_{ij} / \sum_{j=1}^{M} w_j \qquad i = 1, 2, ….., N \tag{3.33}$$

In addition to the weighted algebraic means used in the above aggregations, weighted geometric means can be used. The best alternative of group decision is the one associated with the highest value of P_i.

The MADM methods described in this chapter can efficiently deal with objective attributes. But most of the real-world MADM problems involve objective (*i.e.*, crisp) as well as subjective (*i.e.*, fuzzy and/or linguistic) attributes. Hence fuzzy MADM methods have been proposed by different researchers that can deal with fuzzy as well as crisp data of the attributes. However, Table 2.1, as explained in Chapter 2, may be used for the purpose of assigning a crisp value to the subjective attribute. Once a subjective attribute is represented by a crisp value, then the decision table contains only crisp data, and any MADM method can be applied.

The next chapter presents a logical approach to solve fuzzy MADM problems.

References

Barzilai J, Lootsma FA (1994) Power relations and group aggregation in the multiplicative AHP and SMART. In: Proc. of the 3rd International Symposium on the AHP, George Washington University, Washington

Belton V, Gear T (1983) On a short-coming of Saaty's method of analytic hierarchies. Omega 11:228–230

Chen CT (2000) Extension of the TOPSIS for group decision making under fuzzy environment. Fuzzy Sets and Systems 114:1–9

Chen SJ, Hwang CL (1992) Fuzzy multiple attribute decision making-methods and applications. Lecture Notes in Economics and Mathematical Systems, Springer, New York

Csáki P, Rapcsák T, Turchányi P, Vermes M (1995) Research and development for group decision aid in Hungary by WINGDSS, a Microsoft Windows based group decision support system. Decision Support Systems 14:205–221

Deng H, Yeh CH, Willis RJ (2000) Inter-company comparison using modified TOPSIS with objective weights. Computers & Operations Research 27: 963–973

Dyer RF, Forman EH (1992) Group decision support with the analytic hierarchy process. Decision Support Systems 8:99–124

Edwards W, Newman JR, Snapper K, Seaver D (1982) Multiattribute Evaluation. SAGE Publications, Newbury Park, California

Figueira J, Greco S, Ehrgott M (2004) Multiple criteria decision analysis: state of the art surveys. Springer, New York

Fishburn PC (1967) Additive utilities with incomplete product set: applications to priorities and assignments. Operations Research Society of America, Baltimore

Forman E, Penewati K (1998) Aggregating individual judgements and priorities with the analytic hierarchy process. European Journal of Operational Research 108:165–169

Gal T (1999) Multicriteria decision making:advances in MCDM models algorithms, theory and applications. Kluwer Academic Publishers, Dordrecht

Hwang CL, Yoon KP (1981) Multiple attribute decision making: methods and applications. Springer, New York

Jaganathan S, Erinjeri JJ, Ker JI (2006) Fuzzy analytic hierarchy process based group decision support system to select and evaluate new manufacturing technologies. International Journal of Advanced Manufacturing Technology doi:10.1007/s00170-006-0446-1

Kwiesielewicz M, Uden EV (2004) Inconsistent and contradictory judgements in pairwise comparison method in the AHP. Computers & Operations Research 31:713–719

Lai VS, Bo KW, Cheung W (2002) Group decision making in a multiple criteria environment: a case using the AHP in software selection. European Journal of Operational Research 137:134–144

Lai YJ, Liu TY, Hwang CL (1994) TOPSIS for MODM. European Journal of Operational Research 76:486–500

Lootsma FA (1999) Multi-criteria decision analysis via ratio and difference judgement. Applied Optimization Series, Kluwer Academic Publishers, Dordrecht

Miller DW, Starr MK (1969) Executive decisions with operations research. Prentice Hall, Englewood Cliffs, New Jersey

Olson DL (1996) Decision aids for selection problems. Springer, New York

Opricovic S, Tzeng GH (2002) Multicriteria planning of post-earthquake sustainable reconstruction. Computer-Aided Civil and Infrastructure Engineering 17:211–220

Opricovic S, Tzeng GH (2003) Fuzzy multicriteria model for post-earthquake land-use planning. Natural Hazards Review 4:59–64

Opricovic S, Tzeng GH (2004) Compromise solution by MCDM methods: a comparative analysis of VIKOR and TOPSIS. European Journal of Operational Research 156:445–455

Opricovic S, Tzeng GH (2007) Extended VIKOR method in comparison with outranking methods. European Journal of Operational Research 178:514–529

Saaty TL (1980) The analytic hierarchy process. McGraw Hill, New York
Saaty TL (2000) Fundamentals of decision making and priority theory with the AHP. RWS Publications, Pittsburg
Shannon CE, Weaver W (1947) The mathematical theory of communication. The University of Illinois Press, Urbana
Triantaphyllou E (2000) Multi-criteria decision making methods: a comparative study. Kluwer Academic Publishers, Dordrecht
Triantaphyllou E and Sanchez (1997) A sensitivity analysis approach for some deterministic multi-criteria decision-making methods. Decision Sciences 28:151–194
Tzeng GH, Tsaur SH, Laiw YD, Opricovic S (2002) Multicriteria analysis of environmental quality in Taipei:public preferences and improvement strategies. Journal of Environmental Management 65:109–120
Tzeng GH, Teng MH, Chen JJ, Opricovic S (2002) Multicriteria selection for a restaurant location in Taipei. International Journal of Hospitality Management 21:171–187
Tzeng GH, Lin CW, Opricovic S (2005) Multi-criteria analysis of alternative-fuel buses for public transportation. Energy Policy 33:1373–1383
Yoon KP, Hwang CL (1995) Multiple attribute decision making. SAGE Publications, Beverly Hills, CA
Yu PL (1973) A class of solutions for group decision problems. Management Science 19:936–946
Zanakis S, Solomon A, Wishart N, Dublish S (1998) Multi-attribute decision making: a comparison of select methods. European Journal of Operational Research, 107:507–529
Zeleny M (1982) Multiple criteria decision making. McGraw Hill, New York

4

A Logical Approach to Fuzzy MADM Problems

4.1 Introduction

The classical MADM methods assume all measures of performance of alternatives (m_{ij}) and weights of attributes (w_j) values are crisp numbers. The alternatives with higher overall or composite performance scores are considered to be preferred by the decision maker. Since the final scores are real numbers, the preferred alternatives are those with higher overall or composite performance scores. In reality, measure of performance (m_{ij}) can be crisp, fuzzy and/or linguistic. For example, let a material be chosen for making an engineering component and the attributes considered are: cost of material, tensile strength, hardness, density, and corrosion resistance. The last attribute, corrosion resistance, is not quantifiable; rather, it is represented by linguistic terms such as 'low', 'average', 'high', *etc.* The other attributes can be crisp numbers. This MADM problem contains a mixture of fuzzy and crisp data. Most of the real-world MADM problems are of this type.

Fuzzy MADM methods are proposed to solve problems that involve fuzzy data. Bellman and Zadeh (1970) were the first to relate fuzzy set theory to decision-making problems. Yager and Basson (1975) proposed fuzzy sets for decision making. Bass and Kwakernaak (1977) proposed a fuzzy MADM method that is widely regarded as the classic work of fuzzy MADM methods. During the last three decades, several fuzzy MADM methods have been proposed and reviewed (Chen and Hwang, 1992; Triantaphyllou and Lin, 1996; Triantaphyllou, 2000; Figueira *et al.*, 2004). After a systematic and critical study of the existing fuzzy MADM methods, it has been found that the majority of the approaches require cumbersome computations. As a result, none of them are suitable for solving problems with more than 10 alternatives associated with more than 10 attributes. That drawback certainly limits their applicability to real-world problems. Further, most approaches require that the elements in the decision matrix be presented in a fuzzy format, though they are crisp in nature. Such an assumption violates the original intent of fuzzy set theory. If the data is precisely known, there is no subjectivity involved in the decision problem. Such data should never be represented in any fuzzy format. The conversion of crisp data into fuzzy format will increase the computational requirements. This, in turn, makes these fuzzy methods cumbersome to use, and incapable of solving problems that contain more than 10 alternatives and 10 attributes.

Chen and Hwang (1992) proposed an approach to avoid the abovementioned difficulties, so that MADM problems can be meaningfully and efficiently solved in a fuzzy environment. The approach is composed of two phases. The first phase converts fuzzy data into crisp scores. The result of the first phase is a decision matrix that contains only crisp data. In the second phase, MADM methods, described in Chapter 3, can be utilized to determine the ranking order of alternatives. The easy-to-use and easy-to-understand characteristics of this approach make it valuable to management and system analysts.

4.2 Method Proposed by Chen and Hwang (1992)

The method proposed by Chen and Hwang (1992) first converts linguistic terms into fuzzy numbers and then the fuzzy numbers into crisp scores. The method is described below.

4.2.1 Converting Linguistic Terms to Fuzzy Numbers

This method systematically converts linguistic terms into their corresponding fuzzy numbers. It contains eight conversion scales. The conversion scales were proposed by synthesizing and modifying the works of Wenstop (1976), Bass and Kwakernaak (1977), Efstathiou and Rajkovic (1979), Bonissone (1982), Efstathiou and Tong (1982), Kerre (1982), and Chen (1988),

To demonstrate the method, a 5-point scale having the linguistic terms low, fairly low, medium, fairly high, and high, as shown in Figure 4.1 (Chen and Hwang, 1992), is considered. These linguistic terms can be equated to other terms like low, below average, average, above average, and high.

4.2.2 Converting Fuzzy Numbers to Crisp Scores

The method uses a fuzzy scoring approach that is a modification of the fuzzy ranking approaches proposed by Jain (1976, 1977), and Chen (1985). The crisp score of fuzzy number 'M' is obtained as follows:

$$\mu_{max}(x) = \begin{cases} x, 0 \le x \le 1 \\ 0, \text{otherwise} \end{cases} \tag{4.1}$$

$$\mu_{min}(x) = \begin{cases} 1 - x, 0 \le x \le 1 \\ 0, \text{otherwise} \end{cases} \tag{4.2}$$

The fuzzy max and fuzzy min of fuzzy numbers are defined in a manner such that absolute locations of fuzzy numbers can be automatically incorporated in the comparison cases. The left score of each fuzzy number 'M_i' is defined as

$$\mu_L(M_i) = \underset{x}{\text{Sup}}[\mu_{min}(x) \wedge \mu_{Mi}(x)] \tag{4.3}$$

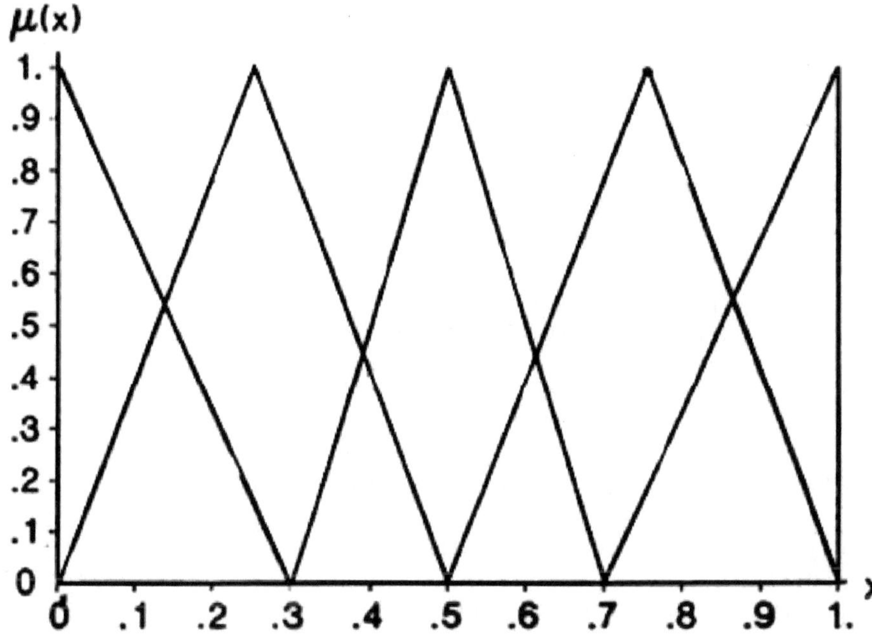

Figure 4.1. Linguistic terms to fuzzy numbers conversion (5-point scale) (from Chen and Hwang 1992; with kind permission of Springer Science and Business Media)

The $\mu_L(M_i)$ score is a unique, crisp, real number in (0, 1). It is the maximum membership value of the intersection of fuzzy number M_i and the fuzzy min. The right score may be obtained in a similar manner:

$$\mu_R(M_i) = \sup_x[\mu_{max}(x) \wedge \mu_{Mi}(x)] \quad (4.4)$$

Again, $\mu_R(M_i)$ is a crisp number [0,1]. Given the left and right scores, the total score of a fuzzy number M_i is defined as:

$$\mu_T(M_i) = [\mu_R(M_i) + 1 - \mu_L(M_i)] / 2 \quad (4.5)$$

4.3 Demonstration of the Method

Now, the 5-point scale is considered to demonstrate the conversion of fuzzy numbers into crisp scores (Figure 4.1).

Linguistic term	Fuzzy number
Low	M_1
Below average	M_2
Average	M_3
Above average	M_4
High	M_5

The maximizing and minimizing sets are defined as Equations 4.1 and 4.2. From Figure 4.1, membership functions of M_1, M_2, M_3, M_4, and M_5 are written as:

$$\mu_{M1}(x) = \begin{cases} 1, x = 0 \\ (0.3\text{-}x) / (0.3), 0 \le x \le 0.3 \end{cases}$$

$$\mu_{M2}(x) = \begin{cases} (x\text{-}0)/ (0.25), 0 \le x \le 0.3 \\ (0.5\text{-}x) / (0.25), 0.25 \le x \le 0.5 \end{cases}$$

$$\mu_{M3}(x) = \begin{cases} (x\text{-}0.3)/ (0.2), 0.3 \le x \le 0.5 \\ (0.7\text{-}x)/ (0.2), 0.5 \le x \le 0.7 \end{cases}$$

$$\mu_{M4}(x) = \begin{cases} (x\text{-}0.5)/ (0.25), 0.5 \le x \le 0.75 \\ (1.0\text{-}x)/ (0.25), 0.75 \le x \le 1.0 \end{cases}$$

$$\mu_{M5}(x) = \begin{cases} (x\text{-}0.7)/ (0.3), 0.7 \le x \le 1.0 \\ 1, x = 1 \end{cases}$$

The right, left, and total scores are computed as follows for M_1:

$$\mu_R (M_1) = \sup_x [\mu_{max} (x) \wedge \mu_{M1}(x)] = 0.23$$

$$\mu_L(M_1) = \sup_x [\mu_{min} (x) \wedge \mu_{M1}(x)] = 1.0$$

$$\mu_T (M_1) = [\mu_R (M_1) + 1 - \mu_L(M_1)] / 2 = 0.115$$

Similarly, the right, left, and total scores are computed for M_2, M_3, M_4, and M_5 and are tabulated as follows:

i	$\mu_R (M_i)$	$\mu_L (M_i)$	$\mu_T (M_i)$
1	0.23	1.0	0.115
2	0.39	0.8	0.295
3	0.58	0.59	0.495
4	0.79	0.4	0.695
5	1.0	0.23	0.895

Hence, the linguistic terms with their corresponding crisp scores are given in Table 4.1. Instead of assigning arbitrary values for various attributes, this fuzzy method reflects the exact linguistic descriptions in terms of crisp scores. Hence, it gives better approximation of linguistic descriptions that are widely used.

It may be added here that this method can be used not only for assigning values to the attributes, but also for deciding the relative importance between the attributes. For example, using the same 5-point scale, the relative importance between two attributes can be described as given in Table 4.2.

Table 4.1. Conversion of linguistic terms into fuzzy scores (5-point scale)

Linguistic term	Fuzzy number	Crisp score
Low	M_1	0.115
Below average	M_2	0.295
Average	M_3	0.495
Above average	M_4	0.695
High	M_5	0.895

Table 4.2. Conversion of linguistic terms into fuzzy scores (relative importance value on a 5-point scale)

Linguistic term	Fuzzy number	Crisp score
One attribute is very less important than the other	M_1	0.115
One attribute is less important than the other	M_2	0.295
Two attributes are equally important	M_3	0.495
One attribute is more important than the other	M_4	0.695
One attribute is much more important than the other	M_5	0.895

The decision makers can appropriately make use of any of the eight scales suggested by Chen and Hwang (1992). For example, an 11-point scale is shown in Figure 4.2, and the corresponding crisp scores of the fuzzy numbers are given in Table 4.3.

Table 4.3. Conversion of linguistic terms into fuzzy scores (11-point scale)

Linguistic term	Fuzzy number	Crisp score
Exceptionally low	M_1	0.045
Extremely low	M_2	0.135
Very low	M_3	0.255
Low	M_4	0.335
Below average	M_5	0.410
Average	M_6	0.500
Above average	M_7	0.590
High	M_8	0.665
Very high	M_9	0.745
Extremely high	M_{10}	0.865
Exceptionally high	M_{11}	0.955

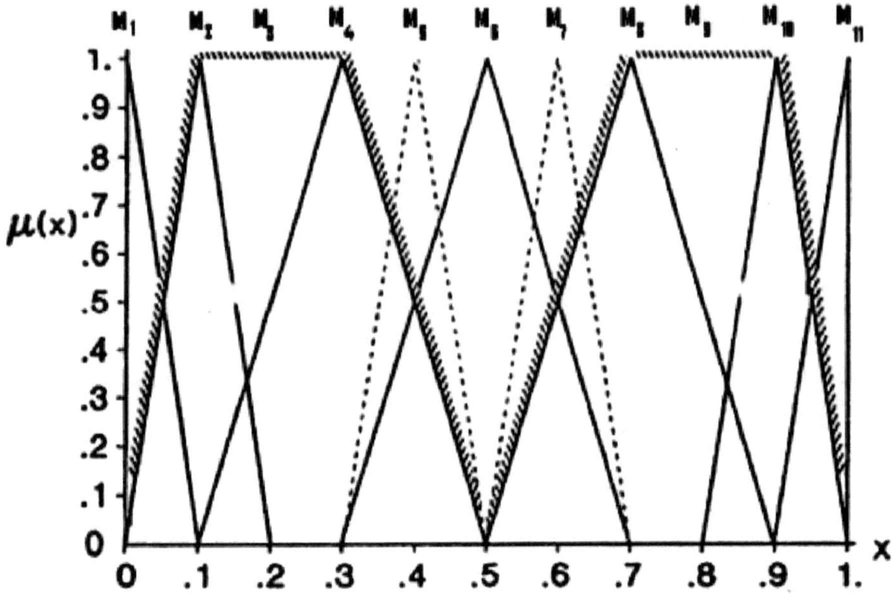

Figure 4.2. Linguistic terms to fuzzy numbers conversion (11-point scale) (from Chen and Hwang 1992; with kind permission of Springer Science and Business Media)

Using the same 11-point scale, the relative importance between two attributes can be described as given in Table 4.4.

Table 4.4. Conversion of linguistic terms into fuzzy scores (relative importance value on an 11-point scale)

Linguistic term	Fuzzy number	Crisp score
One attribute is exceptionally less important than the other	M_1	0.045
One attribute is extremely less important than the other	M_2	0.135
One attribute is very less important than the other	M_3	0.255
One attribute is less important than the other	M_4	0.335
One attribute is slightly less important than the other	M_5	0.410
Two attributes are equally important than the other	M_6	0.500
One attribute is slightly more important than the other	M_7	0.590
One attribute is more important than the other	M_8	0.665
One attribute is much more important than the other	M_9	0.745
One attribute is extremely more important than the other	M_{10}	0.865
One attribute is exceptionally more important than the other	M_{11}	0.955

It may be remembered that Tables 2.1 and 2.2 are suggested in Chapter 2 for assigning the objective values to the subjective attributes, and for assigning the relative importance between the attributes, respectively. Now, Tables 4.1 (or 4.3) and 4.2 (or 4.4) may be used for the same purpose, as these give better approximation of the linguistic terms. The case studies presented in Chapters 5–30 of this book utilize Tables 4.3 and 4.4.

References

Bass SJ, Kwakernaak H (1977) Rating and ranking of multi-aspects alternatives using fuzzy sets. Automatica 13:47–58

Bellman RE, Zadeh LE (1970) Decision-making in a fuzzy environment. Management Science 17:212–223

Bonissone PP (1982) A fuzzy sets based linguistic approach: theory and applications. In: Gupta MM, Sanchez E (eds) Approximate reasoning in decision analysis. North Holland, pp 329–339

Chen SH (1985) Ranking fuzzy numbers with maximizing set and minimizing set. Fuzzy Sets and Systems 17:113–129

Chen SM (1988) A new approach to handling fuzzy decision-making problems. In: Proc. 18th International Symposium on Multiple-Valued Logic, Computer Society Press, Palma de Mallorca, Spain

Chen SJ, Hwang CL (1992) Fuzzy multiple attribute decision making-methods and applications. Lecture Notes in Economics and Mathematical Systems, Springer, New York

Efstathiou J, Rajkovic V (1979) Multiattribute decision making using a fuzzy hewristic approach. IEEE Transactions on Systems, Man, Cybernetics, SMC-9:326–333

Efstathiou J, Tong R (1980) Ranking fuzzy sets using linguistic preference relations. In: Proc. 10th International Symposium on Multiple-Valued Logic, Northwestern University, Evanston

Figueira J, Greco S, Ehrgott M (2004) Multiple criteria decision analysis: state of the art surveys. Springer, New York

Jain R (1976) Decision making in the presence of fuzzy variables. IEEE Transactions on Systems, Man and Cybernetics, SMC 6:698–703

Jain R (1977) A procedure for multi-aspect decision making using fuzzy sets. International Journal of System Science 8:1–7

Kerre EE (1982) The use of fuzzy set theory in electrocardiological diagnostics. In: Gupta MM, Sanchez E (eds) Approximate reasoning in decision analysis. North Holland, pp 277–282

Triantaphyllou E (2000) Multi-criteria decision making methods: a comparative study. Kluwer Academic Publishers, Dordrecht

Triantaphyllou E, Lin CT (1996) Development and evaluation of five fuzzy multi-attribute decision making methods. International Journal of Approximate Reasoning 14:281–310

Wenstop F (1976) Fuzzy set simulation models in a systems dynamic perspective. Kybernetes 6:209–218

Yager RR, Basson D (1975) Decision-making with fuzzy sets. Decision Sciences 6:590–600

Part 2

Applications of GTMA and Fuzzy MADM Methods in the Manufacturing Environment

5

Material Selection for a Given Engineering Application

5.1 Introduction

An ever increasing variety of materials is available today, each having its own characteristics, applications, advantages, and limitations. When selecting materials for engineering designs, a clear understanding of the functional requirements for each individual component is required, and various important criteria or attributes need to be considered. Material selection attribute is defined as a factor that influences the selection of a material for a given application. These attributes include: physical properties, electrical properties, magnetic properties, mechanical properties, chemical properties, manufacturing properties (machinability, formability, weldability, castability, heat treatability, *etc.*), material cost, product shape, material impact on environment, performance characteristics, availability, fashion, market trends, cultural aspects, esthetics, recycling, target group, *etc.*

The selection of an optimal material for an engineering design from among two or more alternative materials on the basis of two or more attributes is a multiple attribute decision-making problem. Various approaches have been proposed in the past to help address the issue of material selection. Liao (1996) presented a fuzzy multicriteria decision-making method for material selection. However, the method is complicated and requires much more computation. Farag (1997) proposed a simple mathematics-based weighted properties method that can be used when several properties should be taken into consideration. Giachetti (1998) described a prototype material and manufacturing process selection system that integrates a formal multiple attribute decision model with a relational database. The decision model enables the representation of the designer's preferences over the decision attributes. A compatibility rating between the product profile requirements and the alternatives stored in the database for each decision attribute was generated using possibility theory. The vectors of compatibility ratings were aggregated into a single rating of that alternative's compatibility. A ranked set of compatible material and manufacturing process alternatives was the output by the system.

Ashby (2000) proposed multi-objective optimization in materials design and selection, using 'utility' functions. Ashby *et al.* (2004) provided a comprehensive review of the strategies or methods for materials selection, from which three types of materials selection methodology were identified: (i) free searching based on quantitative analysis, (ii) checklist/questionnaire based on expertise capture, and (iii) inductive reasoning and analog procedure. All of these methods use materials data in either a non-computerized or computerized form.

For the free-searching method, there are already a number of well-documented methods, the best known being the graphical engineering selection method or the ranking method (Ashby, 1992; Ashby and Johnson, 2002). A checklist/questionnaire method has been proposed by a number of researchers, the most recent described by Edwards (2005), where the author developed a structured set of questions to improve the likelihood of achieving an optimal design solution. The inductive reasoning and analog procedure resulted from the rapid development of information technology tools, and the application of artificial intelligence. Some of representative examples include a knowledge-based system for materials management that involves materials selection (Trethewey *et al.*, 1998), a knowledge-based system for materials selection (Sapuan, 2001), integrated information technology approach (Jalham, 2006), fuzzy knowledge-based decision support system for selection of manufacturing processes and materials (Zha, 2005) and a case-based reasoning method (Amen and Vomacka, 2001). However, these systems and methods are complex and necessitate knowledge extensive.

A framework to represent and deal with the relationships between design variables of both materials parameters and system-level parameters was proposed by Raj (2000) and Raj *et al.* (2000). The idea of an integrated approach for materials selection and structural design had been advocated by Edwards (2002). The materials parameters could be material properties, or they could be parameters describing the micro/nanostructure of the materials. Ermolaeva *et al.* (2002) studied materials selection combined with structural optimization. However, the elaborate materials selection method proposed by these authors was limited to selecting from a limited number of specific materials. Lin and Lin (2003) discussed state-of-art research on environmentally conscious material selection methodologies. Ljungberg (2005) presented guidelines for sustainable product development with special regard to materials, design and ecology. Giudice *et al.* (2005) proposed a method to integrate mechanical and environmental performances for materials selection in the life-cycle design process. Kuo *et al.* (2006) presented an innovative method, namely, green fuzzy design analysis (GFDA), which involves simple and efficient procedures to evaluate product design alternatives based on environmental consideration using fuzzy logic. The hierarchical structure of environmentally conscious design indices was constructed using the analytical hierarchy process (AHP), which includes five aspects: (1) energy, (2) recycling, (3) toxicity, (4) cost, and (5) material. After weighting factors for the environmental attributes are determined, the most desirable design alternative can be selected using a fuzzy MADM method.

Edwards and Deng (2006) discussed the aspects of supporting design decision making when applying materials in combination. Deng and Edwards (2007) presented an overview of recent research in materials identification and materials selection. Shanian and Savadogo (2006a) had presented a material selection model using an MADM method known as ELECTRE. However, the ELECTRE method uses the concept of outranking relationship, and the procedure is rather lengthy. Only a partial prioritization of alternative materials is computed in ELECTRE models. As the number of alternatives increases, the amount of calculations rises quite rapidly, and the computational procedures are elaborate. In their other works (Shanian and Savadogo, 2006b, 2006c), the authors had proposed ELECTRE IV and TOPSIS methods for material selection of metallic bipolar plates for polymer electrolyte fuel cell.

Matos and Simplicio (2006) presented a practical example concerning the selection of materials to substitute polyvinyl chloride in automobile interiors. Bovea and Gallardo (2006) tested five life-cycle impact assessment methods, and applied to different polymer materials used for packaging purposes. The aim of the study was to demonstrate the need to perform a sensitivity analysis when a single environmental score is applied during the process of selecting materials, in order to enhance the environmental performance of products. Chan and Tong (2006) proposed a multicriteria weighted average method using gray relational analysis to rank the materials. Rao (2006) presented a material selection model using graph theory and the matrix approach. A 'material suitability index' was proposed that evaluates and ranks the materials for a given engineering component. Kumar and Singh (2006) presented an intelligent system for selection of materials for progressive die components. Cheng *et al.* (2006) used the fuzzy AHP method for selection of technological forecasting methods for predictuion of new materials development.

Manshadi *et al.* (2007) proposed a numerical method for materials selection combining non-linear normalization with a modified digital logic method. Guisbiers and Wautelet (2007) presented the details of materials selection for thin films for radio frequency micro-electromechanical systems (MEMS). Rao and Davim (2007) used the TOPSIS method for selection of materials for a given application.

A good amount of research work has been carried out in the past on materials selection. However, there is a need for a simple, systematic, and logical scientific method or mathematical tool to guide user organizations in taking a proper material selection decision. The objective of a material selection procedure is to identify the material selection attributes, and obtain the most appropriate combination of material selection attributes in conjunction with the real requirement. Thus, efforts need to be extended to determine attributes that influence material selection, using a simple logical approach, to eliminate unsuitable materials and to select a proper material to strengthen the existing material selection procedure. This is considered in this chapter using graph theory and the matrix approach (GTMA) and fuzzy MADM methods described in Chapters 2–4 of the book.

5.2 Examples

To demonstrate and validate the application of decision-making methods, two examples are considered. In both, GTMA is applied first, and subsequently a few MADM methods are applied to rank and select the materials for the given applications.

5.2.1 Example 1

Manshadi *et al.* (2007) proposed a numerical method for materials selection combining nonlinear normalization with a modified digital logic method. This example problem is related with selection of a suitable material for a cryogenic storage tank for transportation of liquid nitrogen. The material selection problem considers seven alternative materials and seven attributes, and the data are given in Table 5.1.

Table 5.1. Objective data of the attributes of example 5.2.1(from Manshadi *et al.*, 2007; reprinted with permission from Elsevier)

Material	Material selection attributes						
	TI	YS	YM	D	TE	TC	SH
1	75.5	420	74.2	2.8	21.4	0.37	0.16
2	95	91	70	2.68	22.1	0.33	0.16
3	770	1,365	189	7.9	16.9	0.04	0.08
4	187	1,120	210	7.9	14.4	0.03	0.08
5	179	875	112	4.43	9.4	0.016	0.09
6	239	1,190	217	8.51	11.5	0.31	0.07
7	273	200	112	8.53	19.9	0.29	0.06

TI = Toughness index (based on UTS, yield strength YS, and ductility e at -196°C) = (UTS + YS)e/2; YS = Yield strength (MPa); YM = Young's modulus (GPa); D= Density (g/cm^3); TE = Thermal expansion (given in 10^{-6}/°C); TC = Thermal conductivity (cal/cm^2/cm/°C/s); SH = Specific heat (cal/g/°C)
Material 1:Al 2024-T6; Material 2:Al 5052-O; Material 3:SS 301-FH; Material 4:SS310-3AH; Material 5:Ti-6Al-4V; Material 6:Inconel 718; Material 7:70Cu-30Zn

5.2.1.1 Application of Graph Theory and Matrix Approach (GTMA)

Various steps of the methodology, proposed in Section 2.6, are carried out:

Step 1: In the present work, the attributes considered are the same as those of Manshadi *et al.* (2007) and these are: toughness index (TI), yield strength (YS), Young's modulus (YM), density (D), thermal expansion (TE), thermal conductivity (TC) and specific heat (SH). The quantitative values of the material selection attributes, which are given in Table 5.1, are normalized. TI, YS, and YM are considered as beneficial attributes, and the remaining attributes as non-beneficial. Values of these seven attributes are normalized, as explained in Section 2.4, and are given in Table 5.2 in the respective columns.

Table 5.2. Normalized data of the attributes of example 5.2.1

Material	Normalized values of material selection attributes						
	TI	YS	YM	D	TE	TC	SH
1	0.0981	0.3077	0.3419	0.9571	0.4393	0.0432	0.375
2	0.1234	0.0667	0.3226	1	0.4253	0.0485	0.375
3	1	1	0.8709	0.3392	0.5562	0.4	0.75
4	0.2429	0.8205	0.9677	0.3392	0.6528	0.5333	0.75
5	0.2325	0.6410	0.5161	0.6049	1	1	0.6667
6	0.3104	0.8718	1	0.3149	0.8174	0.0516	0.8571
7	0.3546	0.1465	0.5161	0.3142	0.4724	0.0552	1

Relative importance of attributes (a_{ij}) is also assigned the values, as explained in Section 2.4, using Table 4.4. Let the decision maker (*i.e.*, designer) select the following assignments:

$$A1_{7x7} = \begin{array}{c} \\ TI \\ YS \\ YM \\ D \\ TE \\ TC \\ SH \end{array} \begin{array}{c} \begin{array}{ccccccc} TI & YS & YM & D & TE & TC & SH \end{array} \\ \begin{bmatrix} - & 0.745 & 0.865 & 0.590 & 0.665 & 0.865 & 0.865 \\ 0.255 & - & 0.665 & 0.335 & 0.335 & 0.665 & 0.665 \\ 0.135 & 0.335 & - & 0.745 & 0.255 & 0.500 & 0.500 \\ 0.410 & 0.665 & 0.255 & - & 0.590 & 0.745 & 0.745 \\ 0.335 & 0.665 & 0.745 & 0.410 & - & 0.665 & 0.665 \\ 0.135 & 0.335 & 0.500 & 0.255 & 0.335 & - & 0.500 \\ 0.135 & 0.335 & 0.500 & 0.255 & 0.335 & 0.500 & - \end{bmatrix} \end{array} \quad (5.1)$$

However, it may be added that, in actual practice, the designer can judiciously decide these values of relative importance depending on the requirements. The assigned values are for demonstration purpose only.

Step 2:

1. The material selection attributes graph, showing the presence as well as relative importance of the above attributes, is similar to Figure 2.2 but with seven attributes is drawn. It is not shown here for obvious reasons.
2. The material selection attributes matrix of this graph can be written based on Equation 2.10. It is similar to matrix Equation 5.1 but also with the presence of diagonal elements A_i.
3. The material selection attributes function is written. However, it may be added that as a computer program is developed for calculating the permanent function value of a matrix, this step can be skipped.

4 & 5. The material selection index is calculated using the values of A_i and a_{ij} for each alternative material. The material selection index values of different materials are given below in descending order:

Material 3: SS 301-FH	39.1123
Material 5: Ti-6Al-4V	34.0554
Material 4: SS 310-3AH	30.6316
Material 6: Inconel 718	29.0377

Material 7: 70Cu-30Zn 20.0377
Material 1: Al 2024-T6 17.2897
Material 2: Al 5052-O 16.2634

From the above values of the material selection index, it is understood that the material designated as 3, *i.e.*, SS 301-FH, is the right choice for the given problem of selection of a suitable material for a cryogenic storage tank for transportation of liquid nitrogen. The second choice is Ti-6Al-4V, and the last choice is Al 5052-O. These results match those suggested by Manshadi *et al.* (2007) using nonlinear normalization and a modified digital logic method. However, it may be mentioned that the ranking depends upon the judgements of relative importance made by the designer. The ranking presented may change if the designer assigns different relative importance values to the attributes. The same is true for the approach proposed by Manshadi *et al.* (2007). However, the GTMA method is superior to the method used by Manshadi *et al.* (2007) in that it enables a more critical analysis than the digital logic method, since any number of quantitative and qualitative attributes can be considered. Also, the proposed method can deal with material selection attributes on a qualitative scale using fuzzy logic. Such a provision is missing in the method suggested by Manshadi *et al.* (2007). Further, the proposed method assigns the values of relative importance based on a fuzzy scale, whereas the weights assigned to various attributes by Manshadi *et al.* (2007) were rather arbitrary and too simplistic. The use of permanent concept helps in better appreciation of the attributes, and it characterizes the considered material selection problem, as it contains all possible structural components of the attributes and their relative importance (from a combinatorial point of view). The coefficients of similarity are calculated and are given in Table 5.3.

Table 5.3. Values of coefficient of similarity for the alternative materials of example 5.2.1

Material	2	3	4	5	6	7
1	0.94	0.442	0.564	0.508	0.595	0.863
2		0.416	0.531	0.531	0.56	0.812
3			0.783	0.871	0.742	0.512
4				0.899	0.948	0.654
5					0.853	0.588
6						0.69

5.2.1.2 SAW Method

Using the same weights of the attributes as those of Manshadi *et al.* (2007), the overall performance score (*i.e.*, material selection index, in this example) for each material is calculated using the normalized data of the attributes given in Table 5.2, and Equation 3.2. For example, the value of P_i for the material designated as 1 is calculated as: 0.28x0.0981 + 0.14x0.3077 + 0.05x0.3419 + 0.24x0.9571 + 0.19x0.4393 + 0.05x0.0432 + 0.05x0.0432 = 0.421722. The values of P_i are arranged in descending order as given below:

Material 3: SS 301-FH	0.4217
Material 5: Ti-6Al-4V	0.5991
Material 6: Inconel 718	0.5352
Material 4: SS 310-3AH	0.5008
Material 1: Al 2024-T6	0.4217
Material 2: Al 5052-O	0.4020
Material 7: 70Cu-30Zn	0.3635

The SAW method also suggests the material designated as 3, *i.e.*, SS 301-FH, as the right choice for the given problem of selection of a suitable material for a cryogenic storage tank for transportation of liquid nitrogen. The second choice is Ti-6Al-4V, and the last choice is the material designated as 7, *i.e.*, 70Cu-30Zn. However, comparing the attribute data of the materials of the last two choices, *i.e.*, materials 2 and 7, it may not be logical to propose 7 as the last choice.

5.2.1.3 WPM

The overall performance score (*i.e.*, material selection index, in this example) for each material is calculated using the normalized data of the attributes given in Table 5.2 for the given weights of the attributes, and Equation 3.4. For example, the value of P_i for the material designated as 1 is calculated as: $0.0981^{0.28} + 0.3077^{0.14} + 0.3419^{0.05} + 0.9571^{0.24} + 0.4393^{0.19} + 0.0432^{0.05} + 0.0432^{0.05} = 0.2889$. The values of P_i are arranged in the descending order as given below:

Material 3: SS 301-FH	0.6843
Material 5: Ti-6Al-4V	0.5248
Material 4: SS 310-3AH	0.4440
Material 6: Inconel 718	0.4412
Material 1: Al 2024-T6	0.2889
Material 2: Al 5052-O	0.2505
Material 7: 70Cu-30Zn	0.1809

WPM also suggests the material designated as 3, *i.e.*, SS 301-FH, as the right choice for the given material selection problem. The second choice is Ti-6Al-4V, and the last choice is material designated as 7, *i.e.*, 70Cu-30Zn.

5.2.1.4 AHP and its Versions

As the weights of the attributes are already available, the alternatives are compared pair-wise with respect to how much better they are in satisfying each of the attributes. This means ascertaining how well each alternative serves each attribute. In this example, as there are seven alternatives and seven attributes, there will be seven numbers of 7 x 7 matrices of judgements.

The absolute mode is used, as data of the attributes for different alternatives to be evaluated are readily available. Comparison of alternative materials is shown in Table 5.4 with respect to TI (a beneficial attribute), and D (a non-beneficial attribute) only for demonstration purpose. Similar comparisons can be shown with respect to the other five attributes. Since the exact values are used in these comparison matrices, CI is always equal to 0, as there is complete consistency in judgements.

Table 5.4. Pair-wise comparison matrices for the alternative materials of example 5.2.1

	1	2	3	4	5	6	7	R	I
TI									
1	1	0.7947	0.098	0.4037	0.4218	0.316	0.2765	0.0415	0.0981
2	1.258	1	0.123	0.508	0.531	0.397	0.348	0.0522	0.1234
3	10.2	8.13	1	4.12	4.3	3.22	2.82	0.4235	1
4	2.477	1.97	0.243	1	1.045	0.782	0.685	0.1028	0.2429
5	2.37	1.883	0.233	0.957	1	0.75	0.656	0.0984	0.2325
6	3.1645	2.52	0.31	1.28	1.33	1	0.875	0.1314	0.3104
7	3.62	2.87	0.355	1.46	1.524	1.143	1	0.1501	0.3546
D									
1	1	0.957	2.82	2.82	1.58	3.04	0.046	0.2473	0.9571
2	1.045	1	2.95	2.95	1.653	3.17	3.18	0.2584	1
3	0.355	0.339	1	1	0.561	1.077	1.079	0.0877	0.3392
4	0.355	0.339	1	1	0.561	1.077	1.079	0.0877	0.3392
5	0.633	0.605	1.78	1.78	1	1.92	1.926	0.1563	0.6049
6	0.329	0.315	0.929	0.929	0.521	1	1.002	0.0815	0.3149
7	0.329	0.315	0.929	0.929	0.521	1	1.002	0.0812	0.3142

R: Relative weight I: Ideal weight

In the above table, both relative (R) and idealized (I) weight vectors of the seven alternatives are given. The idealized vector is obtained by dividing each element of the relative weight vector by its largest element. The advantage of using idealized weights is that the ranking of the existing alternatives does not change even if a new alternative, identical to a non-optimal alternative, is introduced.

It may be observed that the idealized weights of the alternatives obtained for the attributes in Table 5.4 are nothing but the normalized data given in Table 5.2. This means that whenever quantitative data of the attributes are available, the data can be normalized directly as explained in Section 3.2.1.

The overall or composite performance scores (*i.e.,* material selection indexes, in this example) for the alternatives are obtained by multiplying the relative normalized weight (w_j) of each attribute with its corresponding normalized weight value (relative weight or ideal weight) for each alternative, and summing over all the attributes for each alternative. This step is similar to the SAW method. The alternative materials are arranged in the descending order of the material selection index. The results of the revised AHP and relative AHP are shown below:

Material	Ideal mode	Relative mode
Material 3: SS 301-FH	0.4217	0.2246
Material 5: Ti-6Al-4V	0.5991	0.1691
Material 6: Inconel 718	0.5352	0.1449
Material 4: SS 310-3AH	0.5008	0.1397
Material 1: Al 2024-T6	0.4217	0.1105
Material 2: Al 5052-O	0.4020	0.1062
Material 7: 70Cu-30Zn	0.3635	0.1049

Both the AHP and revised AHP methods give the same material rankings in this example.

The application of the multiplicative AHP method gives the ranking of materials in the sequence of 3-5-4-6-1-2-7 for the given weights of the attributes. This ranking is the same as that obtained using WPM.

5.2.1.5 TOPSIS Method

Step 1: The objective is to evaluate the seven alternative materials, and the attributes are: toughness index (TI), yield strength (YS), Young's modulus (YM), density (D), thermal expansion (TE), thermal conductivity (TC), and specific heat (SH). For this particular material selection problem, TI, YS, and YM are considered as beneficial attributes, and remaining attributes as non-beneficial.

Step 2: The next step is to represent all the information available for the attributes in the form of a decision matrix. The data given in Table 5.1 are represented as matrix D_{7x7}. However, the matrix is not shown here, as it is simply the repetition of data given in Table 5.1 but represented in a matrix form.

Step 3: The quantitative values of the material selection attributes, which are given in Table 5.1, are normalized as explained in Section 3.2.6 and the normalized matrix R_{7x7} is shown below:

$$\begin{bmatrix} 0.0843 & 0.1787 & 0.1842 & 0.1604 & 0.4719 & 0.5660 & 0.5640 \\ 0.1058 & 0.0387 & 0.1738 & 0.1535 & 0.4873 & 0.5040 & 0.5640 \\ 0.8575 & 0.5808 & 0.4690 & 0.4526 & 0.3727 & 0.0610 & 0.2820 \\ 0.2083 & 0.4765 & 0.5212 & 0.4526 & 0.3176 & 0.0458 & 0.2820 \\ 0.1994 & 0.3723 & 0.2780 & 0.2538 & 0.2073 & 0.0244 & 0.3170 \\ 0.2662 & 0.5064 & 0.5384 & 0.4875 & 0.2537 & 0.4734 & 0.2466 \\ 0.3040 & 0.0851 & 0.2780 & 0.4888 & 0.4389 & 0.4430 & 0.2114 \end{bmatrix}$$

Step 4: Relative importance of attributes (a_{ij}) can be assigned the values as explained in Section 3.2.6. However, to make a comparison of the proposed method with that of Manshadi *et al.* (2007), the same weights considered by those authors are assigned in the present work. These are: $W_{TI} = 0.28$, $W_{YS} = 0.14$, $W_{YM} = 0.05$, $W_D = 0.24$, $W_{TE} = 0.19$, $W_{TC} = 0.05$, and $W_{SH} = 0.05$.

Step 5: The weighted normalized matrix, V_{7x7}, is calculated.

$$\begin{bmatrix} 0.0236 & 0.0250 & 0.0092 & 0.0385 & 0.0896 & 0.0283 & 0.0282 \\ 0.0296 & 0.0054 & 0.0087 & 0.0368 & 0.0926 & 0.0252 & 0.0282 \\ 0.2401 & 0.0813 & 0.0234 & 0.1086 & 0.0708 & 0.00305 & 0.0141 \\ 0.0583 & 0.0667 & 0.0260 & 0.1086 & 0.0603 & 0.0022 & 0.0141 \\ 0.0558 & 0.0521 & 0.0139 & 0.0609 & 0.0393 & 0.0012 & 0.0158 \\ 0.0745 & 0.0708 & 0.0269 & 0.1170 & 0.0482 & 0.0236 & 0.0123 \\ 0.0851 & 0.0119 & 0.0139 & 0.1173 & 0.0834 & 0.0221 & 0.0105 \end{bmatrix}$$

Step 6: The next step is to obtain the ideal (best) and negative ideal (worst) solution. These are calculated as:

$V_{TI}^{+} = 0.24011 \qquad V_{TI}^{-} = 0.02362$

$V_{YS}^{+} = 0.08131 \qquad V_{YS}^{-} = 0.00542$

V_{YM}^{+}	=	0.02692	V_{YM}^{-}	=	0.00869
V_{D}^{+}	=	0.03685	V_{D}^{-}	=	0.11730
V_{TE}^{+}	=	0.03938	V_{TE}^{-}	=	0.09260
V_{TC}^{+}	=	0.00122	V_{TC}^{-}	=	0.02830
V_{SH}^{+}	=	0.01057	V_{SH}^{-}	=	0.02820

Step 7: The next step is to obtain the separation measures, and these are calculated as:

S_1^{+}	=	0.23222	S_1^{-}	=	0.08126
S_2^{+}	=	0.23263	S_2^{-}	=	0.08073
S_3^{+}	=	0.07854	S_3^{-}	=	0.23287
S_4^{+}	=	0.19715	S_4^{-}	=	0.08516
S_5^{+}	=	0.18865	S_5^{-}	=	0.10071
S_6^{+}	=	0.18583	S_6^{-}	=	0.09724
S_7^{+}	=	0.19456	S_7^{-}	=	0.06547

Step 8: The relative closeness of a particular alternative to the ideal solution is calculated and these are:

$P_1 = 0.25922$ $P_2 = 0.25763$ $P_3 = 0.74779$
$P_4 = 0.30165$ $P_5 = 0.34804$ $P_6 = 0.34352$
$P_7 = 0.25178$

This relative closeness to ideal solution can be considered as the 'material selection index '.

Step 9: The alternative materials are arranged in descending order of their material selection index. This can be arranged as: 3-5-6-4-1-2-7. From these values of index, it is understood that the material designated as 3 is the first right choice, material 5 the second choice, and material 7 the last choice for the given application under the given conditions. These results match with those suggested by Manshadi *et al.* (2007) regarding the first five choices. Manshadi *et al.* (2007) proposed a preference order of 3-5-6-4-1-7-2 for the same weights assigned to the attributes. A close look at the data presented in Table 5.1 suggests that material 7 is better than 2 with respect to six attributes (it may be remembered that the total number of attributes is seven). Thus, it is not logical to propose material 7 as the last choice in the TOPSIS method. Rather, proposing material 2 as the last choice is the right decision. Thus, the proposal of material 2 as the last choice by Manshadi *et al.* (2007) is logical.

5.2.1.6 Modified TOPSIS Method

In this process, the positive ideal solution (R^{+}) and the negative ideal solution (R^{-}), which are not dependent on the weighted decision matrix, are given by using Equations 3.19 and 3.20.

R_{TI}^{+}	=	0.8575	R_{TI}^{-}	=	0.0843
R_{YS}^{+}	=	0.5808	R_{YS}^{-}	=	0.0387
R_{YM}^{+}	=	0.5384	R_{YM}^{-}	=	0.1738
R_{D}^{+}	=	0.1535	R_{D}^{-}	=	0.4888
R_{TE}^{+}	=	0.2073	R_{TE}^{-}	=	0.4873
R_{TC}^{+}	=	0.0244	R_{TC}^{-}	=	0.5660
R_{SH}^{+}	=	0.2114	R_{SH}^{-}	=	0.5640

The weighted Euclidean distances are calculated as

D_1^+	=	0.4801	D_1^-	=	0.1693
D_2^+	=	0.4885	D_2^-	=	0.1652
D_3^+	=	0.1650	D_3^-	=	0.4821
D_4^+	=	0.3789	D_4^-	=	0.2459
D_5^+	=	0.3657	D_5^-	=	0.2563
D_6^+	=	0.3688	D_6^-	=	0.2497
D_7^+	=	0.4117	D_7^-	=	0.1476

The relative closeness of a particular alternative to the ideal solution is calculated (*i.e.*, material selection index) and these are:

$P_{1\text{-mod}} = 0.26067$ $P_{2\text{-mod}} = 0.25268$ $P_{3\text{-mod}} = 0.74505$
$P_{4\text{-mod}} = 0.39345$ $P_{5\text{-mod}} = 0.41196$ $P_{6\text{-mod}} = 0.40376$
$P_{7\text{-mod}} = 0.26392$

The alternative materials are arranged in descending order of their material selection index. This can be arranged as: 3-5-6-4-7-1-2. It can be observed that material 2 can be proposed as the last choice. Thus, the modified TOPSIS method has provided a more logical selection procedure, compared to the simple TOPSIS method.

5.2.1.7 Compromise Ranking Method (VIKOR)

Step 1: The objective is to evaluate the seven alternative materials, and the attributes are: toughness index (TI), yield strength (YS), Young's modulus (YM), density (D), thermal expansion (TE), thermal conductivity (TC), and specific heat (SH). For this particular material selection problem, TI, YS, and YM are considered as beneficial attributes, and remaining attributes as non-beneficial. The best, *i.e.*, $(m_{ij})_{max}$, and the worst, *i.e.*, $(m_{ij})_{min}$ values of all attributes are also determined.

Step 2: The values of E_i and F_i are calculated using Equations 3.26 and 3.27, and are given below. The same weights as those considered by Manshadi *et al.* (2007) are assigned in the present work. The weights are: $w_{TI} = 0.28$, $w_{YS} = 0.14$, $w_{YM} = 0.05$, $w_D = 0.24$, $w_{TE} = 0.19$, $w_{TC} = 0.05$, and $w_{SH} = 0.05$.

$E_1 = 0.28 + 0.1039 + 0.0486 + 0.0049 + 0.1795 + 0.05 + 0.05 = 0.7169$
$E_2 = 0.2721 + 0.14 + 0.05 + 0 + 0.19 + 0.0443 + 0.05 = 0.7464$
$E_3 = 0 + 0 + 0.0095 + 0.2142 + 0.1122 + 0.0034 + 0.001 = 0.3403$
$E_4 = 0.2351 + 0.0269 + 0.0024 + 0.2142 + 0.0748 + 0.00198 + 0.001 = 0.5564$
$E_5 = 0.2383 + 0.0538 + 0.0357 + 0.0718 + 0 + 0 + 0.015 = 0.4146$
$E_6 = 0.2141 + 0.0192 + 0 + 0.2392 + 0.0314 + 0.0415 + 0.005 = 0.5504$
$E_7 = 0.2004 + 0.1280 + 0.0357 + 0.24 + 0.1571 + 0.0387 + 0 = 0.7999$
$E_{i\text{-min}} = 0.3403$ $E_{i\text{-max}} = 0.7999$
$R_1 = 0.28$ $R_2 = 0.2721$ $R_3 = 0.2142$ $R_4 = 0.2351$
$R_5 = 0.2383$ $R_6 = 0.2392$ $R_7 = 0.24$
$F_{i\text{-min}} = 0.2142$ $F_{i\text{-max}} = 0.28$

Step 3: The values of P_i are calculated using Equation 3.28 and for v = 0.5.

$P_1 = 0.9095$ $P_2 = 0.8817$ $P_3 = 0$ $P_4 = 0.3938$
$P_5 = 0.2639$ $P_6 = 0.4185$ $P_7 = 0.696$

Step 4: The alternatives are arranged in ascending order, according to the values of P_i. Similarly, the alternatives are arranged according to the values of E_i and F_i separately. Thus, three ranking lists are obtained. The best alternative, ranked by P_i, is the one with the minimum value of P_i.

$P_3 = 0$	$E_3 = 0.3403$	$F_3 = 0.2142$
$P_5 = 0.2639$	$E_5 = 0.4146$	$F_4 = 0.2351$
$P_4 = 0.3938$	$E_6 = 0.5504$	$F_5 = 0.2383$
$P_6 = 0.4185$	$E_4 = 0.5564$	$F_6 = 0.2392$
$P_7 = 0.696$	$E_1 = 0.7169$	$F_7 = 0.24$
$P_2 = 0.8817$	$E_2 = 0.7464$	$F_2 = 0.2721$
$P_1 = 0.9095$	$E_7 = 0.7999$	$F_1 = 0.28$

Step 5: For the given attribute weights, the compromise solution, alternative material 3, which is the best ranked by the measure P is suggested, as it satisfies both conditions discussed in Section 3.2.7.

It may be noted here that for the same weights of importance of the attributes, all decision-making methods described in the example suggest material 3 as the first right choice. The choice may change when different weights are used.

5.2.2 Example 2

Now, another example is considered to demonstrate the application of the GTMA and fuzzy MADM methods. This example problem is related with selection of a suitable work material for a product that needs to be designed for operating in a high-temperature oxygen-rich environment. This selection problem considers six alternative materials and four attributes and the data are shown in Table 5.5.

Table 5.5. Quantitative data of the attributes of example 5.2.2

Material	Material selection attributes			
	Hardness (HB)	MR (%)	Cost ($/lb)	Corrosion resistance
1	420	25	5	Extremely high (0.865)
2	350	40	3	High (0.665)
3	390	30	3	Very high (0.745)
4	250	35	1.3	High (0.665)
5	600	30	2.2	High (0.665)
6	230	55	4	Average (0.5)

MR: Machinability rating is based upon machining AISI 1112 steel with a rating of 100%

5.2.2.1 Application of Graph Theory and Matrix Approach (GTMA)

Step 1: In this example, the attributes considered are: hardness (H), machinability rating of work material based on cutting speed (MR), cost of the material (C), and corrosion resistance (CR). The quantitative values of the material selection attributes, which are given in Table 5.5, are to be normalized. For the given material selection problem, H, M, and CR are considered as beneficial attributes and C as a non-beneficial attribute. Cost is not considered that important in the present example. Corrosion resistance (CR) is expressed qualitatively, and hence ranked value judgements on fuzzy conversion scale, as shown in Table 2.3, are made and given in parentheses in Table 5.5. Values of the four attributes are normalized, and are given in Table 5.6 in the respective columns.

Table 5.6. Normalized data of example 5.2.2

Material	Normalized values of material selection attributes			
	H	M	C	R
1	0.7	0.4545	0.26	1
2	0.5833	0.7273	0.4333	0.7688
3	0.65	0.5454	0.4333	0.8613
4	0.4167	0.6364	1	0.7688
5	1	0.5454	0.5909	0.7688
6	0.3833	1	0.325	0.578

Step 2: The relative importance of attributes (a_{ij}) is also assigned. Let the decision maker (*i.e.*, designer) select the following assignments:

$$
A2_{4x4} = \begin{array}{c} \\ H \\ M \\ C \\ CR \end{array}
\begin{array}{cccc} H & M & C & CR \end{array}
\left[\begin{array}{cccc}
- & 0.335 & 0.665 & 0.665 \\
0.665 & - & 0.745 & 0.745 \\
0.335 & 0.255 & - & 0.335 \\
0.335 & 0.255 & 0.335 & -
\end{array}\right]
$$

It may be added once again that the assigned values in this example are for demonstration purpose only. Following the remaining steps given in the methodology, the material selection index is calculated using the values of A_i and a_{ij} for each alternative material. The material selection index values of different materials are given below in descending order:

Material 5: 1.635251
Material 4: 1.616897
Material 2: 1.335821
Material 3: 1.316612
Material 1: 1.201707
Material 6: 1.125037

From the above values of the material selection index, it is understood that the material designated as 5 is the right choice for the given material selection problem. The second choice is material 4, and the last choice is material 6.

5.2.2.2 SAW Method

The procedure suggested by Edwards *et al.* (1982) to assess weights for each of the attributes to reflect its relative importance in the material selection decision is followed here. First, the attributes are ranked in order of importance and 10 points are assigned to the least important attribute CR. Cost C is also considered least important, and equal to CR in this example. Then, the next-least important attribute H is chosen, 20 points are assigned to it, and the attribute M is given 30 points to reflect the relative importance. The final weights are obtained by normalizing the sum of the points to one. For example, the weight for attribute M is calculated by 30/(30+20+10+10) = 0.4286. The weights of H, C, and CR are calculated as 0.2857, 0.1428, and 0.1428 respectively. Using these weights and the normalized data of the attributes for different alternative materials, the material selection index values are calculated, and are arranged in descending order of the index.

Material 5:	0.7786
Material 2:	0.6295
Material 3:	0.6193
Material 4:	0.6130
Material 1:	0.6098
Material 6:	0.5789

This method also suggests material 5 as the first choice.

5.2.2.3 WPM

The weights used in the SAW method are used in this method and the values of P_i are calculated. The values of P_i are arranged in descending order as given below:

Material 5:	0.7514
Material 2:	0.6194
Material 3:	0.6074
Material 4:	0.5817
Material 1:	0.5652
Material 6:	0.5222

The ranking suggested by both the SAW method and WPM is the same for this example.

5.2.2.4 AHP and its Versions

The AHP method may use the same weights as those in the SAW method. In that case, the ranking of the materials will be same as that suggested by the SAW method. However, if the decision maker decides to use the AHP method rather than SAW and the weights used in it, then he or she has to make pair-wise comparisons of the attributes to determine the weights (w_j) of the attributes. Let the decision maker prepares the following matrix:

$$A3_{4x4} = \begin{array}{c} \\ H \\ M \\ C \\ CR \end{array} \begin{array}{c} \begin{array}{cccc} H & M & C & CR \end{array} \\ \begin{bmatrix} 1 & 1/3 & 2 & 2 \\ 3 & 1 & 4 & 4 \\ 1/2 & 1/4 & 1 & 1 \\ 1/2 & 1/4 & 1 & 1 \end{bmatrix} \end{array}$$

Following the procedure given in Section 3.2.3 of Chapter 3, the relative normalized weights (w_j) of the attributes are calculated, and these are $W_H = 0.2195$, $W_M = 0.5376$, $W_C = 0.1214$, and $W_{CR} = 0.1214$. The value of λ_{max} is 4.0206, and CR is 0.0077. As the calculated value of CR is less than the allowed CR value of 0.1, there is good consistency in the judgements made. Also there is no contradiction in the judgements.

The value of the material selection index is now calculated using the above weights, and the normalized data of the attributes given in Table 5.6. This leads to the ranking given by the revised AHP or ideal mode of AHP. The materials are arranged in the descending order of the material selection index.

Material 6:	0.7314
Material 5:	0.6777
Material 2:	0.6650
Material 3:	0.5931
Material 4:	0.6483
Material 1:	0.5509

It may be observed that the above ranking is for the given preferences of the decision maker. The ranking depends upon the judgements of relative importance of attributes made by the decision maker.

For the weights of attributes used in this method, the simple AHP as well as the multiplicative AHP methods give the same ranking of materials, *i.e.*, 6-5-2-4-3-1.

5.2.2.5 TOPSIS Method

The quantitative values of the material selection attributes, which are given in Table 5.5, are normalized as explained in Section 3.2.6 and the normalized matrix R_{6x4} is shown below:

$$R_{6x4} = \begin{bmatrix} 0.1868 & 0.0786 & 0.0883 & 0.0730 \\ 0.1556 & 0.1258 & 0.0530 & 0.0560 \\ 0.1734 & 0.0940 & 0.0530 & 0.0628 \\ 0.1112 & 0.1100 & 0.0229 & 0.0560 \\ 0.2668 & 0.0943 & 0.0389 & 0.0560 \\ 0.1023 & 0.1729 & 0.0707 & 0.0421 \end{bmatrix}$$

The relative importance of attributes (a_{ij}) can be assigned values as explained in Section 3.2.6. Using the same weights as those used in SAW method, the ranking of materials is 5-3-2-1-6-4. Using the same weights as those used in the AHP method, the ranking of materials is 6-2-5-4-3-1.

5.2.2.6 Modified TOPSIS Method

Using the same weights as those used in the SAW method, the ranking of materials is 5-2-3-4-6-1. Using the same weights as those used in the AHP method, the ranking of materials is 6-5-2-4-3-1.

It may be noted here that for the same weights of importance of the attributes, all decision-making methods described in this example suggest material 5 as the first right choice. The choice may change when different weights are used.

It is observed, from the application of GTMA and various MADM methods for material selection problems, that the relative importance (*i.e.*, weights) of the material selection attributes decides the ranking of the alternative materials to a significant extent. The decision maker has to be clear about his or her preferences, and choose a particular decision-making method to select the best material for the given engineering application.

References

Amen R, Vomacka P (2001) Case-based reasoning as a tool for materials selection. Materials & Design 22:353–358

Ashby MF (1992) Materials selection in mechanical design. Pergamon Press, New York

Ashby MF (2000) Multi-objective optimization in material design and selection. Materials & Design 48:359–369

Ashby MF, Johnson K (2002) Materials and design: the art and science of materials selection in product design. Butterworth Heinemann, Oxford

Ashby MF, Brechet YJM, Cebon D, Salvo L (2004) Selection strategies for materials and processes. Materials & Design 25:51–67

Bovea MD, Gallardo A (2006) The influence of impact assessment methods on materials selection for eco-design. Materials & Design 27:209–215

Chan JWK, Tong TKL (2006) Multi-criteria material selections and end-of-life product strategy: grey relational analysis approach. Materials & Design doi:10.1016/j.matdes.2006.02.016

Cheng AC, Chen CJ, Chen CY (2006) A fuzzy multiple criteria comparison of technology forecasting methods for predicting the new materials development. Technological Forecasting and Social Change doi:10.1016/j.techfore.2006.08.002

Deng YM, Edwards KL (2007) The role of materials identification and selection in engineering design. Materials & Design 28:131–139

Edwards KL (2002) Linking materials and design: an assessment of purpose and progress. Materials & Design 23:255–264

Edwards KL (2005) Selecting materials for optimum use in engineering components. Materials & Design 26:469–473

Edwards KL, Deng M (2006), Supporting design decision-making when applying materials in combination. Materials & Design doi: 10.1016/ j.matdes.2005.12.009

Edwards W, Newman JR, Snapper K, Seaver D (1982) Multiattribute Evaluation. SAGE Publications, Newbury Park, California

Ermolaeva NS, Kaveline KG, Spoormaker JL (2002) Material selection combined with optimal structural design: concept and some results. Materials & Design 23:459–470

Farag M (1997) Materials selection for engineering design. Prentice-Hall, New York

Giachetti RE (1998) A decision support system for material and manufacturing process selection. Journal of Intelligent Manufacturing 9:265–276

Giudice F, La Rosa G, Risitano A (2005) Materials selection in the lifecycle design process: a method to integrate mechanical and environmental performances in optimal choice. Materials & Design 26:9–20

Guisbiers G, Wautelet M (2007) Materials selection for micro-electromechanical systems. Materials & Design 28:246–248

Jalham IS (2006) Decision-making integrated information technology (IIT) approach for material selection. International Journal of Computer Applications in Technology 25:65–71

Kumar S, Singh R (2006) A short note on an intelligent system for selection of materials for progressive die components. Journal of Materials Processing Technology doi:10.1016/j.jmatprotec.2006.09.004

Kuo TC, Chang SH, Huang SH (2006) Environmentally conscious design by using fuzzy multi-attribute decision-making. International Journal of Advanced Manufacturing Technology 29:209–215

Liao TW (1996) A fuzzy multicriteria decision-making method for material selection. Journal of Manufacturing Systems 15:1–12

Lin F, Lin L (2003) A discussion of the state-of-art research on environmentally conscious material selection methodologies for the reduction of products' toxic impact. The Journal of Sustainable Product Design 3:119–134

Ljungberg LY (2005) Materials selection and design for development of sustainable products. Materials & Design doi:10.1016/j.matdes.2005.09.006

Manshadi BD, Mahmudi H, Abedian A, Mahmudi R (2007) A novel method for materials selection in mechanical design: combination of non-linear normalization and a modified digital logic method. Materials & Design 28:8–15

Matos MJ, Simplicio MH (2006) Innovation and sustainability in mechanical design through materials selection. Materials & Design 27:74–78

Raj R (2000) An interdisciplinary framework for the design and life prediction of engineering systems. Trans ASME, Journal of Engineering Materials Technology 122:348–354

Raj R, Enright MP, Frangopol DM (2000) A system level partitioning approach for analyzing the origins of variability in life prediction of tungsten filaments for incandescent lamps. Materials & Design 21:9–18

Rao RV (2006) A material selection model using graph theory and matrix approach. Materials Science and Engineering A 431:248–255

Rao RV, Davim JP (2007) A decision making framework model for material selection using a combined multiple attribute decision making method. International Journal of Advanced Manufacturing Technology (in press)

Sapuan SM (2001) A knowledge-based system for materials selection in mechanical engineering design. Materials & Design 22:687–695

Shanian A, Savadogo O (2006a) A material selection model based on the concept of multiple factor decision making. Materials & Design 27:329–337

Shanian A, Savadogo O (2006b) TOPSIS multiple-criteria decision support analysis for material selection of metallic bipolar plates for polymer electrolyte fuel cell. Journal of Power Sources 159:1095–1104

Shanian A, Savadogo O (2006c) A non-compensatory compromised solution for material selection of bipolar plates for polymer electrolyte membrane fuel cell (PEMFC) using ELECTRE IV. Electrochimica Acta 51:5307–5315

Trethewey KR, Wood RJK, Puget Y, Roberge PR (1998) Development of a knowledge-based system for materials management. Materials & Design 19:39–56

Zha XF (2005) A web-based advisory system for process and material selection in concurrent product design for a manufacturing environment. International Journal of Advanced Manufacturing Technology 25:233–243

6

Evaluation of Product Designs

6.1 Introduction

Today's world is characterized by major changes in market and economic conditions coupled with rapid advances in technology. As a natural result of this, companies have been forced to develop new products for current markets, principally technology-driven or high-tech markets. The changing economic conditions and technologies combined with increased domestic and global competition, changing customer needs, rapid product obsolescence and emergence of new markets require a very fast innovation process (Ayag, 2005). The final decision to select a particular design for a given product is perhaps the most critical stage in product design development. Obviously, such a decision is influenced by many factors, the specifics of which are not known *a priori* during the design stage. As such, a quantitative basis for comparison and selection of the best design solution among a host of alternatives could greatly impact on the eventual success or failure of a product in the market. The importance of this issue calls for more sophisticated design selection criteria, and methods to incorporate all important factors of interest into the selection of a single final design (Besharati *et al.*, 2006).

Thurston (1991) presented a more formal theory and methodology for design by mathematically modeling the functional relationships between design decisions and the ultimate overall worth of a design. A formal methodology for the evaluation of design alternatives (MEDA) was presented which could be used to evaluate design alternatives in the iterative design/redesign process. Multi-attribute utility analysis was employed to compare the overall utility or value of alternative designs as a function of the levels of several performance characteristics of a manufactured system. The evaluation function reflected the designer's preferences for sets of multiple attributes. A case study of materials selection and design in the automotive industry was presented, which illustrated the steps followed in application of the method.

Hsiao (1998) proposed a fuzzy decision-making method for selecting an optimum design from various design alternatives. The development of a juicer was taken as an example in the study. The evaluation objectives were arranged in a hierarchical structure with several levels. The relative contribution of each objective to the overall value of the solution, and the rating or degree of approximation of a solution with respect to a given objective were quantified with the membership functions of a fuzzy set. A computer program based on a weighted generalized mean method was used to calculate the fuzzy probability level by level from the lowest-level objectives. After the fuzzy expected values of the top-level objectives were calculated, they were used to make a decision quantitatively on selecting the optimal design alternative.

Matsatsinis and Siskos (1999) presented a new methodology for the development of new products, and an intelligent decision support system, named MARKEX, which was an implementation of the methodology. The system acted as a consultant for marketers, providing visual support to enhance understanding and to overcome lack of expertise. The databases of the system were the results of consumer surveys, as well as financial information of the enterprises involved in the decision-making process.

Calantone *et al.* (1999) illustrated the use of the analytic hierarchy process (AHP) as a decision support model to aid managers in selecting new product ideas to pursue. The authors then presented an in-depth example of an actual application of AHP in new product screening, and discussed the usefulness of this process in gathering and processing knowledge for making new product screening decisions. Ozer (1999) conducted a survey on new product evaluation models.

Several market-based decision support methodologies have been reported in the literature to aid product selection (Parameswaran *et al.*, 2001; Choi *et al.*, 2004), single product design selection (Balakrishnan and Jacob, 1995, 1996), and product line design (Alexouda, 2005). The selection criteria in these methods were mostly based on either maximization of the market share, or of the seller's return or minimization of job completion time.

Haque *et al.* (2000) described the development and application of case-based reasoning (CBR) to provide decision support for project managers and engineers during the early phases of new product development in a concurrent engineering (CE) environment. Suh (2001) introduced a metric known as a probability of success in product design which combined the uncertainty in each attribute level with a customer's acceptable range. Edwards (2002) discussed the priorities for concurrent engineering towards more strategic product design for manufacture and assembly.

Hsiao and Chou (2004) presented a creativity-based design process for innovative product design. Gulcin and Orhan (2004a) identified the decision points in the NPD process, and the uncertainty factors affecting those points. Next, the necessary decision models and techniques were determined to help the decision makers to reduce their risks. Finally, an integrated approach based on fuzzy logic to shape the decisions was presented, with an application in software development. In another work, Gulcin and Orhan (2004b) presented the uncertainty factors related to NPD, and proposed an integrated approach based on fuzzy logic, neural networks, and multi-criteria decision making to enable the most appropriate decision making. A case study in a toy manufacturing firm served to demonstrate the potential of the methodology.

Petrick and Echols (2004) proposed that firms adopt a broader heuristic for making new product development choices. The heuristic approach required moving beyond traditional finance-based thinking, and suggested that firms concentrate on technological trajectories by combining technology roadmapping, information technology (IT), and supply chain management to make more sustainable new product development decisions.

Pan and Santner (2004) considered applications where the product design or process design is considered to be seriously flawed if its performance is inferior at any level of the environmental factor. The authors developed a theory for a class of subset selection procedures that identify product designs maximizing the worst-case performance over environmental conditions for general combined array experiments.

Ozer (2005) presented an integrated framework for understanding how various factors affect decision making in new product evaluation, and provided guidelines for reducing their negative impacts on new product decisions. The results indicated that the quality of new product evaluation decisions was affected by four major sets of factors, namely, the nature of the task, the type of individuals who are involved in the decisions, the way the individuals' opinions are elicited, and the way the opinions are aggregated.

Lo *et al.* (2006) reported a new idea-screening method for new product development (NPD), with a group of decision makers having imprecise, inconsistent and uncertain preferences. The authors presented a new method for new product screening in the NPD process by relaxing a number of assumptions, so that imprecise, inconsistent and uncertain ratings could be considered. In addition, a new similarity measure for vague sets was introduced to produce a ratings aggregation for a group of decision makers. The method was able to provide decision makers with consistent information and to model situations where vague and ill-defined information exists in the decision process.

Maddulapalli *et al.* (2006) conducted sensitivity analysis for product design selection with an implicit value function. Besharati *et al.* (2006) proposed a generalized purchase modeling approach that considered three important factors (anticipated market demand for the design, designers' preferences, and uncertainty in achieving predicted design attribute levels under different usage conditions and situations), and developed a customer-based expected utility metric that formed the basis for a decision support system for supporting the selection in product design.

The objective of a product design selection procedure is to identify the product design selection attributes, and obtain the most appropriate combination of the attributes in conjunction with the real requirements. A product design selection attribute is defined as a factor that influences the selection of a product design for a given application. Efforts need to be made to determine attributes which influence product design selection for a given industrial application, using a logical approach, to eliminate unsuitable product designs, and to select a proper product design to strengthen the existing product design selection procedure. Pertinent attributes and the alternative product designs involved are to be identified. Values of the attributes and their relative importance are to be obtained. An objective or subjective value, or its range, may be assigned to each identified attribute as a limiting value, or threshold value, for its acceptance for the considered product design selection problem. An alternative product design with each of its selection attributes, meeting the acceptance value, may be short-listed. After short-listing the alternative product designs, the main task in choosing the alternative product design is to see how it serves the attributes considered.

The next section presents the applications of the GTMA and fuzzy MADM methods for product design selection for a given application.

6.2 Example

Now, to demonstrate and validate the application of proposed decision-making methods, the case study presented by Besharati *et al.* (2006) is considered. Besharati *et al.* (2006) generated a number of product alternatives within the design process. The product attributes are both performance- and market- related, and were obtained using design simulation tools and marketing models. The objective of their work was to present a decision support system (DSS) that aggregates the three factors (market demand, uncertainty in achieving nominal attribute levels and designer's preferences) into a single valued metric. The authors considered the problem of design and selection of a power electronic device based on three performance attributes. The attributes were: manufacturing cost, junction temperature, and thermal cycles to failure. Ten design alternatives were considered that had tradeoffs with respect to one another. Table 6.1 presents the data of the design alternatives.

Table 6.1. Description of design alternatives (from Besharati *et al.*, 2006; reprinted with permission from Elsevier)

Design no.	Junction temperature (°C)	Cycles to failure	Manufacturing cost($)
1	126	22,000	85
2	105	38,000	99
3	138	14,000	65
4	140	13,000	60
5	147	10,600	52
6	116	27,000	88
7	112	32,000	92
8	132	17,000	75
9	122	23,500	85
10	135	15,000	62

6.2.1 Graph Theory and Matrix Approach (GTMA)

The attributes considered are the same as those of Besharati *et al.* (2006), and these are: manufacturing cost (MC), junction temperature (JT), and thermal cycles to failure (CF). The quantitative values of the product design selection attributes, given in Table 6.1, are to be normalized. In this example, CF is a beneficial attribute, and MC and JT are non-beneficial attributes. The values of these attributes are normalized, and are given in Table 6.2 in the respective columns.

Table 6.2. Normalized values of the product design selection attributes

Design no.	JT	CF	MC
1	0.8333	0.5789	0.6118
2	1	1	0.5253
3	0.7609	0.3684	0.8
4	0.75	0.3421	0.8667
5	0.7143	0.2789	1
6	0.9052	0.7105	0.5909
7	0.9375	0.8421	0.5652
8	0.7955	0.4474	0.6933
9	0.8607	0.6184	0.6118
10	0.7778	0.3947	0.8397

Let the designer chooses the following preferences (*i.e.*, relative importance):

	JT	CF	MC
JT	---	0.335	0.255
CF	0.665	---	0.335
MC	0.745	0.665	---

Manufacturing cost (MC) is considered more important to the designer than cycles to failure (CF), than junction temperature (JT). The product design selection attributes digraph, product design selection attributes matrix of the digraph and product design selection function for the matrix can be prepared. However, these are not shown here.

The value of the product design selection index (PDSI) is calculated using the values of A_i and a_{ij} for each product design. The product design selection index values of different product designs are given below in descending order:

2	1.2514
7	1.1373
6	1.0447
9	0.9675
1	0.9234
10	0.8890
8	0.8598
4	0.8439
3	0.8383
5	0.8305

From the values of the product design selection index, it is understood that the product design designated as 2 is the best choice among the considered ten product designs for the given power electronic device. The next choice is 7, and the last choice is 5. However, the ranking obtained using GTMA differs from that obtained by Besharati *et al.* (2006), according to which the first choice was 5. The ranking proposed by Besharati *et al.* (2006) was 5-10-4-3-7-6-2-8-9-1.

A close look at the values of the attributes of the alternative product designs 2 and 5 reveal that 2 is much better than 5 in the case of JT and CF attributes, and 5 is better in the case of the MC attribute. In fact, the values of JT and CF for 5 are worst among all alternative product designs. At the same time, the values are best for 2 among all the alternative product designs considered. It appears that the results obtained by Besharati *et al.* (2006) are biased towards the MC attribute.

In the above case, only the designer's preferences are accounted. Besharati *et al.* (2006) presented a second scenario in which the market information was also accounted for in terms of customer's purchase decision. The specific conditions are given below:

- The device has to endure at least 25,000 cycles, or its junction temperature must remain less than 130°C.

or

- The customer is willing to purchase the device if the price is less than \$170 (*i.e.*, manufacturing cost less than \$70), and it lasts at least 20,000 cycles.

Applying the above conditions to the data in Table 6.2 gives only five product designs, 1, 2, 6, 7, and 9. These designs can be ranked as 2-7-6-9-1 using graph theory and the matrix approach. However, the ranking proposed by Besharati *et al.* (2006) was 7-6-9-1-2. Again, it appears that Besharati *et al.* (2006) had given much more importance to the MC attribute than to CF and JT.

Besharati *et al.* (2006) presented a third scenario 3 with multiple segments. Four customer segments were assumed. The purchase decisions were defined as follows:

- Segment I: The device needs to tolerate at least 20,000 cycles. Its junction temperature should not exceed 130°C. The available budget for this purchase is no more than \$185 per product item.
- Segment II: The desired device needs to have one of the following criteria: endure more than 35,000 cycles, junction temperature lower than 110°C, price less than \$160.
- Segment III: The budget does not exceed \$185 per product item, and the eligible device needs to satisfy either one of the following criteria: lasting more than 20,000 cycles, junction temperature lower than 130°C.
- Segment IV: The desired device should tolerate at least 30,000 cycles, and its junction temperature should not exceed 110°C.

From Table 6.1, it can be understood that product design alternatives 1 and 9 lie within the customer's range of segments I and III; 2 lies within the customer's range of segments II and IV, and 4 and 5 lie within the customer's range of segment II. Of these, product design 2 obtains the higher PDSI.

6.2.2 AHP Method

Let the decision maker prepares the following relative importance matrix:

$$
\begin{array}{c} \\ JT \\ CF \\ MC \end{array}
\begin{bmatrix}
JT & CF & MC \\
1 & 1/3 & 1/5 \\
3 & 1 & 1/3 \\
5 & 3 & 1
\end{bmatrix}
$$

In this example, MC is given comparatively higher importance, and CF is given high importance. The normalized weights of each attribute are calculated, and these are: W_{JT} = 0.1047, W_{CF} = 0.2582, and W_{MC} = 0.6371. The value of λ_{max} is 3.0387 and CR = 0.0372, which is much less than the allowed CR value of 0.1. Thus, there is good consistency in the judgements made.

The value of PDSI is now calculated using the above weights and the normalized data of the attributes given in Table 6.2. This leads to the ranking given by the revised AHP or ideal mode of AHP methods. The alternative product designs are arranged in descending order of the PDSI:

5	0.7837
4	0.7186
10	0.7172
2	0.6962
3	0.6839
7	0.6744
6	0.6534
8	0.6396
9	0.6384
1	0.6254

The AHP method suggests product design 5 as the preferred design when no market information is considered. When market information is also considered, then the ranking will be 2-7-6-9-1 (which is the same as that given by GTMA).

It may be noted that the ranking depends upon the judgements of relative importance of attributes made by the decision maker.

6.2.3 TOPSIS Method

Various steps of TOPSIS methodology using the AHP method for assigning the relative importance of attributes are described below:

Step 1: The objective is to evaluate the alternative product designs for the power electronic device. The attributes considered are the same as those of Besharati *et al.* (2006), and these are: manufacturing cost (MC), junction temperature (JT), and thermal cycles to failure (CF).

Step 2: The next step is to represent all the information available on attributes in the form of a decision matrix. The data are shown in Table 6.1.

Step 3: The quantitative values of the product design selection attributes, which are given in Table 6.1, are to be normalized. CF is a beneficial attribute, and higher values are desirable. MC and JT are non-beneficial attributes, and lower values are desirable. The values of these attributes for different product designs are normalized but are not shown here.

Step 4: Let the decision maker assigns the relative importance weights using the AHP method described in Section 6.2.2. The normalized weights of each attribute are calculated, and these are: W_{JT} = 0.1047, W_{CF} = 0.2582, and W_{MC} = 0.6371. The value of λ_{max} is 3.0387 and CR = 0.0333, which is much less than the allowed CR value of 0.1. Thus, there is good consistency in the judgements made.

Step 5: The weighted normalized matrix is calculated.

$$\begin{bmatrix} 0.0326 & 0.0786 & 0.2202 \\ 0.0272 & 0.1357 & 0.2565 \\ 0.0357 & 0.0500 & 0.1684 \\ 0.0362 & 0.0464 & 0.1555 \\ 0.0380 & 0.0373 & 0.1347 \\ 0.0300 & 0.0964 & 0.2280 \\ 0.0289 & 0.1143 & 0.2384 \\ 0.0342 & 0.0607 & 0.1943 \\ 0.0316 & 0.0839 & 0.2202 \\ 0.0349 & 0.0536 & 0.1607 \end{bmatrix}$$

Step 6: The next step is to obtain the ideal (best) and negative ideal (worst) solutions. These are given as:

$V_1^+ = 0.0272$ $V_1^- = 0.0380$
$V_2^+ = 0.1357$ $V_2^- = 0.0379$
$V_3^+ = 0.1347$ $V_3^- = 0.2565$

Step 7: The next step is to obtain the separation measures, and these are:

$S_1^+ = 0.1030$ $S_1^- = 0.0548$
$S_2^+ = 0.1218$ $S_2^- = 0.0985$
$S_3^+ = 0.0925$ $S_3^- = 0.0889$
$S_4^+ = 0.0921$ $S_4^- = 0.1014$
$S_5^+ = 0.0985$ $S_5^- = 0.1218$
$S_6^+ = 0.1013$ $S_6^- = 0.0656$
$S_7^+ = 0.1059$ $S_7^- = 0.0791$
$S_8^+ = 0.0961$ $S_8^- = 0.0664$
$S_9^+ = 0.1001$ $S_9^- = 0.0590$
$S_{10}^+ = 0.0865$ $S_{10}^- = 0.0972$

Step 8: The relative closeness of a particular alternative to the ideal solution is calculated, and these are:

$P_1 = 0.3473$; $P_2 = 0.4471$; $P_3 = 0.4902$; $P_4 = 0.5241$;
$P_5 = 0.5529$; $P_6 = 0.3933$; $P_7 = 0.4276$; $P_8 = 0.4086$;
$P_9 = 0.3709$; $P_{10} = 0.5292$

This relative closeness to the ideal solution is named as the 'product design selection index (PDSI)' in the present example.

Step 9: The scenarios are arranged in descending order of their PDSI. This can be arranged as 5-10-4-3-2-7-8-6-9-1.

From the above values of PDSI, it is understood that product design 5 is the first choice when no market information is considered. When market information is also considered, then the ranking will be 2-7-6-9-11 (which is the same as that given by the GTMA and AHP methods).

6.2.4 Modified TOPSIS Method

For using the same weights of attributes as those used in the AHP and TOPSIS methods, the modified TOPSIS method leads to the following ranking order:

2	0.5311
7	0.4104
6	0.4154
5	0.4272
10	0.3522
4	0.4382
9	0.3747
3	0.4673
1	0.5618
8	0.3913

The modified TOPSIS method suggests product design 2 as the first choice for the given power electronic device when no market information is considered. When market information is also considered, then the ranking will be 2-7-6-9-1 (which is the same as that given by the GTMA, AHP, and TOPSIS methods).

References

Alexouda G (2005) A user-friendly marketing decision support system for product line design using evolutionary algorithms. Decision Support Systems 38:495–509

Ayag Z (2005) An integrated approach to evaluating conceptual design alternatives in a new product development environment. International Journal of Operations Research 43:687–713

Balakrishnan PV, Jacob VS (1995) Triangulation in decision support systems: algorithms for product design. Decision Support Systems 14:313–327

Balakrishnan PV, Jacob VS (1996) Genetic algorithms for product design. Management Science 42:289–298

Besharati B, Azarm S, Kannan PK (2006) A decision support system for product design selection: a generalized purchase modeling approach. Decision Support Systems 42:333–350

Calantone RJ, Benedetto CAD, Schmidt JB (1999) Using the analytic hierarchy process in new producr screening. Journal of Product Innovation Management 16:65–76

Choi HR, Kim HS, Park BJ, Park YJ, Whinston AB (2004) An agent for selecting optimal order set in EC marketplace. Decision Support Systems 36:371–383

Edwards KL (2002) Towards more strategic product design for manufacture and assembly: priorities for concurrent engineering. Materials & Design 23:651–656

Gulcin GB, Orhan F (2004a) A new approach based on soft computing to accelerate the selection of new product ideas. Computers in Industry 54:151–167

Gulcin GB, Orhan F (2004b) A fuzzy-logic-based decision making approach for new product development. International Journal of Production Economics 90: 27–45

Haque BU, Belecheanu RA, Barson RJ, Pawar KS (2000) Towards the application of case based reasoning to decision-making environment in concurrent product development (concurrent engineering). Knowledge-Based Systems 13:101–112

Hsiao SW (1998) Fuzzy logic based decision model for product design. International Journal of Industrial Ergonomics 21:103–116

Hsiao SW, Chou JR (2004) A creativity-based design process for innovative product design. International Journal of Industrial Ergonomics 34:421–443

Lo CC, Wang P, Chao KM (2006) A fuzzy group-preferences analysis method for new-product development. Expert Systems with Applications 31:826–834

Maddulapalli AK, Azarm S, Boyars A (2006) Sensitivity analysis for product design selection with an implicit value function. European Journal of Operational Research doi:10.1016/j.ejor.2006.03.055

Matsatsinis NF, Siskos Y (1999) MARKEX: An intelligent decision support system for product development decisions. European Journal of Operational Research 113:336–354

Ozer M (1999) A survey of new product evaluation models. International Journal of Innovation Management 16:77–94

Ozer M (2005) Factors which influence decision making in new product evaluation. European Journal of Operational Research 163:784–801

Pan G, Santner J (2004) Theory of screening procedures to identify robust product designs using fractional factorial experiments. Journal of Statistical Planning and Inference 125:59–84

Parameswaran M, Stallaert J, Whinston AB (2001) A market based allocation mechanism for the DiffServ framework. Decision Support Systems 31:351–361

Petrick IJ, Echols (2004) Technology roadmapping in review: a tool for making sustainable new product development decisions. Technological Forecasting & Social Change 71:81–100

Suh NP (2001) Axiomatic design: advances and applications. Oxford University Press, New York

Thurston DL (1991) A formal method for subjective design evaluation with multiple attributes. Research in Engineering Design 3:105–122

7

Machinability Evaluation of Work Materials

7.1 Introduction

Machining operations have been the core of the manufacturing industry since the industrial revolution. Machining is a process of material removal using cutting tools and machine tools to accurately obtain the required product dimensions with good surface finish. The manufacturing industries strive to achieve either a minimum cost of production or a maximum production rate, or an optimum combination of both, along with better product quality in machining. Appropriate selection of work piece and tool materials, machine tools, cutting fluids, cutting conditions, and sequences is a key factor in achieving these goals. Moreover, these goals have gained importance within the context of economic liberalization and globalization.

In general, a manufacturing process for a product consists of several phases such as product design, process planning, machining operations, and quality control. The machinability aspect is related to all phases of manufacturing, especially to process planning and machining operations. The general objective of current research on machinability is to improve all phases of manufacturing by optimizing cost, productivity, and quality. Machinability is a measure of ease with which a work material can satisfactorily be machined. The machinability aspect is of considerable importance for production engineers to know in advance the machinability of work materials so that the processing can be planned in an efficient manner. The study can also be a basis for cutting tool and cutting fluid performance evaluation, and machining parameter optimization.

In the process of product design, material selection is important for realizing the design objective and for reducing production costs. The machinability of engineering materials, owing to the marked influence on the production cost, needs to be taken into account in the product design, although it will not always be a criterion considered top priority in the process of material selection. If there are a finite number of work materials from among which the best material is to be chosen, and if each work material satisfies the required design and functionality of the product, then the main criterion to choose the work material is its operational performance during machining, *i.e.*, machinability.

The basis of machinability evaluation depends on the manufacturer's interest, and many other aspects. For instance, some manufacturers consider tool life as the most important criterion to evaluate machinability, while others consider quality of surface cut the dominant factor. The solution to these difficulties has eluded practicing engineers for decades. Since there is no universally accepted methodology for evaluating machinability, and numerous new materials enter the market every year, many manufacturers are encountering difficulties in selecting the most appropriate material for their products.

Machinability is influenced by the machining process input variables, X_r ($r = 1, 2, \ldots\ldots, k$), and the output variables, Y_q ($q = 1, 2, \ldots, n$), and the output variables are the functions of the input variables.

$$Y_q = f(X_1, X_2, \ldots, X_k) \tag{7.1}$$

The machining process is influenced by a number of variables. One may consider any number of machining process input or output variables for the purpose of machinability evaluation of work materials. Table 7.1 presents the most common machining process input and output variables.

Table 7.1. Machining process variables

Machining process input and output variables
Machining process input variables (process-independent variables): 1. Machine tool (rigidity, capacity, accuracy, *etc.*) 2. Cutting tool (material, coating, geometry, nature of engagement with the work material, tool rigidity, *etc.*) 3. Cutting conditions (speed, feed, and depth of cut) 4. Work material properties (hardness, tensile strength, chemical composition, microstructure, method of production, thermal conductivity, ductility, shape and dimensions of the job, work piece rigidity, *etc.*) 5. Cutting fluid properties and characteristics
Machining process output variables (process-dependent variables): 1. Cutting tool life/tool wear/tool wear rate 2. Cutting forces/specific cutting forces 3. Power consumption/specific power consumption 4. Processed surface finish 5. Processed dimensional accuracy 6. Metal removal rate 7. Noise 8. Vibrations 9. Cutting temperature 10. Chip characteristics

However, it may be added that the machining process input variables may not precisely represent machinability. For example, materials of same composition but different metallographic structure may have different machinability characteristics. Machinability evaluation is based on the evaluation of certain economic and technical objectives (such as higher production rate, low operational cost, good product quality, *etc.*), which are the consequences of the machining operation on a given work material. Machining process output variables are nothing but the behavioral properties of the work materials during machining operations in terms of economic and technical consequences, and are directly related to machining operations, and hence to machinability. These machining process output variables are expressed in quantitative terms for the purpose of comparison. As the machining process output variables are directly related to the machining operations, it is quite appropriate to consider the output variables as the pertinent representatives of the machinability of work materials. Moreover, as the machining process output variables are functions of machining process input variables, the majority of researchers have preferred the machining process output variables for the machinability evaluation of work materials. Thus, the machining process output variables are the pertinent, and most commonly accepted measures of machinability.

A machinability attribute is defined as a machining process variable. It can be any machining process input or output variable that affects the machinability. Machinability evaluation of work materials can be carried out using both types of variables. However, as mentioned above, machining process output variables are the pertinent machinability attributes (Table 7.1). These attributes are common to all machining operations, and only the terminology may vary for cutting tools and cutting forces in the machining operations. For example, the cutting tool is called single point tool in turning/shaping/planning/boring operations, drill in drilling operations, reamer in reaming operations, tap in tapping operations, milling cutter in milling operations, grinding wheel in grinding operations, *etc.* Similarly, the cutting forces are named differently in different machining operations: main cutting force, feed force, and thrust force in turning/shaping/planning/boring operations, torque and thrust in drilling/reaming/tapping operations, tangential force and axial force in milling operations, normal force and tangential force in grinding operations, *etc.*

It is also noted from a literature review on machinability evaluation (Bech, 1963; Konig and Erinski, 1983; Mills and Redford, 1983; Ostafev *et al.* 1989; Malakooti *et al.* 1990; Notoya *et al.* 1990; Trent, 1991; Eyada, 1992; Kato *et al.* 1992; Shanmugam and Krishnamurthy, 1992; Jin and Sandstrom, 1994; Yoshikawa *et al.* 1994; Enache *et al.* 1995; Hung *et al.* 1995; Liao, 1996; Arunachalam and Mannan, 2000; Ong and Chew, 2000; Dravid and Utpat, 2001; Kim *et al.* 2002; Rao and Gandhi, 2002; Boubekri *et al.* 2003; Davim and Reis, 2004; Rech *et al.* 2004; Davim and Mata, 2005; Manna and Bhattacharya, 2005; Rao, 2005; Stoić *et al.* 2005; Özdemir and Özek, 2006; Šalak *et al.* 2006; Şeker and Hasirci, 2006; Morehead *et al.* 2007) that the criteria, in general, for the machinability assessment of different work materials include tool life, tool wear/tool wear rate, cutting forces/specific cutting forces, power consumption/specific energy consumption, processed surface finish, dimensional accuracy of the processed surface, *etc.* So far, research has been based mainly on experimental work to characterize the machinability of work materials. Some researchers have evaluated the machinability of different work materials considering 'any one' of the above criteria only (Ostafev *et al.* 1989; Notoya *et al.*

190; Eyada, 1992; Kato *et al.* 1992; Jin and Sandstrom, 1994; Yoshikawa *et al.* 1994; Hung *et al.* 1995; Dravid and Utpat, 2001). Depending on the techno-needs of a process, some criteria may play a primary or secondary role in the machinability evaluation. However, realistic estimation of the machinability can be carried out only by considering all the criteria and their interrelations. The selection procedures suggested by other researchers considered a number of (*i.e.*, more than one) machining process output variables, and these output variables were examined with respect to the work material properties and characteristics. So far, work materials have been evaluated by researchers considering their performance with respect to each machining process output variable separately, and then the final decision regarding the selection of work material (*i.e.*, machinability evaluation) was taken, in a subjective manner, keeping in mind the overall performance.

It is clear that there is a need to develop a scientific/mathematical tool for machinability evaluation that is capable of considering the requirements of a given machining operation. Considerable work in this direction, *i.e.*, simultaneous consideration of machinability attributes using mathematical models, has been reported by a few researchers (Malakooti *et al.* 1990; Enache *et al.* 1995; Liao, 1996; Ong and Chew, 2000; Rao and Gandhi, 2002; Rao, 2005).

It was recommended by Rao (2005) to short-list various work materials on the basis of satisfying the required design and functionality of the product. Machining process input variables such as work material variables play an important role in short-listing. After short-listing the materials, the main criterion to choose the work material is its operational performance while being machined, *i.e.*, the resulting machining process output variables.

7.2 Examples

Now, to demonstrate and validate the application of decision making-methods, two examples are considered. In both examples, GTMA is applied first, and subsequently a few MADM methods are applied to rank and select the work materials from a machinability point of view.

7.2.1 Example 1

Konig and Erinski (1983) listed and discussed the general machining characteristics of aluminum pressure die-cast and die-cast alloys under various machining conditions for turning, face milling, and drilling operations. The authors used the results of turning data (Bech, 1963) of non-ferrous and ferrous alloys machined with high-speed machining tools. The results are given in Table 7.2. One-hour cutting speeds determined from machining tests on aluminum-magnesium die-cast alloy (GK-A1Mg5) and magnesium-aluminum die-cast alloy (GK-MgA19Zn) are the highest compared with the corresponding values for aluminum-silicon die-cast alloys, gray cast iron (GG26) and carbon steel (C35). The one-hour cutting speeds for aluminum-silicon die-cast alloys are higher than for GG26 and C35. The specific cutting forces (*i.e.*, cutting force per unit area of the material removed) when machining aluminum-silicon die-cast alloys and GK-

A1Mg5 are very low compared to those for GG26 and C35, and for GK-MgA19Zn, the specific cutting force is the lowest. Machining of aluminum-silicon die-casting alloys required power 3 to 4.5 times higher than for GG26, and 1.5 to 2 times higher than C35. The alloys GK-A1Mg5 and GK-MgA19Zn required comparatively very high power.

This example is considered to demonstrate the application of the GTAM and MADM methods.

Table 7.2. Objective data of the attributes of example 7.2.1

Work material	VC (m/min)	CF (N/m^2)	PI (kW)
W1	710	400	28
W2	900	415	38
W3	1630	440	59
W4	1720	235	43
W5	120	1150	8
W6	160	1750	19

W1: GK-AlSi10Mg (aluminum-silicon die-cast alloy)
W2: GK-AlSi6Cu4 (aluminum-silicon die-cast alloy)
W3: GK-AlMg5 (aluminum-magnesium die-cast alloy)
W4: GK-MgAl9Zn (magnesium-aluminum die-cast alloy)
W5: GG26 (gray cast iron with lamellar graphite); W6: C35 (low-carbon steel)
VC: One-hour cutting speed; CF = Specific cutting force; CI = Cutting power input
Cutting conditions: dry, tool material–K10, feed–0.175 mm/rev, and depth of cut–2 mm

7.2.1.1 Application of Graph Theory and the Matrix Approach (GTMA)

Pertinent machinability attributes are identified. The attributes considered are: one-hour cutting speed (VC), specific cutting force (CF), and cutting power input (PI). The quantitative values of these attributes are given in Table 7.2, and these are to be normalized. One-hour cutting speed (VC) is a beneficial attribute. A work material is said to possess higher machinability if it allows very high cutting speeds for a specified tool life. So, higher values are desired. Specific cutting force (CF) and cutting power input (PI) are non-beneficial attributes, and lower values are desirable. The values of the three attributes are normalized, and are given in Table 7.3 in their respective columns. Table 7.3 shows the values of A_i for different work materials.

Table 7.3. Normalized data of the attributes of example 7.2.1

Work material	VC	CF	PI
W1	0.4128	0.5875	0.2857
W2	0.5233	0.5663	0.2105
W3	0.9477	0.5341	0.1356
W4	1	1	0.1860
W5	0.0698	0.2043	1
W6	0.0932	0.1343	0.4211

The relative importance of attributes (*i.e.*, a_{ij}) is assigned values as explained in Section 2.4. Let the decision maker select the following assignments:

$$\begin{array}{c} \\ VC \\ CF \\ PI \end{array} \begin{array}{c} \begin{array}{ccc} VC & CF & PI \end{array} \\ \left[\begin{array}{ccc} - & 0.745 & 0.745 \\ 0.255 & - & 0.5 \\ 0.255 & 0.5 & - \end{array}\right] \end{array}$$

For example, one-hour cutting speed is considered much more important than the specific cutting force in turning operations. This is because the one-hour cutting speed is related to high cutting speeds for a specified tool life of one-hour. If a work material permits high cutting speeds for a specified tool life then production time will be reduced, and production costs will also be reduced. Thus, a one-hour cutting speed is considered very important, compared to the other attributes, *i.e.*, specific cutting force and the cutting power input, and thus a relative importance value of 0.745 is assigned to a one-hour cutting speed over specific cutting force and cutting power input (*i.e.*, a_{12} = 0.745 and a_{13} = 0.745), and a relative importance value of 0.255 is assigned to the specific cutting force (*i.e.*, a_{21} = 0.255) and cutting power input (*i.e.*, a_{31} = 0.255). The specific cutting force and cutting power input are considered as equally important in turning operations and thus equal relative importance is assigned to these attributes (*i.e.*, $a_{23} = a_{32}$ = 0.5). It may be added that these values can be decided by the decision maker, depending on the requirements.

The machinability attributes digraph, machinability attributes matrix of the digraph, and machinability function for the matrix can be prepared. The value of the machinability index is calculated using the values of A_i and a_{ij} for each work material.

The machinability index values of different work materials are given below in descending order:

W4: 0.8513
W3: 0.6228
W2: 0.5308
W1: 0.5284
W5: 0.4504
W6: 0.3241

From the above values of the machinability index, it is clear that the work material W4 (*i.e.*, GK – Mg A19Zn: magnesium-aluminum die-cast alloy) is the best choice among the considered materials for the turning operation under the given conditions. The next choice is W3 (*i.e.*, GK-AlMg5: aluminum-magnesium die-cast alloy), and W6 (*i.e.*, low-carbon steel) is the last choice.

Following graph theory and the matrix approach, the coefficients of similarity/dissimilarity are also calculated for different work materials using Equations 2.15 and 2.16. The coefficient of similarity values are given in Table 7.4. These are useful for work materials documentation, and for easy storage and retrieval of the work materials data for turning operations under the given conditions.

Table 7.4. Values of the coefficient of similarity for the work materials of example 7.2.1

Work material	W2	W3	W4	W5	W6
W1	0.9955	0.8483	0.6206	0.8524	0.6134
W2		0.8522	0.6234	0.8486	0.6106
W3			0.7316	0.7232	0.5204
W4				0.5291	0.3807
W5					0.7196

7.2.1.2 SAW Method

The procedure suggested by Edwards *et al.* 1982) to assess weights for each of the attributes to reflect their relative importance to the work material selection decision is as follows. The attributes are ranked in order of importance, and 10 points are assigned to the least important attribute PI. CF is also considered least important and equal to PI in this example. The attribute VC is given 50 points to reflect its relative importance. The final weights are obtained by normalizing the sum of the points to one. For example, the weight for attribute VC is calculated by 50/(50+10+10) = 0.7142. The weights of CF and PI are calculated as 0.1429 each. Using these weights, and the normalized data of the attributes for different work materials, the machinability index values are calculated, and are arranged in descending order of the index.

W4: 0.8838
W3: 0.7726
W2: 0.4848
W1: 0.4196
W5: 0.2219
W6: 0.1458

The SAW method also suggests W4 as the best machinable work material.

7.2.1.3 WPM

The same weights as were used in the SAW method are selected for this method and the values of machinability index are calculated. The values are arranged in descending order.

W4: 0.7863
W3: 0.5693
W2: 0.4646
W1: 0.4119
W5: 0.1190
W6: 0.1216

The ranking of work materials suggested by both the SAW and WPM methods is the same in this example.

7.2.1.4 AHP and its Versions

If the same weights as were used in the SAW method are selected for in this method, then the ranking of work materials obtained by using the relative as well as ideal mode AHP will be same. The multiplicative AHP method also yields the same ranking.

7.2.1.5 TOPSIS Method

Rao (2005) applied the TOPSIS and AHP methods together for machinability evaluation of work materials. The AHP method was used for finding the weights of importance of the attributes. The procedure is given below:

Step 1: The objective is to evaluate the machinability of different non-ferrous and ferrous alloys. The attributes considered are: VC, CF, and PI. VC is the beneficial attribute, and CF and PI are non-beneficial attributes.

Step 2: The next step is to represent all the information available on attributes in the form of a decision matrix. The data given in Table 7.2 are represented as a matrix $D1_{6x3}$, but not shown here.

Step 3: The quantitative values of the machinability attributes, which are given in Table 7.2, are normalized as explained in Section 3.2.6.

Step 4: The relative importance of attributes (a_{ij}) are assigned values using the AHP method as explained in Section 7.2.4. Let the decision maker select the following assignments:

	VC	CF	PI
VC	1	5	5
CF	1/5	1	1
PI	1/5	1	1

Once again, it may be added that, in actual practice, these values of relative importance can be judiciously decided upon by the user/experts, depending on the requirements. The assigned values in this chapter are for demonstration purposes only.

The normalized weight for each attribute is: $W_{VC} = 0.7142$, $W_{CF} = 0.1429$, and $W_{PI} = 0.1429$. The value of λ_{max} is 3.0 and CR = 0.0, and there exists absolute consistency in the judgements made.

Step 5: The weighted normalized matrix $V1_{6x3}$ is calculated.

$$\begin{bmatrix} 0.1921 & 0.0256 & 0.0448 \\ 0.2435 & 0.0266 & 0.0608 \\ 0.4410 & 0.0282 & 0.0943 \\ 0.4654 & 0.0151 & 0.0688 \\ 0.0325 & 0.0737 & 0.0128 \\ 0.0433 & 0.1122 & 0.0304 \end{bmatrix}$$

Step 6: The next step is to obtain the ideal (best) and the negative ideal (worst) solutions, and these are given as:

$V_{VC}^{+} = 0.4654$ $V_{VC}^{-} = 0.0325$
$V_{CF}^{+} = 0.0151$ $V_{CF}^{-} = 0.1122$
$V_{PI}^{+} = 0.0128$ $V_{PI}^{-} = 0.0943$

Step 7: Here, the separation measures are obtained as:

$S_{W1}^{+} = 0.2753$ $S_{W1}^{-} = 0.1882$
$S_{W2}^{+} = 0.2273$ $S_{W2}^{-} = 0.2302$
$S_{W3}^{+} = 0.0861$ $S_{W3}^{-} = 0.4171$
$S_{W4}^{+} = 0.0559$ $S_{W4}^{-} = 0.4444$
$S_{W5}^{+} = 0.4368$ $S_{W5}^{-} = 0.0902$
$S_{W6}^{+} = 0.4335$ $S_{W6}^{-} = 0.0649$

Step 8: The relative closeness of a particular alternative to the ideal solution (*i.e.*, machinability index) is calculated, and these are:

$P_{W1} = 0.4060$ $P_{W2} = 0.5032$ $P_{W3} = 0.8289$ $P_{W4} = 0.8882$
$P_{W5} = 0.1711$ $P_{W6} = 0.1302$

Step 9: The work materials are arranged in descending order of their machinability index, and this can be arranged as W4-W3-W2-W1-W5-W6.

Thus, the TOPSIS method also suggests W4 as the first right-choice work material from the machinability point of view.

7.2.1.6 Modified TOPSIS Method

In this process, the positive ideal solution (R^{+}) and the negative ideal solution (R^{-}), which are not dependent on the weighted decision matrix, are given by using Equations 3.19 and 3.20.

R_{VC}^{+} = 0.6515 R_{VC}^{-} = 0.0455
R_{CF}^{+} = 0.1055 R_{CF}^{-} = 0.7853
R_{PI}^{+} = 0.0895 R_{PI}^{-} = 0.6603

The weighted Euclidean distances are calculated as

D_{W1}^{+} = 0.3354 D_{W1}^{-} = 0.3245
D_{W2}^{+} = 0.2932 D_{W2}^{-} = 0.3486
D_{W3}^{+} = 0.2205 D_{W3}^{-} = 0.5320
D_{W4}^{+} = 0.1481 D_{W4}^{-} = 0.5770
D_{W5}^{+} = 0.5352 D_{W5}^{-} = 0.2386
D_{W6}^{+} = 0.5636 D_{W6}^{-} = 0.1697

The relative closeness of a particular alternative to the ideal solution is calculated (*i.e.*, machinability index), and these are:

$P_{W1\text{-}mod} = 0.4918$ $P_{W2\text{-}mod} = 0.5432$ $P_{W3\text{-}mod} = 0.7070$
$P_{W4\text{-}mod} = 0.7958$ $P_{W5\text{-}mod} = 0.3083$ $P_{W6\text{-}mod} = 0.2315$

The alternative materials are arranged in descending order of their machinability index. This can be arranged as W4-W3-W2-W1-W5-W6.

7.2.2 Example 2

Enache *et al.* (1995) conducted turning experiments on titanium alloys using different cutting tools of different geometries, and presented a mathematical model for assessing the machinability of various work-tool combinations. The work-tool combinations, experimental conditions, and test results are given in Table 7.5. The various steps of the methodology are carried out as described below.

Table 7.5. Objective data of the machinability attributes of example 7.2.2 (from Enache *et al.,* 1995; with permission from CIRP)

Work-tool combination	Tool wear rate (m/min)	Specific energy consumed (N)	Surface roughness (μm)
1	0.061	219.74	5.8
2	0.093	3,523.72	6.3
3	0.064	2,693.21	6.8
4	0.028	761.46	5.8
5	0.034	1,593.48	5.8
6	0.013	2,849.15	6.2

1: TiAl6V4-P20; 2: TiMo32-P20; 3: TiAl5Fe2.5-P20; 4: TiAl6V4-P20 (TiN); 5: TiAl6V4-K20; 6: TiAl6V4-K20* (K20* is a special form of tool without top in contrast with other tools). Cutting conditions: dry, cutting speed–150 m/min, feed–0.15 mm/rev, and depth of cut–0.5 mm

7.2.2.1 Application of SAW Method

Pertinent machinability attributes are identified. The attributes considered are the same as those of Enache *et al.* (1995), and these are tool wear rate (TW), specific energy consumed (SE), and processed surface roughness (SR). These three attributes are non-beneficial, and low values are most desired. In other words, a work material is said to possess higher machinability if it produces very low values of tool wear rate, specific energy consumption, and surface roughness in the turning operation. The objective values of these attributes, given in Table 7.5, are to be normalized. The values of the three attributes are normalized, and are given in Table 7.6 in their respective columns.

Table 7.6. Normalized data of the attributes of example 7.2.2

Work-tool combination	TW	SE	SR
1	0.2131	1	1
2	0.1398	0.0624	0.9206
3	0.2031	0.0816	0.8529
4	0.4643	0.2886	1
5	0.3824	0.1379	1
6	1	0.0771	0.9355

The same weights as those selected by Enache *et al.* (1995) are used in this method, and the values of machinability index are calculated. The values are arranged in descending order as given below:

6: 0.8213
4: 0.4747
1: 0.4248
5: 0.3865
3: 0.2329
2: 0.1885

The above ranking obtained using the SAW method matches very well with the results presented by Enache *et al.* (1995). These show that TiAl6V4 possesses better machinability than TiMo32 and TiAl5Fe2.5. Comparing the machinability of the same work material with different tools, the maximum machinability is obtained with tool K20*, followed by P20 (TiN), P20, and K20.

7.2.2.2 WPM

The same weights as those selected by Enache *et al.* (1995) are used in this method, and the values of the machinability index are calculated. The values are arranged in descending order as given below:

6: 0.6143
4: 0.4518
5: 0.3412
1: 0.3230
3: 0.1922
2: 0.1399

WPM also suggests that work-tool combination designated by 6 possesses better machinability than the other work-tool combinations in this example.

7.2.2.3 AHP and its Versions

If the same weights as those used in the SAW method are selected in this method, then the ranking of work-tool combinations obtained by using the relative as well as ideal mode AHP will be same. The multiplicative AHP method also yields the same ranking as that given by WPM.

7.2.2.4 TOPSIS Method

Rao (2005) applied the TOPSIS and AHP methods together for machinability evaluation of work materials. The AHP method was used to determine the weights of importance of the attributes. The procedure is given below:

Step 1: The objective is to evaluate the machinability of different titanium work materials and work-tool combinations. The attributes considered are the same as those of Enache *et al.* (1995), namely, tool wear rate (TW), specific energy consumed (SE), and processed surface roughness (SR).

Step 2: The next step is to represent all the information available on attributes in the form of a decision matrix. The data given in Table 7.5 are represented as matrix $D2_{6x3}$, but not shown here.

Step 3: The quantitative values of the machinability attributes, given in Table 7.5, are normalized as explained in Section 3.2.6.

$$\begin{bmatrix} 0.4476 & 0.0395 & 0.3862 \\ 0.6824 & 0.6333 & 0.4195 \\ 0.4696 & 0.4841 & 0.4528 \\ 0.2054 & 0.1369 & 0.3862 \\ 0.2495 & 0.2864 & 0.3862 \\ 0.0954 & 0.5121 & 0.4129 \end{bmatrix}$$

Step 4: The relative importance of attributes (a_{ij}) is assigned. Let the decision maker select the following assignments:

$$\begin{array}{c} \\ TW \\ SE \\ SR \end{array} \begin{array}{c} \begin{array}{ccc} TW & SE & SR \end{array} \\ \begin{bmatrix} 1 & 5 & 7 \\ 1/5 & 1 & 3 \\ 1/7 & 1/3 & 1 \end{bmatrix} \end{array}$$

Once again, it may be added that the assigned values in this example are for demonstration purposes only.The normalized weights of each attribute are W_{TW} = 0.7306, W_{SE} = 0.1884, and W_{SR} = 0.0810. The value of λ_{max} is 3.0649 and CR = 0.0624, which is much less than the allowed CR value of 0.1. Thus, there is good consistency in the judgements made.

Step 5: The weighted normalized matrix $V2_{6x3}$ is calculated.

$$\begin{bmatrix} 0.3270 & 0.0074 & 0.0313 \\ 0.4986 & 0.1193 & 0.0339 \\ 0.3431 & 0.0912 & 0.0367 \\ 0.1501 & 0.0258 & 0.0313 \\ 0.1823 & 0.0539 & 0.0313 \\ 0.0697 & 0.0965 & 0.0334 \end{bmatrix}$$

Step 6: The next step is to obtain the ideal (best) and negative ideal (worst) solutions, and these are given as:

$V_{TW}^{+} = 0.0697$ $V_{TW}^{-} = 0.4986$
$V_{SE}^{+} = 0.0074$ $V_{SE}^{-} = 0.1193$
$V_{SR}^{+} = 0.0313$ $V_{SR}^{-} = 0.0367$

Step 7: Here, the separation measures are obtained as:

$S_{Nr1}^{+} = 0.2573$ $S_{Nr1}^{-} = 0.2049$
$S_{Nr2}^{+} = 0.4432$ $S_{Nr2}^{-} = 0.0027$
$S_{Nr3}^{+} = 0.2860$ $S_{Nr3}^{-} = 0.1579$
$S_{Nr4}^{+} = 0.0825$ $S_{Nr4}^{-} = 0.3608$
$S_{Nr5}^{+} = 0.1218$ $S_{Nr5}^{-} = 0.0891$
$S_{Nr6}^{+} = 0.3230$ $S_{Nr6}^{-} = 0.4295$

Step 8: The relative closeness of a particular alternative to the ideal solution (*i.e.*, machinability index) is calculated, and these are:

$P_{Nr1} = 0.4433$ $P_{Nr2} = 0.0060$ $P_{Nr3} = 0.3558$ $P_{Nr4} = 0.8139$
$P_{Nr5} = 0.7262$ $P_{Nr6} = 0.8252$

Step 9: The work-tool combinations are arranged in descending order of their machinability index, and this can be arranged as 6-4-5-1-3-2.

Thus, the TOPSIS method also suggests Nr6 as the best work-tool combination from the machinability point of view.

7.2.2.5 Modified TOPSIS Method

In this process, the positive ideal solution (R^{+}) and the negative ideal solution (R^{-}), which are not dependent on the weighted decision matrix, are used.

R_{TW}^{+} = 0.0954 R_{TW}^{-} = 0.6824
R_{SE}^{+} = 0.0395 R_{SE}^{-} = 0.6333
R_{SR}^{+} = 0.3862 R_{SR}^{-} = 0.4528

The weighted Euclidean distances are calculated as:

D_{Nr1}^{+} = 0.3009 D_{Nr1}^{-} = 0.3280
D_{Nr2}^{+} = 0.5645 D_{Nr2}^{-} = 0.0094
D_{Nr3}^{+} = 0.3743 D_{Nr3}^{-} = 0.1931
D_{Nr4}^{+} = 0.1032 D_{Nr4}^{-} = 0.4618
D_{Nr5}^{+} = 0.1700 D_{Nr5}^{-} = 0.4000
D_{Nr6}^{+} = 0.2061 D_{Nr6}^{-} = 0.5044

The relative closeness of a particular alternative to the ideal solution is calculated (*i.e.*, machinability index), and these are:

$P_{Nr1\text{-}mod} = 0.5216$ $P_{Nr2\text{-}mod} = 0.01642$ $P_{Nr3\text{-}mod} = 0.3403$
$P_{Nr4\text{-}mod} = 0.8174$ $P_{Nr5\text{-}mod} = 0.7017$ $P_{Nr6\text{-}mod} = 0.7099$

The alternative work-tool combinations are arranged in descending order of their machinability index. This can be arranged as 4-6-5-1-3-2. Thus, the modified TOPSIS method suggests 4 as the first right choice, and 6 as the second choice.

References

Arunachalam R, Mannan MA (2000) Machinability of nickel-based high temperature alloys. Machining Science and Technology 4:127–168

Bech HG (1963) Untersuchung derZerspanbarkeit von Leichtmetallegierungen. Dissertation, RWTH, Aachen

Boubekri N, Rodriguez J, Asfour S (2003) Development of an aggregate indicator to assess the machinability of steels. Journal of Materials Processing Technology 134:159–165

Davim JP, Mata F (2005) A new machinability index in turning fiber reinforced plastics. Journal of Materials Processing Technology 170:436–440

Davim JP, Reis P (2004) Machinability study on composite (polyetheretherketone reinforced with 30% glass fibre–PEEK GF 30) using polycrystalline diamond (PCD) and cemented carbide (K20) tools. International Journal of Advanced Manufacturing Technology 23:412–418

Dravid SV, Utpat LS (2001) Machinability evaluation based on the surface finish criterion. Journal of the Institution of Engineers (India), Production Engineering Division 81:47–51

Edwards W, Newman JR, Snapper K, Seaver D (1982) Multiattribute Evaluation. SAGE Publications, Newbury Park, California

Enache S, Strajescu E, Opran C, Minciu C, Zamfirache M (1995) Mathematical model for the establishment of the materials machinability. CIRP Annals 44:79–82

Eyada OK (1992) Reliability of cutting forces in machinability evaluation. Flexible Automation and Information Management 20:937–946

Hung NP, Boey FYC, Khor KA, Oh CA, Lee HF (1995) Machinability of cast and powder-formed aluminum alloys reinforced with SiC particles. Journal of Materials Processing Technology 48:291–297

Jin L, Sandstrom R (1994) Machinability data applied to materials selection. Materials & Design 15:339–346

Kato K, Tokisue H, Chiba I (1992) Effect of side cutting edge angle of tools on the turning machinability of magnesium alloy castings MC2. Journal of Japan Institute of Light Metals 42:453–458

Kim KK, Kang MC, Kim JS, Jung YH, Kim NK (2002) A study on the precision machinability of ball end milling by cutting speed optimization. Journal of Materials Processing Technology 130-131:357–362

Konig W, Erinski D (1983) Machining and machinability of aluminum cast alloys. CIRP Annals 32:535–540

Liao TW (1996) A fuzzy multi criteria decision making method for material selection. Journal of Manufacturing Systems 15:1–12

Malakooti B, Wang J, Tandler EC (1990) Sensor-based accelerated approach for multiattribute machinability and tool life evaluation. International Journal of Production Research 28:2373–2392

Manna A, Bhattacharya B (2003) A study on machinability of Al/SiC-MMC. Journal of Materials Processing Technology 140:711–716

Mills B, Redford AH (1983) Machinability of engineering materials. Applied Science Publishers, London

Morehead M, Huang Y, Hartwig KT (2007) Machinability of ultrafine-grained copper using tungsten carbide and polycrystalline diamond tools. International Journal of Machine Tools and Manufacture 47:286–293

Notoya H, Yamada S, Yoshikawa K, Takatsuji Y (1990) Effects of tool materials on machinability of commercially pure titanium. Journal of the Japan Institute of Metals 54:596–602

Ong SK, Chew LC (2000) Evaluating the machinability of machined parts and their setup plans. International Journal of Production Research 38:2397–2410

Ostafev VA, Minaev AA, Kokarovtsev VV (1989) Fast method for determining machinability of materials. Soviet Engineering Research 9:113–114

Özdemir N, Özek C (2006) An investigation on machinability of nodular cast iron by WEDM. International Journal of Advanced Manufacturing Technology doi:10.1007/s00170-004-2446-3

Rao RV (2005) Machinability evaluation of work materials using a combined multiple attribute decision making method. International Journal of Advanced Manufacturing Technology 28:221–227

Rao RV, Gandhi OP (2002) Digraph and matrix methods for machinability evaluation of work materials. International Journal of Machine Tools and Manufacture 42:321–330

Rech J, Calvez CL, Dessoly M (2004) A new approach for the characterization of machinability-application to steels for plastic injection molds. Journal of Materials Processing Technology 152:66–70

Šalak A, Vasilko K, Selecká M, Danninger H (2006) New short time face turning method for testing the machinability of PM steels. Journal of Materials Processing Technology 176:62–69

Şeker U, Hasirci H (2006) Evaluation of machinability of austempered ductile irons in terms of cutting forces and surface quality. Journal of Materials Processing Technology 173:260–268

Shanmugam S, Krishnamurthy R (1992) Machinability study on pearlitic spheroidal graphite cast iron. International Journal of Production Research 30:189–197

Stoić A, Kopač J, Cukor G (2005) Testing of machinability of mould steel 40CrMnMo7 using genetic algorithm. Journal of Materials Processing Technology 164-165:1624–1630

Trent EM (1991) Metal cutting. Butterworth-Heinemann, London

Yoshikawa T, Miyazawa S, Mori K (1994) Machinability of Ni3Al-based intermetallic compounds. Journal of Mechanical Engineering Laboratory 48:190–196

8

Cutting Fluid Selection for a Given Machining Application

8.1 Introduction

Much heat is generated in metal cutting operations due to plastic deformation of work materials, friction at the tool-chip interface, and friction between the clearance face of the tool and the work piece. The heat generation increases the temperature of both the work piece and the tool point, resulting in decrease in hardness, and hence tool life. The machined surface will also be less smooth, and the possibility of built-up edge increases. So, the use of a cutting fluid during a machining operation very essential. The major factors that govern the selection of cutting fluids are: (i) the machining process, (ii) cutting tool material, and (iii) work piece material. Other factors, such as compatibility with the machine tool, performance requirements, operator interaction, environment friendliness, and economy must also be looked into.

Nowadays, ever increasing environmental problems are becoming a serious threat to the survival and development of society. After the publishing of ISO 9000 quality management standards, the ISO 14000 environmental management system standards, and the OHSAS 18001 occupational health and safety assessment series, one of our greatest strategic challenges is to apply the three series integrated into a management system in enterprises, not only from an engineering but also from a business and marketing perspective. The manufacturing industry is one of the main roots of environmental pollution. Therefore, minimizing the environmental impact of the manufacturing industry has become an important topic for all manufacturers. During these critical times, an advanced manufacturing mode - green manufacturing - suitable for a sustainable development strategy has been presented. Green manufacturing is a modern manufacturing strategy, essential for 21st century manufacturing industries, integrating all issues of manufacturing, its ultimate goal being to reduce and minimize environmental impact and resource consumption during a product's life cycle, which includes design, synthesis, processing, packaging, transportation, and the use of products in continuous or discrete manufacturing industries.

As cutting fluids are widely used in industrial machining operations, and because of their negative effects on health, safety, and environment, legislation and public environmental concerns now have great impacts on their development. Dry machining and minimum quantity lubrication (MQL) machining have been successfully applied in some kinds of machining processes. However, in others, such as grinding, it is very difficult to obtain good results without the help of cutting fluids, because of the high amount of heat generated during grinding. As for MQL machining, although progress is being made, we have a long way to go before this problem is solved in applications workshops. Therefore, research on the composition, supply techniques, selection, cleaning, and maintenance of cutting fluids is still active at present.

The selection of cutting fluids is more an art, than a science, because there is almost no standardized method available for this purpose. Numerous methods have been proposed in the past, yet very few of these gave reasonably satisfactory results. Different metal cutting operations have been used to evaluate cutting fluids. Nagpal and Sharma (1973) presented the results of a series of short- and long-run cylindrical turning tests for the evaluation of most common, commercially available metal cutting fluids, namely, water soluble, straight mineral, chlorinated and sulfo-chlorinated oils. Peters and Aerens (1976) made an attempt to evaluate grinding fluids based on grinding charts obtained in cylindrical plunge grinding. The authors considered performance parameters, roughness, tangential force, normal force, grinding ratio, specific energy, metal removal rate, tool life, and cost for grinding conditions in the middle of the practical usable range. From the comparison, it appeared that the large variety of grinding fluids offered by the market was not justified commercially or technologically. The use of oils led to a significantly lower cost price, and an increased surface quality in external as well as internal grinding, especially when high wheel speed was used.

de Chiffre (1978) studied a series of hole-making operations (drilling, boring, reaming, and tapping) in order to evaluate different types of cutting fluids. After measuring performance parameters such as number of holes to failure, cutting force, and surface finish, the author concluded that the effectiveness of a coolant greatly depended on the machining process and on the performance measures. Sutcliffe *et al.* (1979) used the criterion of catastrophic drill failure, or a maximum of 120 holes. Different feeds, speeds, and types of cutting fluids were tested, including a nitrite-free synthetic coolant that performed very well.

Rowe (1982) performed cutting fluids testing for cylindrical grinding operations, involving the coordination of various chemical and physical properties of the grinding fluids, their physiological actions and their mechanical performance. A simplified databank was also proposed, allocating each result under one of seven categories, and combining these by means of a software program. Various weighting factors were also applied to the practical requirements of specific grinding processes. Rapp (1984) discussed the general criteria for the selection of cutting fluids for machine tools, and identified the advantages offered by an appropriate selection of cutting fluids (*e.g.*, cost reduction, higher productivity, better safety, lower rate of rejects, and less frequent sharpening of tools).

Yuhta *et al.* (1984) carried out experiments on the grinding abilities of grinding fluids (water miscible and insoluble in water). The effects of each component of the grinding fluid on the grindability were discussed. The experimental results showed that when the commercial grinding fluid with the highest grindability was used, oxy-ferric hydroxide formed in the surface layer of the work piece, and the surface layer was thinner than that produced by means of other grinding fluids. This suggested that when grinding steel with a diamond wheel, grindability was improved under conditions in which oxy-ferric hydroxide was produced in the surface layer of the work piece. Lorenz (1985) compared various cutting fluids by tapping test, taking into consideration also the tapping speed. It was shown that the statistical treatment of torque measurements as a function of cutting speed provides a good comparative basis for assessing cutting fluids when machining a particular material, or a particular group of materials. The author discussed the selection of testing tools and test pieces, and also proposed the standardization of test procedure.

Ghio (1986) studied cutting fluids for operations on metal with flexible abrasive belts. Compared with dry grinding, the use of a cutting oil ensured high production, economy, increased belt life, and a better finish. The author suggested the use of very low-viscosity straight mineral oils for materials that tend to clog the belt, and highly fluid compounded oils for carbon steels and non-ferrous metals. Bennett (1987) also presented results of testing of cutting fluids. The coolants tested were synthetic, semi-synthetic, soluble, and straight grinding oils. The three parameters monitored were grinding ratio, surface finish, and load on the wheel head. Narheim and Kendig (1987) evaluated the cutting fluid effectiveness in machining using electrochemical techniques. A correlation was found between the degree of electro-absorption of surfactants from cutting fluids at metal surfaces, and cutting forces in machining. The effectiveness of cutting fluids, as characterized by cutting forces, was assessed using rapid electrochemical techniques, thereby reducing the need for time-consuming and costly machinability tests.

Wakabayashi and Ogura (1990) evaluated cutting fluids in terms of consumption energy in tapping tests. The consumption energy was estimated by integration of the total torque-time curve as an alternative to the tapping torque commonly used to evaluate cutting fluids. It was pointed out that cutting fluids influence the overall cutting process, rather than causing only a reduction of friction on the interface. Based on consumption energy, it was possible to account for the overall cutting process. Cholakov *et al.* (1992) compared lubricating properties of 16 oil- and water-based fluids, tap water, and air in surface grinding of En9 steel specimens. The oils showed an overall better lubricity, which was less affected by changes of operation parameters. Some water-based fluids, under particular operating conditions, were equal to or better than the oils in lowering forces and in wheel protection, but none achieved the surface quality obtained with oil.

Okuyama *et al.* (1993) studied the cooling action of grinding fluid in shallow grinding. A new method was proposed for measuring the heat transfer coefficient in the vicinity of the wheel-work piece contact zone. The experiments were performed under a variety of conditions during which grinding fluid was supplied and the authors recommended certain measures for increasing the cooling efficiency (*i.e.*, setting the velocity of the coolant higher than the critical value to penetrate the air flow layer formed around the wheel periphery, using a nozzle with a thin throat and attaching a scrapper plate above the nozzle outlet, choosing a wheel of larger grain size, and setting a higher wheel speed). de Chiffre *et al.* (1994) used a reaming test for cutting fluid evaluation, as an alternative to tapping torque measurement and thread finish evaluation. Davinson (1995) provided some guidelines for choosing the correct cutting fluids, and disposing of the used coolants, including waste minimization and elimination. Research into the effects of a coherent cutting fluid jet as opposed to a dispersed jet, upon exit from the nozzle, was carried out by Webster *et al.* (1995). Compared with a dispersed jet, the authors reported that when a coherent jet was maintained, the grinding temperature was reduced.

Sheng and Oberwalleney (1997) reviewed the basic components, performance and health effects, and post-processing options for non-water-miscible and water mixed fluids. Time-based degradation mechanisms for cutting fluid performance were examined, and disposal pre-treatment options for cutting fluids were discussed. Maekawa (1998) reviewed computational aspects of tribological phenomena in metal machining. Emphasis was laid on the interaction between the mechanical aspects of tribology, and the characteristics of the cutting process. Brinksmeier *et al.* (1999) discussed aspects of cooling lubrication reduction in machining advanced materials, *e.g.*, titanium alloys and extreme low-sulfur steels. The authors focused the research on cutting tool performance and wear mechanism at high cutting speeds, while using different lubricants and cooling supply strategies. The investigations contributed to increasing process stability and tool life, improving of machined surface finish, and avoiding tensile residual stresses.

Yamanaka *et al.* (1996) developed a new, easy, and accurate method to evaluate the performance of grinding fluids by means of a block-on test ring machine and an electro-plated CBN wheel. Based on this, the authors developed a new grinding fluid for the CBN wheel. In another work, Yamanaka *et al.* (1997) had provided reasons for the outstanding properties of extreme-pressure agents based on the analysis of specimen surfaces after the tests. Further, Yamanaka *et al.* (1998) evaluated the grinding performances of 11 typical metal working additives, and found that sulfur-type EP additives and phosphorous-type EP additives showed excellent grinding performance, even at low concentration. Yamanaka *et al.* (2000a), reported on whether or not any synergic effect can be observed in grinding performance when two different types of metal working additives are used together. The results showed that there was no synergic effect on grinding performance in a total of 12 cases. Further, Yamanaka *et al.* (2000b) studied the grinding performance of various types of carboxylic acids, and found that among those tested, straight chain saturated higher fatty acids with carbon atom numbers exceeding that of lauric acid are the best in grinding performance.

Ebbrell *et al.* (2000) studied the effects of cutting fluid application method on the grinding process. Results from three experiments with different quantities of cutting fluid passing through the grinding zone were presented. Michigan Technological University developed a cutting fluid evaluation software test bed. Upton (2000) described a new drilling test for the evaluation of cutting fluids. The technique was based on a procedure that relied on gathering performance data from tests using the same drill with different cutting fluids, or lubricant concentrations, rather than on the life time or wear rate of individual tools. Chen *et al.* (2001) presented an analytical model for the prediction of shop floor aerosol generation rate, and particulate size distribution associated with the spin-off motion of cutting fluid from a rotational work piece in a turning operation. The predictive models developed can be used as a basis for human exposure and health hazard analysis.

Belluco and de Chiffre (2001) presented the results of cutting fluid testing through subsequent hole-making operations. AISI 316L stainless steel specimens were machined with drilling, core drilling, reaming and tapping using HSS-E tools. The effect of different lubricants on cutting force and power was investigated in connection with the development of vegetable-based cutting oils. de Chiffre *et al.* (2001) aimed to ream austenitic stainless steel using water-based fluids, and to evaluate the effect of cutting fluid on cutting forces, surface finish, and hole diameter. Results showed that torque and thrust measurements offer a reliable description of the lubricating properties of cutting fluids, while conventional surface roughness evaluation was associated with a large scatter in the data. Eppert *et al.* (2001) presented a methodology using the cluster analysis in a hierarchical agglomerative form, for the development of a classification scheme based on physical properties of a wide array of cutting fluids. Bartz (2001) described ecological and environmental aspects of cutting fluids, and suggested that all components, base oils and additives, have to be selected very carefully in order to minimize any health problems and any impact to the environment. Rao and Gandhi (2001) presented a cutting fluid selection index using digraph and matrix methods, which can serve for the evaluation and selection of cutting fluids.

Sun *et al.* (2001) presented a two-grade fuzzy synthetic decision-making method using AHP for evaluation of grinding fluids. Varadarajan *et al.* (2002) investigated hard turning operations with MQL, and made a comparison with dry and wet turning. Tan *et al.* (2002) presented a decision-making framework model for cutting fluid selection for green manufacturing, together with a case study. Sokovic and Mijanovic (2001) studied ecological aspects of cutting fluids, and their influence on quantifiable parameters of cutting processes. Rao (2004) presented a combined MADM method for the selection of environmentally conscious cutting fluids using the TOPSIS and AHP methods.

Dhar *et al.* (2006) studied the effect of minimum quantity lubrication (MQL) on tool wear and surface roughness while machining AISI 4340 steel. In another work, Dhar and Kamruzzaman (2006) conducted turning experiments on AISI 4037 steel using cryogenic cooling by liquid nitrogen jets. Haq and Tamizharasan (2005) investigated the effects of cooling in hard turning operations. Reddy and Rao (2006) studied the effects of solid lubricants on cutting forces and surface quality in end milling. The results indicated that there was a considerable improvement in process performance with solid lubricant-assisted machining, compared to that of machining with cutting fluids. Obikawa *et al.* (2006) investigated high-speed grooving operations with minimum quantity lubrication (MQL). Heinemann *et al.* (2006) studied the effect of MQL on the tool life of small twist drills in deep-hole drilling.

It is evident from the above that existing procedures of cutting fluid selection for a given machining application focus mainly on identifying the cutting fluid matching with a tool, work material, and machining operation. Different metal cutting operations have been used to evaluate cutting fluids, and the performance of a cutting fluid judged by the resulting machining process output variables such as: tool life (*i.e.*, life of single point tool in turning/boring, drill in drilling, reamer in reaming, tap in tapping, grinding wheel in grinding), cutting forces (*i.e.*, main cutting force and/or thrust in turning/boring, torque and/or thrust in drilling/reaming/tapping, normal force and/or tangential force in grinding), power consumption, cost per unit volume of material removed, surface finish, cutting temperature, dimensional accuracy, *etc.* The selection procedures suggested by earlier researchers considered either a single machining process output variable, or a number of machining process output variables, and these output variables were examined with respect to cutting fluid properties and characteristics. So far, cutting fluids have been evaluated in terms of their performance with respect to each machining process output variable separately, and then the final decision regarding selection was taken, in a subjective manner, keeping in mind the overall performance. It is clear that there is a need to develop a mathematical tool for cutting fluid selection that is capable of considering the requirements of a given machining application. The objective of a cutting fluid selection procedure is to identify cutting fluid properties, and obtain the most appropriate combination of cutting fluid properties in conjunction with the real requirement of a machining application. Thus, efforts need to be extended to determine attributes that influence cutting fluid selection for a given machining application, using a logical approach, to eliminate unsuitable cutting fluids and to select an appropriate cutting fluid to strengthen the existing cutting fluid selection procedure. A few researchers, such as Rowe (1982), Sun *et al.* (2001), Rao and Gandhi (2001), Tan *et al.* (2002) and Rao (2004), have presented some mathematical models for cutting fluid selection.

A cutting fluid attribute is defined as a property or characteristic of the cutting fluid, or a machining process variable on which the cutting fluid has influence. Cutting fluid attributes can be broadly classified into two types, and are listed below:

1. Cutting fluid properties and characteristics such as viscosity, viscosity index, composition, flash point, specific heat, thermal conductivity, lubricity, durability, film formation, anti-foaming characteristics, anti-contamination characteristics, cooling capacity, evaporation rate, toxicity, degradation, disposability, corrosion resistance, compatibility, cost of cutting fluid, molecular size, thermal stability, emulsion stability, chemical stability, handling qualities, physiological properties, operator acceptability, and ecological and environmental characteristics.

2. Machining process variables on which the cutting fluid has influence such as tool life (*i.e.*, life of single point tool in turning/boring, drill in drilling, reamer in reaming, tap in tapping, grinding wheel in grinding), cutting forces (*i.e.*, main cutting force and/or thrust in turning/boring, torque and/or thrust in drilling/reaming/tapping, normal force and/or tangential force in grinding), power consumption, cost per unit volume of material removed, surface finish, cutting temperature, dimensional accuracy, metal removal rate, *etc.*

Rao (2004) proposed that the cutting fluids be short-listed for a given machining application, on the basis of cutting fluid attributes of first type, *i.e.*, properties or characteristics of the cutting fluid satisfying the machining application requirements. *The machining application involves the machining process, tool, and work materials.* An objective or subjective value, or its range, may be assigned to each identified attribute as a limiting value, or threshold value, for acceptance in the cutting fluid selection problem considered. A cutting fluid with each of its selection attribute, meeting the acceptance value, may be short-listed. After short-listing, the main criterion to choose the cutting fluid for a given machining application is its operational performance during machining. The operational performance of the cutting fluid is indicated by the cutting fluid attributes of second type, *i.e.*, machining process output variables.

The next section describes the application of graph theory and the matrix approach, and fuzzy MADM methods for cutting fluid selection in a given machining application.

8.2 Examples

Now, to demonstrate and validate the application of decision making methods, two examples are considered. For a start, GTMA is applied, and subsequently a few MADM methods are applied to rank and select the cutting fluids for a given machining application.

8.2.1 Example 1

A cylindrical grinding operation is considered in which four grinding fluids are tested. Eight cutting fluid attributes are considered, of which four are the machining process output variables wheel wear (WW), tangential force (TF), grinding temperature (GT), and surface roughness (SR), and four are the cutting fluid properties and characteristics recyclability (R), toxic harm rate (TH), environment pollution tendency (EP), and stability (S). The cutting fluid properties and characteristics are expressed in linguistic terms. Table 8.1 presents the data on cutting fluid selection attributes for the four grinding fluids tested.

Table 8.1. Data of cutting fluid selection attributes of example 8.2.1

Cutting fluid	WW (mm)	TF (N)	GT (°C)	SR (μm)	R	TH	EP	S
1	0.035	34.5	847	1.76	L	A	AA	AA
2	0.027	36.8	834	1.68	L	H	H	H
3	0.037	38.6	808	2.40	AA	AA	BA	A
4	0.028	32.6	821	1.59	A	AA	AA	BA

L: Low; BA: Below average; A: Average; AA: Above average; H: High

The linguistic terms are converted to fuzzy scores as explained in Chapter 4 using Table 4.3. Table 8.2 presents the objective data of cutting fluid selection attributes accordingly.

Table 8.2. Objective data of cutting fluid selection attributes of example 8.2.1

Cf	WW	TF	GT	SR	R	TH	EP	S
1	0.035	34.5	847	1.76	0.335	0.500	0.590	0.590
2	0.027	36.8	834	1.68	0.335	0.665	0.665	0.665
3	0.037	38.6	808	2.40	0.590	0.590	0.410	0.500
4	0.028	32.6	821	1.59	0.500	0.590	0.590	0.410

Cf: Cutting fluid

8.2.1.1 Application of Graph Theory and Matrix Approach (GTMA)

Various steps of the methodology, proposed in Section 2.6, are carried out as described below.

In the present work, the attributes considered are wheel wear (WW), tangential force (TF), grinding temperature (GT), surface roughness (SR), recyclability (R), toxic harm rate (TH), environment pollution tendency (EP), and stability (S). The objective values of the cutting fluid selection attributes, which are given in Table 8.2, are to be normalized. R and S are beneficial attributes, and higher values are desirable. Values of these attributes are normalized, as explained in Section 2.4, and are given in Table 8.3 in the respective columns. WW, TF, GT, SR, TH, and EP are non-beneficial attributes and lower values are desirable. The values of these attributes for different cutting fluids are normalized, and given in Table 8.3 in the respective columns.

Table 8.3. Normalized data of cutting fluid selection attributes of example 8.2.1

Cf	WW	TF	GT	SR	R	TH	EP	S
1	0.7714	0.9449	0.9539	0.9034	0.5678	1	0.6949	0.8872
2	1	0.8859	0.9688	0.9464	0.5678	0.7519	0.6165	1
3	0.7297	0.8445	1	0.6625	1	0.8475	1	0.7519
4	0.9643	1	0.9842	1	0.8475	0.8475	0.6949	0.6165

Cf: Cutting fluid

Relative importance of attributes (a_{ij}) is also assigned the values as explained in Section 2.4. Let the decision maker (*i.e.*, user organization) makes the following assignments:

	WW	TF	GT	SR	R	TH	EP	S
WW	-	0.745	0.665	0.745	0.745	0.665	0.665	0.745
TF	0.255	-	0.335	0.5	0.59	0.41	0.41	0.59
GT	0.335	0.665	-	0.665	0.665	0.59	0.59	0.665
SR	0.255	0.5	0.335	-	0.59	0.41	0.41	0.59
R	0.255	0.41	0.335	0.41	-	0.335	0.335	0.5
TH	0.335	0.59	0.41	0.59	0.665	-	0.5	0.665
EP	0.335	0.59	0.41	0.59	0.665	0.5	-	0.665
S	0.255	0.41	0.335	0.41	0.5	0.335	0.335	-

However, it may be added that the above-assigned values are for demonstration purposes only.

The cutting fluid attributes digraph, cutting fluid attributes matrix of the digraph and cutting fluid function for the matrix can be prepared. The value of the cutting fluid selection index is calculated using the values of A_i and a_{ij} for each cutting fluid. The cutting fluid selection index values of different cutting fluids are given below in descending order:

Cutting fluid 4: 246.8591
Cutting fluid 3: 238.2171
Cutting fluid 2: 233.2670
Cutting fluid 1: 231.1462

From the above values of the cutting fluid selection index, it is clear that the cutting fluid, designated as 4 is the best choice among the cutting fluids considered for the cylindrical grinding operation under the given conditions. The next choice is cutting fluid 3, and cutting fluid 1 is the last choice. It may be observed that this ranking is based upon simultaneous consideration of the machining process output variables on which the cutting fluid has influence, as well as the environmental properties and characteristics of the cutting fluids.

Following graph theory and the matrix approach, the coefficients of similarity/dissimilarity are also calculated for different cutting fluids, using Equations 2.15 and 2.16. The coefficient of similarity values are given in Table 8.4. These are useful for cutting fluids documentation, for easy storage, and for retrieval of cutting fluids data for cylindrical grinding operations under the given conditions.

Table 8.4. Values of coefficient of similarity for the cutting fluids of example 8.2.1

Cutting fluid	2	3	4
1	0.9909	0.9703	0.9363
2		0.9792	0.9449
3			0.9650

8.2.1.2 SAW Method

The procedure suggested by Edwards *et al.* (1982) to assess weights for each of the attributes to reflect relative importance to the cutting fluid selection decision is followed. The attributes are ranked in order of importance and 10 points are assigned to the least important attribute S. R is also considered least important and equal to S in this example. The attribute WW is given 60 points to reflect its relative importance. GT is given 30 points, TH and EP are given 25 points each and TF and SR are given 15 points each. The final weights are obtained by normalizing the sum of the points to one. For example, the weight for attribute WW is calculated by 60/(60+30+25+25+15+15+10+10) = 0.316. The weight for attribute GT is 0.158, the weights for TH and EP are 0.132 each, those for TF and SR are 0.079 each, and those for S and R are 0.053 each. Using these weights and the normalized data of the attributes for different cutting fluids, the cutting fluid selection index values are calculated, and are arranged in descending order.

Cutting fluid 4: 0.8994
Cutting fluid 2: 0.8775
Cutting fluid 3: 0.8443
Cutting fluid 1: 0.8413

From the above values of the cutting fluid selection index, it is clear that the cutting fluid, designated as 4 is the best choice among the cutting fluids considered for the cylindrical grinding operation under the given conditions.

8.2.1.3 WPM

Using the same weights of attributes as selected for the SAW method, the following ranking of cutting fluids is obtained:

Cutting fluid 4: 0.8884
Cutting fluid 2: 0.8603
Cutting fluid 3: 0.8332
Cutting fluid 1: 0.8300

The ranking is the same as that obtained by using the SAW method.

8.2.1.4 AHP and its Versions

If the same weights as those used in the SAW method are selected for this method, then the ranking of cutting fluids obtained by using the relative as well as ideal mode AHP will be the same. The multiplicative AHP method yields the same ranking as that given by WPM. However, if the decision maker decides to use the AHP method, rather than the weights used in the SAW method, then he or she has to make pair-wise comparisons of the attributes to determine the weights (w_j) of the attributes. Let the decision maker prepare the following matrix:

	WW	TF	GT	SR	R	TH	EP	S
WW	1	5	3	5	5	3	3	4
TF	1/5	1	1/3	1	2	1/2	1/2	2
GT	1/3	3	1	3	3	2	2	3
SR	1/5	1	1/3	1	2	1/2	1/2	2
R	1/5	1/2	1/3	1/2	1	1/3	1/3	1
TH	1/3	2	1/2	2	3	1	1	3
EP	1/3	2	1/2	2	3	1	1	3
S	1/5	1/2	1/3	1/2	1	1/3	1/3	1

Wheel wear (WW) is strongly more important than the tangential force (TF) in the grinding operation. Reducing WW is strongly more important than reducing TF. Attention should be paid to reducing the value of WW so as to reduce the machining cost. So, a relative importance value of 5 is assigned to WW over TF (*i.e.*, a_{12} = 5), and a relative importance value of 1/5 is assigned to TF over WW (*i.e.*, a_{21} = 1/5). Wheel wear (WW) is moderately more important than the grinding temperature (GT). So, a relative importance value of 3 is assigned to WW over GT (*i.e.*, a_{13} = 3), and a relative importance value of 1/3 is assigned to GT over WW (*i.e.*, a_{31} = 1/3). Similarly, the relative importance among other attributes can be explained. It may be added that these values are to be arrived at judiciously after careful analysis. The assigned values in this chapter are for demonstration purposes only.

The normalized weights of each attribute are calculated following the procedure presented in Section 3.2.3, and these are Www = 0.3306, W_{TF} = 0.0718, W_{GT} = 0.1808, W_{SR} = 0.0718, W_R = 0.0459, W_{TH} = 0.1260, W_{EP} = 0.1260, and W_S = 0.0472. The value of λ_{max} is 8.19 and CR = 0.0194, which is much less than the allowed CR value of 0.1. Thus, there is good consistency in the judgements made.

The value of the cutting fluid selection index is now calculated using the above weights, and the normalized data of the attributes given in Table 8.3. This leads to the ranking given by the revised AHP or ideal mode of AHP methods. The alternative cutting fluids are arranged in descending order of the cutting fluid selection index.

Cutting fluid 4: 0.9027
Cutting fluid 2: 0.8830
Cutting fluid 3: 0.8444
Cutting fluid 1: 0.8417

From the above values of the cutting fluid selection index, it is clear that the cutting fluid designated as 4 is the best choice among the cutting fluids considered for the cylindrical grinding operation under the given conditions.

For the above weights of importance of attributes, multiplicative AHP also leads to the same ranking order of 4-3-2-1.

It may be observed that the above ranking is for the given preferences of the decision maker. The ranking depends upon the judgements of relative importance of attributes made by the decision maker.

8.2.1.5 TOPSIS Method

Step 1: The objective is to evaluate the four alternative cutting fluids, the pertinent attributes considered being WW, TF, GT, SR, R, TH, EP, and S.

Step 2: The next step is to represent all the information available on attributes in the form of a decision matrix. The data given in Table 8.2 can be represented as matrix D_{4x8}. However, the matrix is not shown here, as it is nothing but the repetition of data given in Table 8.2 but represented in a matrix form.

Step 3: The quantitative values of the flexible manufacturing system selection attributes, which are given in Table 8.2, are normalized as explained in Section 3.2.6.

Step 4: Relative importance of attributes (a_{ij}) is assigned using the AHP method as explained in Section 8.2.1.3, and these are Www = 0.3306, W_{TF} =

0.0718, W_{GT} = 0.1808, W_{SR} = 0.0718, W_R = 0.0459, W_{TH} = 0.1260, W_{EP} = 0.1260, and W_S = 0.0472. The value of λ_{max} is 8.19 and CR = 0.02, which is much less than the allowed CR value of 0.1. Thus, there is good consistency in the judgements made.

Step 5: The weighted normalized matrix, V_{4x8} is calculated, and is shown below:

$$\begin{bmatrix} 0.1806 & 0.0347 & 0.0925 & 0.0335 & 0.0170 & 0.0535 & 0.0650 & 0.0253 \\ 0.1393 & 0.0370 & 0.0911 & 0.0320 & 0.0170 & 0.0710 & 0.0732 & 0.0285 \\ 0.1909 & 0.0388 & 0.0883 & 0.0457 & 0.0298 & 0.0631 & 0.0450 & 0.0215 \\ 0.1445 & 0.0328 & 0.0897 & 0.0303 & 0.0253 & 0.0631 & 0.0650 & 0.0176 \end{bmatrix}$$

Step 6: The next step is to obtain the ideal (best) and negative ideal (worst) solutions, and these are given as:

V_{WW}^+ = 0.1393 V_{WW}^- = 0.1909
V_{TF}^+ = 0.0328 V_{TF}^- = 0.0388
V_{GT}^+ = 0.0883 V_{GT}^- = 0.0925
V_{SR}^+ = 0.0303 V_{SR}^- = 0.0457
V_R^+ = 0.0298 V_R^- = 0.0169
V_{TH}^+ = 0.0535 V_{TH}^- = 0.0711
V_{EP}^+ = 0.0452 V_{EP}^- = 0.0733
V_S^+ = 0.0285 V_S^- = 0.0176

Step 7: The next step is to obtain the separation measures, and these are:

S_1^+ = 0.0480 S_1^- = 0.0267
S_2^+ = 0.0360 S_2^- = 0.0545
S_3^+ = 0.0555 S_3^- = 0.0325
S_4^+ = 0.0256 S_4^- = 0.0514

Step 8: The relative closeness of a particular alternative to the ideal solution is calculated and these are P_1 = 0.3571, P_2 = 0.6024, P_3 = 0.3691, and P_4 = 0.6675.

This relative closeness to ideal solution can be named 'cutting fluid selection index' in the present work.

Step 9: The alternative cutting fluids are arranged in descending order of their cutting fluid selection index. This can be arranged as 4-2-3-1.

8.2.1.6 Modified TOPSIS Method

In this method, the positive ideal solution (R^+) and the negative ideal solution (R^-) are used, and the values are given below:

R_{WW}^+	=	0.4213	R_{WW}^-	=	0.5774
R_{TF}^+	=	0.4566	R_{TF}^-	=	0.5407
R_{GT}^+	=	0.4881	R_{GT}^-	=	0.5117
R_{SR}^+	=	0.4218	R_{SR}^-	=	0.6367
R_R^+	=	0.6505	R_R^-	=	0.3694
R_{TH}^+	=	0.4243	R_{TH}^-	=	0.5644
R_{EP}^+	=	0.3587	R_{EP}^-	=	0.5818
R_S^+	=	0.6049	R_S^-	=	0.3729

The weighted Euclidean distances are calculated as

D_1^+	=	0.1114	D_1^-	=	0.0831
D_2^+	=	0.1127	D_2^-	=	0.1152
D_3^+	=	0.1169	D_3^-	=	0.1041
D_4^+	=	0.0833	D_4^-	=	0.1138

The relative closeness of a particular alternative to the ideal solution is calculated (*i.e.*, cutting fluid selection index), and these are:

$P_{1\text{-mod}} = 0.4272$ $P_{2\text{-mod}} = 0.5054$ $P_{3\text{-mod}} = 0.4709$ $P_{4\text{-mod}} = 0.5774$

The alternative cutting fluids are arranged in the descending order of their cutting fluid selection index. This can be arranged as: 4-2-3-1.

8.2.2 Example 2

The results of a cylindrical turning test are presented in Table 8.5. This test is conducted for the purpose of evaluation of most common, commercially available metal cutting fluids, namely, water soluble, straight mineral, chlorinated and sulfo-chlorinated oils.

Table 8.5. Data of cutting fluid attributes of example 8.2.2

Cutting fluid	F_c (N)	F_t (N)	WL (mm*100)	R_{rms} (μm)
Dry	1,324	725	7	9
Water soluble	1,082	485	16	7
Straight mineral oil	1,098	516	8	4.7
Chlorinated oil	1,158	494	15	4.9
Sulfo-chlorinated oil	962	393	6	8

F_c: Cutting force; F_t: Thrust force; WL: Wear land; R_{rms}: Processed surface roughness expressed in rms value.
Work material: medium-carbon steel; Tool: HSS; Cutting conditions: speed–33.5 m/min, feed–0.24 mm/rev

This example is considered to demonstrate further the application of the GTMA and MADM methods for cutting fluid selection.

8.2.2.1 GTMA

In the present work, the attributes considered are cutting force (FC), thrust force (TF), wear land (WL), and processed surface roughness (R). The objective values of the cutting fluid selection attributes, which are given in Table 8.5, are to be normalized. All four attributes are of non-beneficial type, and lower values are desirable. Values of these attributes are normalized, as explained in Section 2.4, and are given in Table 8.6 in the respective columns.

Table 8.6. Normalized data of cutting fluid attributes of example 8.2.2

Cutting fluid	FC (N)	TF (N)	WL (mm * 100)	R (μm)
Dry	0.7251	0.5489	1	0.5222
Water soluble	0.9074	0.8206	0.4375	0.6714
Straight mineral oil	0.8743	0.7713	0.875	1
Chlorinated oil	0.8290	0.8057	0.4667	0.9592
Sulfo-chlorinated oil	1	1	1	0.5875

Relative importance of attributes (a_{ij}) is assigned values as explained in Section 2.4. Let the decision maker select the following assignments:

	FC	TF	WL	R
FC	---	0.665	0.335	0.590
TF	0.335	---	0.255	0.410
WL	0.665	0.745	---	0.665
R	0.410	0.590	0.335	---

Finally, the cutting fluid selection index values of different cutting fluids are calculated, and are given below in descending order:

Sulfo-chlorinated oil	2.8871
Straight mineral oil	2.8172
Chlorinated oil	2.1483
Water soluble	1.9204
Dry	1.9076

From the above values of the cutting fluid selection index, it is understood that the sulfo-chlorinated oil is the best choice among the cutting fluids considered for the cylindrical turning operation under the given conditions. The last choice is dry cutting.

Following graph theory and the matrix approach, the coefficients of similarity/dissimilarity are also calculated, and are given in Table 8.7.

Table 8.7. Values of coefficient of similarity for the cutting fluids of example 8.2.2

Cutting fluid	Water Soluble	Straight min. oil	Chl. oil	Sulfo-chlorinated oil
Dry	0.9933	0.6671	0.8879	0.6607
Water soluble		0.6817	0.8939	0.6652
Straight mineral oil			0.7626	0.9758
Chlorinated oil				0.7441

8.2.2.2 SAW Method

The procedure suggested by Edwards and Newman (1982) to assess weights for each of the attributes to reflect relative importance to the cutting fluid selection decision is followed here. The attributes are ranked in order of importance, and 10 points are assigned to the least important attribute TF. R is considered the next-least important attribute, and is given 20 points. FC is given 30 points, and WL 40 points. The final weights are obtained by normalizing the sum of the points to one. For example, the weight for attribute WL is calculated by 40/(40+30+20+10) = 0.40. The weight for attribute FC is 0.30, that for R 0.20 and that for TF 0.10. Using these weights, and the normalized data of the attributes for different cutting fluids, the cutting fluid selection index values are calculated, and are arranged in descending order of the index.

Sulfo-chlorinated oil	0.9588
Straight mineral oil	0.8561
Water soluble	0.7638
Chlorinated oil	0.7626
Dry	0.7069

The SAW method also suggests sulfo-chlorinated oil as the first choice for the cylindrical turning operation under the given conditions.

8.2.2.3 WPM

Using the same weights of attributes as those selected for the SAW method, the following ranking of cutting fluids is obtained:

Sulfo-chlorinated oil	0.9482
Straight mineral oil	0.8536
Chlorinated oil	0.7435
Water soluble	0.7383
Dry	0.6883

This method also suggests sulfo-chlorinated oil as the right choice in this example.

8.2.2.4 AHP and its Versions

If the same weights as those used in the SAW method are selected for this method, then the ranking of cutting fluids obtained by using the relative as well as ideal mode AHP methods will be same. The multiplicative AHP method yields the same ranking as that given by WPM.

8.2.2.5 TOPSIS Method

Following the steps of the TOPSIS method, the following ranking is obtained:

Sulfo-chlorinated oil	0.7976
Straight mineral oil	0.7933
Dry	0.6502
Chlorinated oil	0.2979
Water soluble	0.2108

This method also suggests sulfo-chlorinated oil as the right choice. However, the water-soluble fluid is shown as the last choice (unlike dry cutting as given by the other methods). This may be due to TOPSIS being normally biased towards the alternative having a higher value of attribute with higher relative importance. In this example, attribute WL is given maximum weight of importance, and as far as this attribute is concerned, dry cutting is better than the water-soluble fluid.

8.2.2.6 Modified TOPSIS Method

Following the steps of the modified TOPSIS method, the following ranking is obtained:

Sulfo-chlorinated oil	0.7892
Straight mineral oil	0.7285
Chlorinated oil	0.4500
Dry	0.4468
Water soluble	0.4084

This method also suggests sulfo-chlorinated oil as the right choice in this example.

References

Bartz WJ (2001) Ecological and environmental aspects of cutting fluids. Lubrication Engineering 57:13–16

Belluco W, de Chiffre L (2001) Testing of vegetable based cutting fluids by hole making operations. Lubrication Engineering 57:12–16

Bennett BG (1987) Effects of coolants and application in using vitrified CBN grinding wheels. Carbide and Tool Journal 19:23–26.

Brinksmeier E, Waiter A, Janssen R, Diersen P (1999) Aspects of cooling lubrication reduction in machining advanced materials. Proc. IME., Journal of Engineering Manufacture 213:769–778.

Chen Z, Wong K, Li W, Liang SY, Stephenson DA (2001) Cutting fluid aerosol generation due to spin-off in turning operation: analysis for environmentally conscious machining. Journal of Manufacturing Science and Engineering 123:507–512

Cholakov GS, Guest TL, Rowe GW (1992) Lubricating properties of grinding fluids. 1. Comparison of fluids in surface grinding experiments. STLE Transactions 48:155–163

Davinson JF (1995) Cutting fluids and coolants. Tooling and Production 60:4–8

de Chiffre L (1978) Testing the overall performance of cutting fluids. STLE Transactions 34:244–251

de Chiffre L, Lassen S, Pedersen KB, Skade S (1994) Reaming test for cutting fluid evaluation. Journal of Synthetic Lubrication 11:17–34

de Chiffre L, Belluco W, Zeng Z (2001) An investigation of reaming test parameters used for cutting fluid evaluations. Lubrication Engineering 57:24–28

Dhar NR, Kamruzzaman M (2006) Cutting temperature, tool wear, surface roughness and dimensional deviation in turning AISI-4037 steel under cryogenic condition. International Journal of Machine Tools and Manufacture doi:10.1016/j.ijmachtools.2006.09.018

Dhar NR, Kamruzzaman M, Ahmed M (2006) Effect of minimum quantity lubrication (MQL) on tool wear and surface roughness in turning AISI-4340 steel. Journal of Materials Processing Technology 172:299–304

Ebbrell S, Woolley NH, Tridimas YD, Allanson DR, Rowe WB (2000) The effects of cutting fluid application methods on the grinding process. International Journal of Machine Tools & Manufacture 40:209–223

Edwards W, Newman JR, Snapper K, Seaver D (1982) Multiattribute Evaluation. SAGE Publications, Newbury Park, California

Eppert JJ, Gunter KL, Sutherland JW (2001) Development of a cutting fluid classification system using cluster analysis. Tribology Transactions 44:375–382

Ghio F (1986) Cutting fluids for operations on metal with flexible abrasive belts. Australian Machinery and Production Engineering 38:19–20

Haq AN, Tamizharasan T (2005) Investigation of the effects of cooling in hard turning operations. International Journal of Advanced Manufacturing Technology 30:808–816

Heinemann N, Hinduja S, Barrow G, Petuelli G (2006) Effect of MQL on the tool life of small twist drills in deep-hole drilling. International Journal of Machine Tools and Manufacture 46:1–6

Lorenz G (1985) Reliable cutting fluid rating. CIRP Annals 34:95–99

Maekawa K (1998) Computational aspects of tribology in metal machining. Proc. of I.Mech.E., Journal of Engineering Tribology 212:307–318

Nagpal BK, Sharma CS (1973) Evaluation of four common, commercially available cutting fluids used in flooding. Journal of the Institution of Engineers (India) 23:249–253

Narheim Y, Kendig M (1987) Evaluation of the cutting fluid effectiveness in machining using electrochemical techniques. Wear 114:51–57

Obikawa T, Kamata Y, Shinozuka J (2006) High-speed grooving with applying MQL. International Journal of Machine Tools and Manufacture 46:1854–1861

Okuyama S, Nakamura Y, Kawamura S (1993) Cooling action of grinding fluid in shallow grinding. International Journal of Machine Tools & Manufacture 33:13–23

Peters J, Aerens R (1976) An objective method for evaluating grinding coolants. CIRP Annals 25:247–250

Rao RV (2004) Performance evaluation of cutting fluids for green manufacturing using a combined multiple attribute decision making method. International Journal of Environmentally Conscious Design and Manufacturing 12:526–535

Rao RV, Gandhi OP (2001) Digraph and matrix method for selection, identification and comparison of metal cutting fluids. Proc. IME, Journal of Engineering Tribology 212:307–318

Rapp W (1984) Selection of cooling lubricants. VDI - Z 126:213–220

Reddy NSK, Rao PV (2006) Experimental investigation to study the effect of solid lubricants on cutting forces and surface quality in end milling. International Journal of Machine Tools and Manufacture 46:189–198

Rowe GW (1982) Lubricant testing for grinding operations. Wear 77:73–80

Sheng PS, Oberwalleney S (1997) Life-cycle planning of cutting fluids - a review. Journal of Manufacturing Science and Engineering 119:791–800

Sokovic M, Mijanovic K (2001) Ecological aspects of the cutting fluids and its influence on quantifiable parameters of the cutting processes. Journal of Materials Processing Technology 109:181–189

Sun J, Ge P, Zhenchang L (2001) Two-grade fuzzy synthetic decision-making system with use of an analytic hierarchy process for performance evaluation of grinding fluids. Tribology International 34:683–688

Sutcliffe T, Barber SJ, Dycan W (1979) Coolants laboratory evaluation by drill test technique using high speed drills. STLE Transactions 35:145–152

Tan XC, Liu F, Cao HJ, Zhang H (2002) A decision-making framework model of cutting fluid selection for green manufacturing and a case study. Journal of Materials Processing Technology 129:467–470

Upton DP (2000) Optimization of cutting fluid performance. International Journal of Production Research 38:1219–1223

Varadarajan AS, Philip PK, Ramamoorthy B (2002) Investigations on hard turning with minimal cutting fluid application (HTMF) and its comparison with dry and wet turning. International Journal of Machine Tools and Manufacture 42:193–200

Wakabayashi T, Ogura S (1990) Evaluation of cutting fluids by consumption energy in tapping test. STLE Transactions 46:715–720

Webster JA, Cui C, Mindek RB (1995) Grinding fluid application system design. CIRP Annals 4:333–338

Yamanaka Y, Oi T, Sakai Y, Mukai D (1996) Development of a new grinding fluid for the CBN grinding wheel. Part I. STLE Transactions 52:359–364

Yamanaka Y, Hayama M, Oi T, Sakai Y, Satoh M (1997) Development of a new grinding fluid for the CBN grinding wheel. Part II. STLE Transactions 53:20–26

Yamanaka Y, Hayama M, Oi T, Imai J, Satoh M (1998) Development of a new grinding fluid for the CBN grinding wheel. Part III. STLE Transactions 54:24–30

Yamanaka Y, Hayama M, Oi T, Imai J, Satoh M (2000) Development of a new grinding fluid for the CBN grinding wheel. Part IV. STLE Transactions 56:17–24

Yamanaka Y, Oi T, Nanao H, Satoh M (2000) Development of a new grinding fluid for the CBN grinding wheel. Part V. STLE Transactions 56:25–31

Yuhta T, Igarashi S, Hukushima, H, Satoh T (1984) On the influence of grinding fluid in grinding carbon steel with a diamond wheel. Japan Society of Precision Engineering 1:363–368

9

Evaluation and Selection of Modern Machining Methods

9.1 Introduction

Traditional machining processes, such as turning, grinding, drilling, milling, *etc.*, remove material by chip formation, abrasion, or micro-chipping. There are situations, however, where these processes are not satisfactory, economical, or even possible, for the following reasons (Kalpakjian and Schmid, 2000):

1. The hardness and strength of the material is very high (typically above 400 HB) or the material is too brittle.
2. The work piece is too flexible, slender, or delicate to withstand the cutting or grinding forces, or the parts are too difficult to fix.
3. The shape of the part is complex .
4. Surface finish and dimensional tolerance requirements are more rigorous than those obtained by other processes.
5. Temperature rise and residual stresses in the work piece are not desirable or acceptable.

These requirements have led to the development of chemical, electrical, laser and other means of material removal. Beginning in the 1940s, these advanced methods are called non-traditional or unconventional machining processes. Over the last four decades, there has been a large increase in the number of non-traditional machining processes (NTMPs). Today, NTMPs with vastly different capabilities and specifications are available for a wide range of applications. Effective utilization of the capabilities of NTMPs needs careful selection of a suitable process for the application (Benedict, 1987; Yurdakul and Cogun, 2003). The lack of versatility of NTMPs, uncertainties regarding the capabilities of NTMPs, and different cost elements of operating NTMPs make the comparison and ranking of NTMPs a challenging task. An increasing shortage of experienced experts in the field of NTMPs makes the selection of appropriate NTMPs a critical problem.

There is not enough published work on the selection of NTMPs. A few attempts have been made to suggest a systematic procedure for selection of a particular NTMP for a given application. Alder *et al.* (1986) outlined the background, and some of the problems associated with the selection of conventional processes and NTMPs. A range of material types to achieve a given task by means of a knowledge-based expert system was also examined. Cogun (1994) developed a procedure that identifies suitable alternative NTMPs for the user, with a list of suitable processes for parts with relatively slack design requirements. The main objective of the work was to remove unsuitable NTMPs from consideration, but not the ranking of the NTMPs. Jain and Jain (2001) reviewed the modeling of material removal in mechanical-type advanced machining processes, and gave a brief summary of research work on these processes.

Yurdakul and Cogun (2003) developed a multi-attribute selection procedure for NTMP selection using the technique for order preference by similarity to ideal solution (TOPSIS) and the analytic hierarchy process (AHP) methods. AHP was used to assign weights of relative importance to various process selection attributes, and TOPSIS was used to obtain a ranking score for each of the alternative NTMPs. However, the authors had not considered the subjective attributes. Further, the authors had made certain mistakes in applying the basic technique of TOPSIS. Chakraborthy and Dey (2006) suggested a quality function deployment (QFD) based expert system for NTMP selection. The developed expert system employs the use of a house of quality (HOQ) matrix for comparison of relevant product and process characteristics. The weights obtained for various process characteristics were utilized to estimate an overall score for each of the NTMPs, and the process having the maximum score was selected as the optimal choice. However, the procedure is knowledge-intensive and may go beyond the capabilities of the non-expert user.

There is a need for a simple scientific method or mathematical tool to guide users in taking a proper NTMP selection decision. The objective of an NTMP selection procedure is to identify the NTMP selection attributes, and obtain the most appropriate combination of attributes in conjunction with the real requirements of the machining application. Efforts need to be extended to determine attributes that influence NTMP selection for a given machining application, using a logical approach, to eliminate unsuitable NTMPs, and for the selection of a proper NTMP to strengthen the existing NTMP selection procedure. This is considered in this chapter using the GTMA and other fuzzy MADM methods.

An NTMP selection attribute is defined as a factor that influences the selection of an NTMP for a given industrial application. NTMP attributes include work piece material, cost involved, and process capability attributes such as tolerance, surface finish, surface damage, corner radii, taper, hole diameter, depth/diameter ratio for cylindrical holes, depth/width ratio for blind cavities, width of cut, material removal rate, part size, part exterior and interior shape details, *etc.*

As a first step in NTMP selection, the decision maker has to identify the NTMP selection attributes for a given industrial application, and short-list the NTMP processes on the basis of identified attributes satisfying the requirements. A quantitative or qualitative value, or its range, may be assigned to each identified attribute as a limiting value, or threshold value, for its acceptance in the application considered. An NTMP process with each of its attributes, meeting the criterion, may be short-listed.

Now, an example is included to demonstrate and validate the proposed decision making-methods for the selection of an NTMP process for a given industrial application.

9.2 Examples

Two examples of NTMP selection are considered.

9.2.1 Example 1

Yurdakul and Cogun (2003) developed a multi-attribute selection procedure for NTMP selection using the TOPSIS and AHP methods. The authors presented different case studies, one of which is considered here. The details of the case study are given in Table 9.1. The NTMPs eliminated on the basis of the work material were ECM, ECG, ECH, EDM, WEDM, and PAC. The NTMP eliminated on the basis of the shape applications was WJM. No NTMPs were eliminated on the basis of process capabilities. This elimination procedure is similar to the short-listing of alternative NTMPs as described in Section 9.1. Feasible NTMPs to be ranked are AJM, USM, CHM, EBM, and LBM.

Table 9.1. Data of the NTMP selection attributes of example 9.2.1 (from Yurdakul and Cogun 2003; permission of the Council of the Institution of Mechanical Engineers, UK)

NTMP	T	SF	SD	TR	MR	WM	C
AJM	0.05	0.6	2.5	0.005	50	3	4
USM	0.013	0.5	25	0.005	500	3	5
CHM	0.03	2	5	0.3	40	1	2
EBM	0.02	3	100	0.02	2	3	1
LBM	0.02	1	100	0.05	2	3	1

Work material: Ceramic (non-conductive); Shape application: Cylindrical through hole drilling; Process requirements: 930 holes of 0.64 mm diameter, L/D = 5.7

T: Tolerance (mm); SF: Surface finish (μm); SD: Surface damage (μm); TR: Taper (mm/mm); MR: Material removal rate (mm^3/min); WM: Work material (NTMP process suitability is assigned on a scale of 1–3, 1 for poor and 3 for good application); C: Cost (on a scale of 1–9, 1 for low, 5 for medium and 9 for very high)

USM: Ultrasonic machining; AJM: Abrasive jet machining; LBM: Laser beam machining; EBM: Electron beam machining; CHM: Chemical machining

9.2.1.1 Graph Theory and Matrix Approach (GTMA)

Now, various steps of the proposed procedure are carried out as described next:

1. The NTMP selection attributes considered are the same as those of Yurdakul and Cogun (2003) and these are: tolerance (T), surface finish (SF), surface damage (SD), taper (TR), material removal rate (MR), work material (WM), and cost (C).
2. The quantitative values of the NTMP selection attributes, which are given in Table 3, are to be normalized. MR and WM are beneficial attributes, and higher values are desirable. Values of these attributes are normalized, and are given in Table 9.2 in the respective columns. T, SF, SD, TR, and C are non-beneficial attributes, and lower values are desirable. The values of these attributes for different NTMPs are normalized, and are given in Table 9.2 in the respective columns.

Table 9.2. Normalized data of the NTMP selection attributes of example 9.2.1

NTMP	T	SF	SD	TR	MR	WM	C
AJM	0.26	0.83	1	1	0.1	1	0.25
USM	1	1	0.1	1	1	1	0.2
CHM	0.43	0.25	0.5	0.02	0.08	0.33	0.5
EBM	0.65	0.17	0.03	0.25	0.004	1	1
LBM	0.65	0.5	0.03	0.1	0.004	1	1

Relative importance of attributes (a_{ij}) is also assigned values, as explained in Chapter 4. Let the decision maker (*i.e.*, user organization) select the following assignments:

$$\begin{array}{c} \\ T \\ SF \\ SD \\ TR \\ MR \\ WM \\ C \end{array} \begin{array}{c} \begin{array}{ccccccc} T & SF & SD & TR & MR & WM & C \end{array} \\ \begin{bmatrix} - & 0.59 & 0.865 & 0.865 & 0.59 & 0.665 & 0.665 \\ 0.410 & - & 0.745 & 0.745 & 0.5 & 0.590 & 0.590 \\ 0.135 & 0.255 & - & 0.5 & 0.255 & 0.335 & 0.335 \\ 0.135 & 0.255 & 0.500 & - & 0.255 & 0.335 & 0.335 \\ 0.410 & 0.500 & 0.745 & 0.745 & - & 0.590 & 0.590 \\ 0.335 & 0.410 & 0.665 & 0.665 & 0.41 & - & 0.500 \\ 0.335 & 0.410 & 0.665 & 0.665 & 0.41 & 0.500 & - \end{bmatrix} \end{array}$$

The assigned values in this example are for demonstration purposes only.
3. The NTMP selection attributes digraph, showing the presence as well as relative importance of the above attributes, is similar to Fig. 2.2 but with seven attributes. However, it is not shown here.
4. The NTMP selection attributes matrix of this digraph is written. However, it is not shown here.
5. The NTMP selection attributes function is written. However, as a computer program is developed for calculating the permanent function value of a matrix, this step can be skipped.
6. The NTMP selection index (NTMP-SI) is calculated using the values of A_i and a_{ij} for each alternative NTMP and the values are given in descending order.

Ultrasonic machining (USM)	40.50211
Abrasive jet machining (AJM)	31.76501

Laser beam machining (LBM)	21.05954
Electron beam machining (EBM)	20.13952
Chemical machining (CHM)	15.90063

From the above values of the NTMP selection index, USM is understood as the best choice among the alternatives considered for the given hole making operations. The ranking of NTMPs based on the proposed methodology is USM-AJM-LBM-EBM-CHM; by contrast, the ranking presented by Yurdakul and Cogun (2003) was USM-LBM-EBM-CHM-AJM. Both the methods suggest USM as the first right choice. However, the ranking of certain alternative NTMPs obtained by using the proposed procedure is different from that reported by Yurdakul and Cogun (2003). For example, AJM is the second choice based on the proposed procedure, whereas it was LBM in Yurdakul and Cogun (2003), and AJM was proposed as the last choice by these authors. A closer look at the quantitative data of the attributes of LBM and AJM reveals that AJM is better than LBM in the case of four out of seven attributes (*i.e.*, SF, SD, T, and MR), and equal to LBM in the case of attribute WM. LBM is better than AJM only in the case of two attributes (*i.e.*, T and C). Thus, keeping in mind the values of the attributes and the relative importance of the attributes, proposing AJM as the second choice by the proposed method based on the method used here seems to be more appropriate, compared to LBM as proposed by Yurdakul and Cogun (2003). Thereby, the differences in the ranking of alternatives between the procedure proposed here and that suggested by Yurdakul and Cogun (2003) can be explained.

It may be added here, however, that the weights of relative importance used by Yurdakul and Cogun (2003) were different from those used in the present work. Further, it may be mentioned that ranking depends upon the judgements of relative importance made by the decision maker (*i.e.*, user organization). The ranking may change if the decision maker assigns different relative importance values to the attributes. The same is true with the approach proposed by Yurdakul and Cogun (2003). Yurdakul and Cogun (2003) had made certain mistakes in applying the basic technique of TOPSIS in their model (*e.g.*, in normalization of the attributes, and calculation of final ranking scores).

9.2.1.2 SAW Method

To start with, the attributes are ranked in order of importance, and 10 points each are assigned to the least important attributes SD and TR. The attributes WM and C are considered as equally important in the present example, and given 20 points each to reflect their relative importance. SF and MR are considered as equally important, and given 30 points each, and T is given 40 points. The final weights are obtained by normalizing the sum of the points to one. Thus, the weights of T, SF, MR, WM, C, SD, and TR are calculated as 0.25, 0.1875, 0.1875, 0.125, 0.125, 0.0625, and 0.0625, respectively. Using these weights, and the normalized data of the attributes for different NTMPs, the NTMP-SI values are calculated, and are arranged in descending order of the index.

Ultrasonic machining (USM)	0.8438
Abrasive jet machining (AJM)	0.5206
Laser beam machining (LBM)	0.5151

Electron beam machining (EBM)	0.4626
Chemical machining (CHM)	0.3056

The SAW method also suggests USM as the right choice for the given NTMP selection problem.

9.2.1.3 WPM

Using the same weights of attributes as those selected for the SAW method, the NTMP-SI value for each NTMP is calculated, and the values are given below:

Ultrasonic machining (USM)	0.7082
Abrasive jet machining (AJM)	0.4478
Chemical machining (CHM)	0.2328
Laser beam machining (LBM)	0.1948
Electron beam machining (EBM)	0.1685

WPM also suggests USM as the right choice for the given NTMP selection problem. However, CHM is proposed as the third choice, and EBM as the last choice.

9.2.1.4 AHP and its Versions

The AHP method may use the same weights as those selected for the SAW method. In that case, the ranking of the NTMPs will be same. However, if the decision maker decides to use the AHP method for determining the weights, rather than adopting the weights used in SAW method, then he or she has to make pair-wise comparisons of the attributes to determine the weights (w_j) of the attributes. Let the decision maker prepare the following matrix:

	T	SF	SD	TR	MR	WM	C
T	1	2	7	7	2	3	3
SF	1/2	1	5	5	1	2	2
SD	1/7	1/5	1	1	1/5	1/3	1/3
TR	1/7	1/5	1	1	1/5	1/3	1/3
MR	1/2	1	5	5	1	2	2
WM	1/3	1/2	3	3	1/2	1	1
C	1/3	1/2	3	3	1/2	1	1

The normalized weights of each attribute are calculated following the procedure presented in Section 3.2.3 and these are $W_T = 0.3224$, $W_{SF} = W_{MR} = 0.1938$, $W_{WM} = W_C = 0.1063$, and $W_{SD} = W_{TR} = 0.0387$. The value of λ_{max} is 7.028 and CR = 0.003457, which is much less than the allowed CR value of 0.1. Thus, there is good consistency in the judgements made.

The value of NTMP-SI is now calculated using the above weights, and the normalized data of the attributes given in Table 9.3. The alternative NTMPs are arranged in descending order of the NTMP-SI:

Ultrasonic machining (USM)	0.8801
Laser beam machining (LBM)	0.5249
Abrasive jet machining (AJM)	0.4743
Electron beam machining (EBM)	0.4667
Chemical machining (CHM)	0.3109

For the above weights of importance of attributes, the multiplicative AHP method leads to the following ranking order:

Ultrasonic machining (USM)	0.7709
Abrasive jet machining (AJM)	0.3451
Chemical machining (CHM)	0.2466
Laser beam machining (LBM)	0.2084
Electron beam machining (EBM)	0.1752

It may be observed that the ranking order given by multiplicative AHP for the given weights is similar to that given by WPM.

9.2.1.5 TOPSIS Method

Using the same weights as those selected for the AHP method, and following the steps of the methodology given in Section 3.2.6, the TOPSIS method gives the following ranking order of NTMPs:

Ultrasonic machining (USM)	0.7709
Abrasive jet machining (AJM)	0.3451
Chemical machining (CHM)	0.2466
Laser beam machining (LBM)	0.2084
Electron beam machining (EBM)	0.1752

This ranking order is similar to that given by the multiplicative AHP method.

9.2.1.6 Modified TOPSIS Method

For the same weights as those used in the AHP method, the modified TOPSIS method gives the following ranking order:

Ultrasonic machining (USM)	0.7693
Laser beam machining (LBM)	0.4893
Abrasive jet machining (AJM)	0.4231
Electron beam machining (EBM)	0.4180
Chemical machining (CHM)	0.3594

This ranking order is similar to that given by the AHP method.

9.2.2 Example 2

Another case study presented by Yurdakul and Cogun (2003) is considered here. No NTMPs were eliminated on the basis of the work material. The NTMPs eliminated on the basis of the shape applications were WJM, ECM, ECG, ECH, CHM, WEDM, and PAC. The NTMP eliminated on the basis of process capability was AJM. This elimination procedure is similar to the short-listing of alternative NTMPs described in Section 9.1. Feasible NTMPs to be ranked are USM, EDM, EBM, and LBM. The details of the case study are given in Table 9.3.

Table 9.3. Data of the NTMP selection attributes of example 9.2.2 (from Yurdakul and Cogun 2003; permission of the Council of the Institution of Mechanical Engineers, UK)

NTMP	SF	SD	TR	MR	WM	C
USM	0.5	25	0.005	500	2	5
EDM	2	20	0.001	800	3	7
EBM	3	100	0.02	2	2	1
LBM	1	100	0.05	2	2	1

Work material: Hardened 52100 steel; Shape application: Cylindrical through hole drilling; Process requirements: 800 holes of 0.175 mm diameter, L/D = 5.7

SF: Surface finish (μm); SD: Surface damage (μm); TR: Taper (mm/mm); MR: Material removal rate (mm^3/min); WM: Work material (NTMP process suitability is assigned on a scale of 1–3, 1 for poor and 3 for good application); C: Cost (on a scale of 1–9, 1 for low, 5 for medium and 9 for very high)

USM: Ultrasonic machining; EDM: Electric discharge machining; EBM: Electron beam machining; LBM: Laser beam machining

9.2.2.1 Graph Theory and the Matrix Approach (GTMA)

In the present work, the attributes considered are the same as those of Yurdakul and Cogun (2003) and these are: surface finish (SF), surface damage (SD), taper (TR), material removal rate (MR), work material (WM), and cost (C).

The quantitative values of the NTMP selection attributes, which are given in Table 9.3, are to be normalized. SF, SD, TR, and C are non-beneficial attributes, and MR and WM are beneficial attributes. The values of the attributes are normalized, and are shown in Table 9.4.

Table 9.4. Normalized data of the NTMP selection attributes of example 9.2.2

NTMP	SF	SD	TR	MR	WM	C
USM	1	0.8	0.2	0.625	0.6667	0.2
EDM	0.25	1	1	1	1	0.1428
EBM	0.1667	0.2	0.05	0.0025	0.6667	1
LBM	0.5	0.2	0.02	0.0025	0.6667	1

Let the decision maker select the following assignments of relative importance:

	SF	SD	TR	MR	WM	C
SF	-	0.59	0.5	0.41	0.665	0.41
SD	0.41	-	0.41	0.335	0.59	0.335
TR	0.5	0.59	-	0.41	0.665	0.41
MR	0.59	0.665	0.59	-	0.745	0.59
WM	0.335	0.41	0.335	0.255	-	0.255
C	0.59	0.665	0.59	0.41	0.745	-

The NTMP selection attributes digraph, NTMP selection attributes matrix of the digraph, and NTMP selection function for the matrix can be prepared. The value of the NTMP selection index is calculated using the values of A_i and a_{ij} for each NTMP. The NTMP selection index values of different NTMPs are given below in descending order:

EDM	14.0184
USM	10.4897
LBM	7.1186
EBM	6.4465

From the above values of the NTMP selection index, EDM is identified as the best choice among the alternatives considered for the given operations. The ranking of NTMPs based on the methodology proposed is EDM-USM-LBM-EBM.

9.2.2.2 TOPSIS Method

The same relative importance matrix as in Yurdakul and Cogun (2003) is used here.

	SF	SD	TR	MR	WM	C
SF	1	2	1	1/2	4	1/2
SD	1/2	1	1/2	1/3	2	1/3
TR	1	2	1	1/2	3	1/2
MR	2	3	2	1	6	2
WM	1/4	1/2	1/3	1/6	1	1/6
C	2	3	2	1/2	6	1

The normalized weights of each attribute are calculated following the procedure presented in Section 3.2.3, and these are $W_{SF} = 0.155$, $W_{SD} = 0.0853$, $W_{TR} = 0.1478$, $W_{MR} = 0.3162$, $W_{WM} = 0.0447$, and $W_C = 0.251$. The value of λ_{max} is 6.0725 and CR = 0.0116, which is much less than the allowed CR value of 0.1. Thus, there is good consistency in the judgements made.

Following the steps of the methodology given in Section 3.2.6, the TOPSIS method gives the following ranking order of NTMPs:

EDM	0.6250
USM	0.6121
EBM	0.3932
LBM	0.3865

This ranking order also suggests EDM as the first choice. However, Yurdakul and Cogun (2003) who also used the above relative importance matrix and the TOPSIS method, obtained a different ranking order, *i.e.*, EDM-USM-LBM-EBM. As explained in Section 9.2.1.1, Yurdakul and Cogun (2003) had made certain mistakes in applying the basic technique of TOPSIS in their model.

9.2.2.3 Modified TOPSIS Method

For the same weights as those used in the TOPSIS method, the modified TOPSIS method gives the following ranking order:

USM	0.6467
EDM	0.6215
EBM	0.4100
LBM	0.3959

This ranking order suggests USM as the first right choice. However, a closer look at the values of the attributes for USM, and the corresponding values of the attributes for EDM indicates that proposing USM is not logical. Thus, it can be said that modified TOPSIS method does not provide logical results for the example considered here.

References

Alder GM, McGeough JA, Spencer CA (1986) Selection of machining processes by intelligent knowledge base systems. In: Proc. International Conference on Computer Aided Production Engineering, University of Edinburgh, Edinburgh pp 61–66

Benedict GF (1987) Non-traditional manufacturing processes. Marcel Dekker, New York

Chakraborthy S, Dey S (2006) QFD-based expert system for non-traditional machining processes selection. Expert Systems with Applications doi:10.1016/j.eswa.2006.02.010

Cogun C (1994) Computer aided preliminary selection of non traditional machining processes. International Journal of Machine Tools and Manufacture 34:315–326

Jain NK, Jain VK (2001) Modeling of material removal in mechanical type advanced machining processes: a state-of-art review. International Journal of Machine Tools and Manufacture 41:1573–1635

Kalpakjian S, Schmid SR (2000) Manufacturing engineering and technology. Addison Wesley Longman (Singapore), Indian branch, Delhi

Yurdakul M, Cogun C (2003) Development of a multi-attribute selection procedure for non-traditional machining processes. Proc. IME, Journal of Engineering Manufacture 217:993–1009

10

Evaluation of Flexible Manufacturing Systems

10.1 Introduction

A flexible manufacturing system (FMS) consists of a group of processing work stations (usually CNC machine tools) interconnected by an automated material handling and storage system, and controlled by a distributed computer system. The reason the FMS is called 'flexible' is that it is capable of processing a variety of different part styles simultaneously at the various work stations, and the mix of part styles and quantities of production can be adjusted in response to changing demand patterns. The evolution of flexible manufacturing systems offers great potential for increasing flexibility and changing the basis of competition by ensuring both cost-effective and customized manufacturing at the same time.

The decision to invest in FMS and other advanced manufacturing technology has been an issue in the practitioner and academic literature for over two decades. An effective justification process requires the consideration of many quantitative attributes (*e.g.*, costs involved, floor space requirements, *etc.*) and qualitative attributes (*e.g.*, product-mix flexibility, routing flexibility, *etc.*). An FMS selection attribute is defined as a factor that influences the selection of a flexible manufacturing system for a given application. These attributes include: costs involved, floor space requirements, labor requirements, throughput time, work-in-process, setup cost, quality, volume flexibility, product-mix flexibility, process/routing flexibility, expansion flexibility, utilization rate, risk, ease of operation, maintenance aspects, payback period, reconfiguration time, company policy, *etc.*

To help address this issue of effective evaluation and justification of flexible manufacturing systems, various mathematical and systems modeling approaches have been proposed. Kochan (1987) discussed the importance of selection of flexible manufacturing and CAD/CAM systems. Troxler (1990) estimated the cost impact of flexible manufacturing systems. Dhavale (1990) proposed a manufacturing cost model for computer-integrated manufacturing systems. Layek and Wolf (1991) evaluated flexibility of alternative FMS designs using a comparative measure. Sriram and Gupta (1991) discussed the impact of FMS and its implications in terms of information reporting, strategic cost analyses, and control. Suresh and Kaparthi (1992) presented a procedure that combined a general mixed integer goal programming (GP) formulation with the analytic hierarchy process (AHP) for use in deciding upon flexible automation investments. Gerwin and Kolodny (1992) discussed the aspects of management of advanced manufacturing technologies.

Elango and Meinhart (1994) proposed a strategic framework for selecting an FMS. Kuula (1993) presented a risk management model for FMS selection decisions using a multiple criteria decision-making approach. Tabucanon *et al.* (1994) proposed a decision support system for multiple criteria machine selection for flexible manufacturing systems. The approach presented combined the analytic hierarchy process (AHP) technique with the rule-based technique for creating Expert Systems (ES). Myint and Tabucanon (1994) used AHP method and goal programming (GP) model to determine the satisfactory FMS configuration from the short-listed FMS configurations.

Shang and Sueyoshi (1995) proposed a unified framework to facilitate decision-making in the design and planning stage of FMS. The recommended framework contains three individual modules: an analytic hierarchy process (AHP), a simulation module, and an accounting procedure. These modules were unified through an efficiency measurement method called data envelopment analysis (DEA). The AHP model examines the non-monetary criteria associated with corporate goals and long-term objectives, while the simulation model was employed to analyze the tangible benefits. Both the AHP and simulation models were used to generate the necessary outputs for the DEA, whereas the accounting procedure determines the required inputs, such as expenditures and resources for realizing the potential benefits.

Albayrakoglu (1996), and Mohanty and Venkataraman (1996) proposed the application of AHP for justification of new manufacturing technologies. Sarkis (1997) presented an illustrative problem for evaluating flexible manufacturing systems for an industrial application using DEA. The problem considered 24 alternative flexible manufacturing systems, and eight selection attributes. Perego and Rangone (1998) presented a reference framework for the application of three categories of fuzzy MADM techniques to select advanced manufacturing technologies.

Talluri *et al.* (2000) proposed a method based on the combined application of data envelopment analysis (DEA) and nonparametric statistical procedures for FMS evaluation. Chan *et al.* (2000) developed intelligent decision support tools to aid the design of flexible manufacturing systems. Karsak and Tolga (2001) proposed a fuzzy multiple criteria decision-making procedure for evaluating advanced manufacturing system investments. Karsak and Kuzgunkaya (2002) proposed a fuzzy multiple objective programming approach for the selection of a flexible manufacturing system. The model proposed by the authors determines the most appropriate FMS alternative through maximization of objectives such as reduction in labor cost, reduction in setup cost, reduction in work-in-process (WIP), increase in market response and improvement in quality, and minimization of capital and maintenance costs as well as floor space used. These objectives were assigned priorities indicating their importance levels based on linguistic variables.

Sarkis and Talluri (1999) presented a decision model using DEA for evaluation of flexible manufacturing systems in the presence of both cardinal and ordinal factors. Karsak (2002) presented a distance-based fuzzy MCDM approach for evaluating flexible manufacturing system alternatives. The method is similar to the TOPSIS method. Tseng (2004) presented the details of strategic choice of flexible manufacturing technologies. Laosirihongthong *et al.* (2003) presented case studies related to new manufacturing technology implementation. Lloréns *et al.* (2005) described the aspects of flexibility of manufacturing systems, strategic change, and performance. The authors showed that manufacturing flexibility at system level can be a critical factor in the process of strategic change, which means that it can have an impact on the desirability of strategic change, or on the more specific strategic fit.

Bayazit (2005) used AHP to implement FMS in a tractor manufacturing plant. Also a sensitivity analysis was conducted to assess how realistic the final outcome was. Kulak and Kahraman (2005) proposed axiomatic design (AD) principles for multiple attribute comparison of advanced manufacturing systems. The comparison was made for cases of both complete and incomplete information. The crisp AD approach for complete information, and the fuzzy AD approach for incomplete information were developed. Rao (2006) presented a decision-making model for FMS selection using digraph and matrix methods. A 'flexible manufacturing system selection index' was proposed that evaluates and ranks flexible manufacturing systems for a given industrial application. In another work, Rao (2007) used the TOPSIS and AHP methods in combination for evaluating flexible manufacturing systems.

Now, to demonstrate and validate the application of decision-making methods, two examples are considered. In both, GTMA is applied first, and then a few MADM methods are applied to rank and select the flexible manufacturing systems.

10.2 Examples

Two examples are considered to demonstrate the application of the GTMA and fuzzy MADM methods.

10.2.1 Example 1

Karsak and Kuzgunkaya (2002) proposed a fuzzy multiple objective programming approach for the selection of a flexible manufacturing system. The authors had considered eight alternative flexible manufacturing systems and seven attributes. Five attributes were expressed objectively, and two attributes were expressed subjectively. Table 10.1 presents the data.

Table 10.1. Data of attributes of example 10.2.1 (from Karsak and Kuzgunkaya 2002; reprinted with permission from Elsevier)

FMS	RLC	RWP	RSC	IMR	IQ	CMC	FSU
1	30	23	5	Good	Good	1,500	5,000
2	18	13	15	Good	Good	1,300	6,000
3	15	12	10	Fair	Fair	950	7,000
4	25	20	13	Good	Good	1,200	4,000
5	14	18	14	Worst	Good	950	3,500
6	17	15	9	Good	Fair	1,250	5,250
7	23	18	20	Fair	Good	1,100	3,000
8	16	8	14	Worst	Fair	1,500	3,000

RLC: Reduction in labor cost (%) RWP: Reduction in WIP (%)
RSC: Reduction in set up cost (%) IMR: Increase in market response
IQ: Increase in quality CMC: Capital and maintenance cost ($1,000)
FSU: Floor space used (sq. ft.)
The above data for RLC, RWP, RSC, CMC, and FSU are actually the middle values of the range presented by Karsak and Kuzgunkaya (2002)

10.2.1.1 Application of Graph Theory and Matrix Approach (GTMA)

Various steps of the methodology, proposed in Section 2.6, are carried out as described below:

Step 1: In the present work, the attributes considered are the same as those of Karsak and Kuzgunkaya (2002), and these are: reduction in labor cost (RLC), reduction in WIP (RWP), reduction in setup cost (RSC), increase in market response (IMR), increase in quality (IQ), capital and maintenance cost (CMC), and floor space used (FSU). The subjective data of the two attributes IMR and IQ are converted into appropriate objective data using Table 4.3, and the objective data for all seven attributes are given in Table 10.2.

Table 10.2. Objective data of attributes of example 10.2.1

FMS	RLC	RWP	RSC	IMR	IQ	CMC	FSU
1	30	23	5	0.745	0.745	1,500	5,000
2	18	13	15	0.745	0.745	1,300	6,000
3	15	12	10	0.500	0.500	950	7,000
4	25	20	13	0.745	0.745	1,200	4,000
5	14	18	14	0.255	0.745	950	3,500
6	17	15	9	0.745	0.500	1,250	5,250
7	23	18	20	0.500	0.745	1,100	3,000
8	16	8	14	0.255	0.5	1,500	3,000

The objective values of the FMS selection attributes, which are given in Table 10.2, are to be normalized. RLC, RWP, RSC, IMR, and IQ are beneficial attributes, and higher values are desirable. CMC and FSU are non-beneficial attributes, and lower values are desirable. The values of the attributes for different FMSs are normalized, and given in Table 10.3 in the respective columns.

Table 10.3. Normalized data of attributes of example 10.2.1

FMS	RLC	RWP	RSC	IMR	IQ	CMC	FSU
1	1	1	0.25	1	1	0.6333	0.6
2	0.6	0.5652	0.75	1	1	0.7308	0.5
3	0.5	0.5217	0.5	0.6711	0.6711	1	0.4286
4	0.83333	0.8696	0.65	1	1	0.7917	0.75
5	0.4667	0.7826	0.7	0.3423	1	1	0.8571
6	0.5667	0.6527	0.45	1	0.6711	0.76	0.5714
7	0.7667	0.7826	1	0.6711	1	0.8636	1
8	0.5333	0.3478	0.7	0.3423	0.6711	0.6333	1

Relative importance of attributes (a_{ij}) is also assigned values, as explained in Section 2.4. Let the decision maker (*i.e.*, user organization) select the following assignments:

	RLC	RWP	RSC	IMR	IQ	CMC	FSU
RLC	-	0.5	0.665	0.5	0.335	0.335	0.665
RWP	0.5	-	0.665	0.5	0.335	0.335	0.665
RSC	0.335	0.335	-	0.335	0.255	0.255	0.5
IMR	0.5	0.5	0.665	-	0.335	0.335	0.665
IQ	0.665	0.665	0.745	0.665	-	0.5	0.745
CMC	0.665	0.665	0.745	0.665	0.5	-	0.745
FSU	0.335	0.335	0.5	0.335	0.255	0.255	-

As was assigned by Karsak and Kuzgunkaya (2002), more relative importance is given to IQ and CMC, less to RLC, RWP, and IMS, and even lesser to RSC and FSU. However, the above-assigned values are for demonstration purposes only.

Step 2:

1. The FMS selection attributes digraph, showing the presence as well as relative importance of the above attributes is similar to Figure 2.2 but with seven attributes is not shown here due to obvious reasons.
2. The FMS selection attributes matrix of this digraph is written based on Equation 2.10. This is not shown here due to space restriction.
3. The FMS selection attributes function is written but not shown here. However, it may be added that as a computer program is developed for calculating the permanent function value of a matrix, this step can be skipped.

4 & 5. The flexible manufacturing system selection index (FMS-SI) is calculated using the values of A_i and a_{ij} for each alternative flexible manufacturing system. The FMS-SI values of different flexible manufacturing systems are given in descending order:

7	61.2188
4	57.2741
1	48.9012
5	45.6628

6	39.3644
2	45.5043
3	35.2635
8	34.7198

From the above values of FMS-SI, it is understood that the flexible manufacturing system designated as 7 is the right choice for the given industrial application under the given conditions, and the second choice is 4. These results are similar to those suggested by Karsak and Kuzgunkaya (2002) using the fuzzy multiple objective programming approach. However, it may be mentioned that the ranking depends upon the judgments of relative importance made by the user. The ranking may change if the user assigns different relative importance values to the attributes. The same is true with the approach proposed by Karsak and Kuzgunkaya (2002).

The fuzzy method proposed by Karsak and Kuzgunkaya (2002) is cumbersome in terms of the mathematical equations involved, representation of weights of relative importance, fuzzy distributions, *etc.* Further, the authors had converted the available objective values of the attributes (of RLC, RWP, RSC, CMC, and FSU) into fuzzy values which violates the basic rule of fuzzy logic, *i.e.*, the available objective values need not be fuzzified (*i.e.*, the actual objective values of the attributes are to be taken as is). Comparatively, the GTMA proposed here provides a simple, straight-forward and logical procedure for the FMS selection problem.

10.2.1.2 AHP and its Versions

Let the decision maker prepare the following relative importance matrix:

	RLC	RWP	RSC	IMR	IQ	CMC	FSU
RLC	1	1	3	1	1/3	1/3	3
RWP	1	1	3	1	1/3	1/3	3
RSC	1/3	1/3	1	1/3	1/5	1/5	1
IMR	1	1	3	1	1/3	1/3	3
IQ	3	3	5	3	1	1	5
CMC	3	3	5	3	1	1	5
FSU	1/3	1/3	1	1/3	1/5	1/5	1

RLC is considered moderately more important than RSC in FMS selection. So, a relative importance value of 3 is assigned to RLC over RSC (*i.e.*, a_{13} = 3), and a relative importance value of 1/3 is assigned to RSC over RLC (*i.e.*, a_{31} = 1/3). RLC and RWP are considered equally important attributes in FMS selection. So, a relative importance value of 1 is assigned to RLC over RWP (*i.e.*, a_{12} = 1), and a relative importance value of 1/1 is assigned to RWP over RLC (*i.e.*, a_{21} = 1/1=1). Similarly, the relative importance among other attributes can be explained. However, it may be added that, in actual practice, these values of relative importance can be judiciously decided upon by the user/expert depending on the requirements. The normalized weights of each attribute are calculated following the procedure presented in Section 3.2.3, and these are W_{RLC} = 0.1181, W_{RWP} =

0.1181, W_{RSC} = 0.046, W_{IMR} = 0.1181, W_{IQ} = 0.3, W_{CMC} = 0.3, and W_{FSU} = 0.046, and good consistency is found in the judgments made.

The value of the FMS selection index is now calculated using the above weights, and the normalized data of the attributes given in Table 10.2. This leads to the ranking given by the revised AHP or ideal mode of AHP method. The alternative FMS configurations are arranged in descending order of the FMS selection index:

4	0.9211
7	0.9133
1	0.8834
5	0.8596
2	0.8324
3	0.7440
6	0.7384
8	0.6140

For the above weights of importance of attributes, multiplicative AHP leads to the following ranking order:

4	0.8684
7	0.7993
1	0.7990
2	0.7657
5	0.7641
6	0.6826
3	0.6728
8	0.5495

It may be observed that the above ranking is for the given preferences of the decision maker. The ranking depends upon the judgements of relative importance of attributes made by the decision maker.

10.2.2 Example 2

Now another example is considered to further demonstrate the potential of the proposed GTMA and fuzzy MADM methods.

Kulak and Kahraman (2005) proposed axiomatic design (AD) principles for multiple attribute comparison of advanced manufacturing systems. The authors presented the case study of a company manufacturing tractor components that wished to renew the manufacturing system. In order to produce a group of products, the company had to decide and select the most appropriate one among the different alternative FMSs. The attributes considered were: annual depreciation and maintenance costs (ADM), quality of results (Q), ease of use (E), competitiveness (C), adaptability (A), and expandability (X). The linguistic expressions of the attributes are given in Table 10.4.

Table 10.4. System data of the attributes (from Kulak and Kahraman 2005; reprinted with permission from Elsevier)

FMS	ADM	Q	E	C	A	X
FMS-I	High	Excellent	Very good	Excellent	Very good	Very good
FMS-II	Very low	Very good	Good	Very good	Very good	Very good
FMS-III	Medium	Good	Good	Very good	Excellent	Good
FMS-IV	Low	Fair	Good	Very good	Very good	Good

ADM: annual depreciation and maintenance costs Q: quality of results
E: ease of use C: competitiveness A: adaptability X: expandability

10.2.2.1 Application of GTMA

In the present work, the attributes considered are the same as those of Kulak and Kahraman (2005), and these are: annual depreciation and maintenance costs (ADM), quality of results (Q), ease of use (E), competitiveness (C), adaptability (A), and expandability (X). The linguistic terms given in Table 10.4 are converted into appropriate fuzzy scores (using Table 4.3). Table 10.5 gives the fuzzy scores, and these scores are to be normalized. ADM is a non-beneficial attribute, and lower values are desirable. The remaining attributes are beneficial, and higher values are desirable. The fuzzy scores are normalized, as explained in Section 2.4, and are given in Table 10.6 in the respective columns.

Table 10.5. Fuzzy scores of the attributes of example 10.2.2

FMS	ADM	Q	E	C	A	X
FMS-I	0.665	0.955	0.865	0.955	0.865	0.865
FMS-II	0.255	0.865	0.745	0.865	0.865	0.865
FMS-III	0.5	0.745	0.745	0.865	0.955	0.745
FMS-IV	0.335	0.5	0.745	0.865	0.865	0.745

Table 10.6. Normalized values of the attributes of example 10.2.2

FMS	ADM	Q	E	C	A	X
FMS-I	0.3835	1	1	1	0.9058	1
FMS-II	1	0.9058	0.8613	0.9058	0.9058	1
FMS-III	0.51	0.7801	0.8613	0.9058	1	0.8613
FMS-IV	0.7612	0.5236	0.8613	0.9058	0.9058	0.8613

Relative importance of attributes (a_{ij}) is also assigned values, as explained in Section 2.4. Let the decision maker (*i.e.*, user organization) select the following assignments:

	ADM	Q	E	C	A	X
ADM	-	0.665	0.745	0.665	0.745	0.745
Q	0.335	-	0.665	0.5	0.665	0.665
E	0.255	0.335	-	0.5	0.5	0.5
C	0.335	0.5	0.665	-	0.665	0.665
A	0.255	0.335	0.5	0.335	-	0.5
X	0.255	0.335	0.5	0.335	0.5	-

The assigned values are for demonstration purposes only.

Following the procedure of GTMA, the flexible manufacturing system selection index (FMS-SI) is calculated using the values of A_i and a_{ij} for each alternative flexible manufacturing system. The FMS-SI values of different flexible manufacturing systems are given below in descending order:

FMS-II	22.5201
FMS-I	19.2912
FMS-III	17.3958
FMS-IV	17.0109

From the above values of FMS-SI, it is understood that the flexible manufacturing system designated as FMS-II is the right choice under the given conditions. This result matches with that suggested by Kulak and Kahraman (2005). In fact, FMS-I has taken the second position mainly because of its very low normalized fuzzy score for its ADM attribute. Otherwise, it would have become the first choice. In their work, Kulak and Kahraman (2005) had eliminated FMS-I after performing all calculations, reasoning that the value of ADM for this alternative was beyond the acceptable limit. However, this discarding of FMS-I could have been done at the initial short-listing stage itself, as suggested in step 1 of the GTMA methodology presented in Section 2.6. This could be the case for FMS-IV, too.

10.2.2.2 AHP and its Versions

The AHP method may use the same weights as those selected for the SAW method. In that case, the ranking of the materials will be same. However, if the decision maker decides to use the AHP method, rather than SAW method and the weights used in it, then he or she has to make pair-wise comparisons of the attributes to determine the weights (w_j) of the attributes. Let the decision maker prepare the following pair-wise comparison matrix:

	ADM	Q	E	C	A	X
ADM	1	3	5	3	5	5
Q	1/3	1	3	1	3	3
E	1/5	1/3	1	1/3	1	1
C	1/3	1	3	1	3	3
A	1/5	1/3	1	1/3	1	1
X	1/5	1/3	1	1/3	1	1

The assigned values are for demonstration purposes only. The normalized weight of each attribute is calculated following the procedure presented in step 4 of

Section 3.2.3, and these are: W_{ADM} = 0.4188, W_Q = 0.1873, W_E = 0.0688, W_C = 0.1873, W_A = 0.0688, and W_X = 0.0688. The value of λ_{max} is 6.0578 and CR = 0.00933, which is much less than the allowed CR value of 0.1. Thus, there is good consistency in the judgements made.

The value of FMS selection index is now calculated using the above weights and the normalized data of the attributes given in Table 10.5. The alternative FMS configurations are arranged in descending order of the FMS selection index:

FMS-II	0.9485
FMS-I	0.7351
FMS-IV	0.7673
FMS-III	0.7167

Thus, the revised AHP or ideal mode AHP method also suggest FMS-II as the first choice.

For the same weights of importance of attributes, the SAW method also gives the same ranking as that given by AHP method.

For the same weights of importance of attributes, multiplicative AHP leads to the following ranking order:

FMS-II	0.9473
FMS-IV	0.7548
FMS-III	0.6924
FMS-I	0.6648

Thus, multiplicative AHP also suggests FMS-II as the first choice. WPM also suggests the same ranking as that given by the multiplicative AHP method.

It may be observed that the above ranking is for the given preferences of the decision maker. The ranking depends upon the judgements of relative importance of attributes made by the decision maker.

10.2.2.3 TOPSIS & Modified TOPSIS Methods

Application of the TOPSIS and modified TOPSIS methods also suggests FMS-II as the first choice.

10.2.2.4 Compromise Ranking Method (VIKOR)

Step 1: The objective is to evaluate the four flexible manufacturing systems, and the attributes are: annual depreciation and maintenance costs (ADM), quality of results (Q), ease of use (E), competitiveness (C), adaptability (A), and expandability (X). ADM is a non-beneficial attribute, and lower values are desirable. The remaining attributes are beneficial, and higher values are desirable. Table 10.4 gives the fuzzy scores. The best, *i.e.*, $(m_{ij})_{max}$, and the worst, *i.e.*, $(m_{ij})_{min}$, values of all attributes are also determined.

Step 2: The values of E_i and F_i are calculated using Equations 3.26 and 3.27, and are given below. The same weights used in the AHP method are considered, and these are: W_{ADM} = 0.4188, W_Q = 0.1873, W_E = 0.0688, W_C = 0.1873, W_A = 0.0688, and W_X = 0.0688.

$E_1 = 0.42 + 0 + 0 + 0 + 0.07 + 0 = 0.49$

$E_2 = 0 + 0.0376 + 0.07 + 0.19 + 0.07 + 0 = 0.3676$

$E_3 = 0.251 + 0.0877 + 0.07 + 0.19 + 0 + 0.07 = 0.6687$

$E_4 = 0.0819 + 0.19 + 0.07 + 0.19 + 0.07 + 0.07 = 0.6719$
$E_{i\text{-}min} = 0.3676$ $E_{i\text{-}max} = 0.6719$
$R_1 = 0.42$ $R_2 = 0.19$ $R_3 = 0.251$ $R_4 = 0.19$
$F_{i\text{-}min} = 0.19$ $F_{i\text{-}max} = 0.42$

Step 3: The values of P_i are calculated using Equation 3.28, and for v = 0.5.
$P_1 = 0.7011$ $P_2 = 0$ $P_3 = 0.6274$ $P_4 = 0.5$

Step 4: The alternatives are arranged in ascending order, according to the values of P_i. Similarly, the alternatives are arranged according to the values of E_i and F_i separately. Thus, three ranking lists are obtained. The best alternative, ranked by P_i, is the one with the minimum value of P_i.

$P_2 = 0$	$E_2 = 0.3676$	$F_2 = 0.19$
$P_4 = 0.5$	$E_1 = 0.49$	$F_4 = 0.19$
$P_3 = 0.6274$	$E_3 = 0.6687$	$F_3 = 0.251$
$P_1 = 0.7011$	$E_4 = 0.6719$	$F_1 = 0.42$

Step 5: For the given attribute weights, FMS-II is suggested as the compromise solution, as it satisfies both conditions given in Section 3.2.7.

References

Albayrakoglu M (1996) Justification of new manufacturing technology: a strategic approach using the analytical hierarchy process. Production and Inventory Management Journal 37:71–77

Bayazit O (2005) Use of AHP in decision-making for flexible manufacturing systems. Journal of Manufacturing Technology Management 16:808–819

Chan FTS, Jiang B, Tang NKH (2000) The development of intelligent decision support tools to aid the design of flexible manufacturing systems. International Journal of Production Economics 65:73–84

Dhavale DG (1990) A manufacturing cost model for computer-integrated manufacturing systems. International Journal of Operations and Production Management 10:5–18.

Elango B, Meinhart WA (1994) Selecting a flexible manufacturing system - A strategic approach. Long Range Planning 27:118–126

Gerwin D, Kolodny H (1992) Management of advanced manufacturing technology: strategy, organization, and innovation. Wiley, New York

Karsak EE (2002) Distance-based fuzzy MCDM approach for evaluating flexible manufacturing system alternatives. International Journal of Production Research 40:3167–3181

Karsak EE, Kuzgunkaya O (2002) A fuzzy multiple objective programming approach for the selection of a flexible manufacturing system. International Journal of Production Economics 79:101–111

Karsak EE, Tolga E (2001) Fuzzy multi-criteria decision-making procedure for evaluating advanced manufacturing system investments. International Journal of Production Economics 69:49–64

Kochan D (1987) Flexible manufacturing and CAD/CAM-evaluation and selection of systems. Computers in Industry 8:201–207

Kulak O, Kahraman C (2005) Multi-attribute comparison of advanced manufacturing systems using fuzzy vs. crisp axiomatic design approach. International Journal of Production Economics 95:415–424

Kuula M (1993) A risk management model for FMS selection decisions: A multiple-criteria decision-making approach. Computers in Industry 23:99–108

Laosirihongthong T, Paul H, Speece MW (2003) Evaluation of new manufacturing technology implementation: an empirical study in the Thai automotive industry. Technovation 23:321–331

Layek AM, Wolf C (1991) Evaluating flexibility of alternative FMS designs-A comparative measure. International Journal of Production Economics 23:3–10

Lloréns FJ, Molina LM, Verdú AJ (2005) Flexibility of manufacturing systems, strategic change and performance. International Journal of Production Economics 98:273–289

Mohanty RP, Venkataraman S (1996) Use of the analytic hierarchy process for selecting automated manufacturing systems. International Journal of Operations & Production Management 13:45–57

Myint S, Tabucanon MT (1994) A multiple-criteria approach to machine selection for flexible manufacturing systems source. International Journal of Production Economics 33:121–131

Perego A, Rangone A (1998) A reference framework for the application of MADM fuzzy techniques to selecting AMTS. International Journal of Production Research 36:437–458

Rao RV (2006) A decision-making framework model for evaluating flexible manufacturing systems using digraph and matrix methods. International Journal of Advanced Manufacturing Technology 30:1101–1110

Rao RV (2007) Evaluating flexible manufacturing systems using a combined multiple attribute decision-making method. International Journal of Production Research (In Press)

Sarkis J (1997) Evaluating flexible manufacturing systems using data envelopment analysis. The Engineering Economist 43:25–46

Sarkis J, Talluri S (1999) A decision model for evaluation of flexible manufacturing systems in the presence of both cardinal and ordinal factors. International Journal of Production Research 37:2927–2938

Shang J, Sueyoshi T (1995) A unified framework for the selection of a flexible manufacturing system. European Journal of Operational Research 85:297–315

Sriram RS, Gupta YP (1991) Strategic cost measurement for flexible manufacturing systems. Long Range Planning 24:34–40

Suresh NC, Kaparthi S (1992) Flexible automation investments: a synthesis of two multi-objective modeling approaches. Computers & Industrial Engineering 22:257–272

Tabucanon MT, Batanov DN, Verma DK (1994) A decision support system for multicriteria machine selection for flexible manufacturing systems. Computers in Industry 25:131–143

Talluri S, Whiteside MM, Seipel SJ (2000) A nonparametric stochastic procedure for FMS evaluation. European Journal of Operational Research 124:529–538

Troxler JW (1990) Estimating the cost impact of flexible manufacturing. Journal of Cost Management 4:26–35

Tseng MC (2004) Strategic choice of flexible manufacturing technologies. International Journal of Production Economics 91:201–298

11

Machine Selection in a Flexible Manufacturing Cell

11.1 Introduction

Machine selection has been a very important issue for manufacturing companies due to the fact that improperly selected machines can negatively affect the overall performance of a manufacturing system. In addition, the outputs of a manufacturing system (*i.e.*, the rate, quality and cost) depend mostly oan appropriate selection of machines and its implementation (Ayag and Ozdemir, 2006). On the other hand, the selection of a new machine is a time-consuming and difficult process requiring advanced knowledge and experience deeply. So, the process can be a difficult task for engineers and managers. For a proper and effective evaluation, the decision maker may need a large amount of data to be analyzed, and many attributes to be considered. The decision maker should be an expert, or at least be very familiar with the specifications of machines to select the most suitable one. In this chapter, the machine selection problem in a flexible manufacturing cell (FMC) is considered to describe the systematical methods offering the best solution. In this chapter, the word 'machine' in a flexible manufacturing cell may be understood as a group of machines required to form the cell.

Flexible manufacturing cells have been used as a tool to implement flexible manufacturing processes to increase the competitiveness of manufacturing systems. Flexible manufacturing cells represent a class of highly automated systems. The increased importance of these highly automated manufacturing systems to the survival of modern industries has resulted in growing research efforts that address the many issues inherent in flexible manufacturing. One of the key issues is the problem of machine selection in a flexible manufacturing cell, which involves a number of attributes, *e.g.*, purchasing cost, machine type, number of machines in a group, floor space requirement, time needed for production, *etc.* To help address this issue, various mathematical and systems modeling approaches have been proposed to date. Sarkis (1997) used the data envelopment analysis (DEA) method for evaluating flexible manufacturing systems. However, DEA requires more computation, and if the number of factors that the decision maker wishes to consider is very large, and the number of alternatives small, then DEA may be a poor discriminator of good and poor performers. Again, DEA may be at a disadvantage in terms of the method's rationale if the decision maker is unfamiliar with linear programming concepts. Talluri *et al.* (2000) proposed a framework based on the combined application of DEA and nonparametric statistical procedures, for the selection of flexible manufacturing systems.

Wang *et al.* (2000) presented a real case of machine selection in a flexible manufacturing cell using a fuzzy multiple attribute decision-making method. However, the method was cumbersome in terms of the representation of weights of relative importance of the factors, fuzzy distributions, rating and ranking models, computation time, *etc.* Malek and Resare (2000) presented an algorithm-based decision support system for the concerted selection of equipment in machining/assembly cells. Karsak and Tolga (2001) proposed a fuzzy decision algorithm to select the most suitable advanced manufacturing system alternative. Both an economic evaluation criterion and strategic criteria such as flexibility, quality improvement, were considered for selection. Karsak and Kuzgunkaya (2002) proposed a fuzzy multiple objective programming approach for the selection of a flexible manufacturing system. Karsak (2002) presented a distance-based fuzzy multiple criteria decision-making (MCDM) approach based on the concepts of ideal and anti-ideal solutions for the selection of a flexible manufacturing system from a set of mutually exclusive alternatives. Rai *et al.* (2002) proposed a fuzzy goal-programming model using a genetic algorithm for machine tool selection and operation allocation in FMS.

Yurdakul (2004) proposed AHP as a strategic decision-making method for machine tool selection. Tseng (2004) proposed a game theoretical model for selection of flexible manufacturing technologies. Chtourou *et al.* (2005) developed an expert system for manufacturing systems machine selection. Chan and Swarnkar (2005) presented a fuzzy goal-programming approach to model the machine tool selection and operation allocation problem of a flexible manufacturing system. An ant colony optimization (ACO)-based approach was applied to optimize the model. Chan *et al.* (2005) presented a fuzzy goal-programming approach to model a machine tool selection and operation allocation problem; the model was optimized using an approach based on artificial immune systems (AIS).

Djassemi (2005) examined the performance of cellular manufacturing systems in a variable demand and flexible work force environment using simulation modeling. Mishra *et al.* (2006) presented a fuzzy goal-programming model having multiple conflicting objectives and constraints pertaining to a machine tool selection and operation allocation problem, and a new random search optimization methodology termed quick converging simulated annealing (QCSA) was used. Ayag and Ozdemir (2006) proposed a fuzzy AHP method for evaluating machine tool alternatives.

Even though precision-based methods such as expert systems, neural networks, goal programming methods, fuzzy algorithms, genetic algorithms, simulated annealing, *etc.* have been proposed in the past to address the issue of selection of flexible manufacturing technologies, these methods are knowledge-intensive, complicated, require a high level of computation, and may go beyond the capabilities of the real decision maker (*i.e.,* user organization). Also, most research work has concentrated on flexible manufacturing systems (FMS), and only a few authors have considered the problem of machine selection in a flexible manufacturing cell (FMC), once the alternative machines are developed. Thus, there is a need for a simple, systematic and logical method or mathematical tool to guide user organizations in taking a proper decision involving a number of machine selection attributes and their interrelations. This is considered in this chapter using the GTMA and other fuzzy MADM methods.

11.2 Example

Wang *et al.* (2000) presented a real case of a machine group selection in a flexible manufacturing cell including two CNC milling machine groups, a CNC lathe, and a robot for material handling.

The constraints were described as explained below:

Constraint 1: The total purchasing cost should not exceed 600,000 dollars
Constraint 2: for CNC milling machine
Vertical/horizontal: horizontal
Spindle speed : 4,500 rpm
X/Y/Z axis travel : 630/630/500
Feed rate : 5,000 mm/min
Tool capacity : 40
Maximum tool diameter : 130 mm
Constraint 3: for CNC lathe
Maximum swing : 520 mm
Maximum turning diameter : 350 mm
Maximum turning length : 500 mm
Hole through spindle : 70 mm
Chuck size : 8”
Spindle speed : 4,500 rpm
Feed rate : 4,500 mm/min
Constraint 4: for robot
Configuration : arm-like
Max. load capacity at wrist : 60 kg
Allowable load moment of wrist : 36 kg-m
Horizontal reach : 150 cm
Repeatability : 1.0
Drive method : Electrical

Furthermore, in the allowance for the operating procedure, the two milling machines can be replaced with a multifunctional machining center.

After incorporating the above constraints into the total purchasing cost, and into the specifications of the milling machine, lathe machine, and robot, suitable machine groups of FMC were composed into 10 possible alternatives. Table 11.1 presents these 10 short-listed possible alternatives.

Table 11.1. Objective data of attributes of the example considered (Wang *et al.*, 2000; with permission from Taylor & Francis Ltd., http:www.tandf.co.uk/journals)

Alternative	Total purchasing cost ($)	Total floor space (m^2)	MN	Productivity* (mm/min)
1	581,818	54.49	3	5,500
2	595,454	49.73	3	4,500
3	586,060	51.24	3	5,000
4	522,727	45.71	3	5,800
5	561,818	52.66	3	5,200
6	543,030	74.46	4	5,600
7	522,727	75.42	4	5,800
8	486,970	62.62	4	5,600
9	509,394	65.87	4	6,400
10	513,333	70.67	4	6,000

MN: Total number of machines in a machine group of the flexible manufacturing cell
*Productivity (mm/min): the value corresponds to the machine with the slowest feed rate in the machine group

Now, application of the GTMA and other fuzzy MADM methods is carried out as explained below.

11.2.1 Application of GTMA

The machine selection attributes considered are the same as those of Wang *et al.* (2000), and these are total purchasing cost (PC), total floor space (FS), total machine number (MN) and productivity (P). The machines short-listed are also the same as those of Wang *et al.* (2000).

The objective values of the machine selection attributes, which are given in Table 11.1, are to be normalized. Productivity (P) is a beneficial attribute and higher values are desirable. Values of these attributes are normalized, and given in Table 11.2 in the respective column. PC, FS, and MN are non-beneficial attributes, and lower values are desirable. The values of these attributes for different alternative machines are normalized, and given in Table 11.2 in the respective columns.

Relative importance of attributes (a_{ij}) is assigned the values using Table 4.4. Let the decision maker (*i.e.*, user organization) select the following assignments:

$$
\begin{array}{c} \\ PC \\ FS \\ MN \\ P \end{array}
\begin{array}{cccc} PC & FS & MN & P \\ \end{array}
$$

$$
\begin{array}{c} PC \\ FS \\ MN \\ P \end{array}
\begin{bmatrix} \text{---} & 0.665 & 0.745 & 0.590 \\ 0.335 & \text{---} & 0.590 & 0.410 \\ 0.255 & 0.410 & \text{---} & 0.335 \\ 0.410 & 0.590 & 0.665 & \text{---} \end{bmatrix}
$$

Table 11.2. Normalized data of the machine selection attributes of the example considered

Alternative	PC	FS	MN	P
1	0.854	0.839	1.000	0.859
2	0.835	0.919	1.000	0.703
3	0.848	0.892	1.000	0.781
4	0.951	1.000	1.000	0.906
5	0.885	0.868	1.000	0.812
6	0.915	0.614	0.750	0.875
7	0.951	0.606	0.750	0.906
8	1.000	0.730	0.750	0.875
9	0.976	0.694	0.750	1.000
10	0.968	0.647	0.750	0.938

The machine selection attributes digraph, machine selection attributes matrix of the digraph, and machine selection function for the matrix can be prepared. The value of machine selection index is calculated using the values of A_i and a_{ij} for each alternative machine, and these are given below in descending order:

Alternative 4: 3.488808
Alternative 5: 3.002258
Alternative 1: 2.981451
Alternative 3: 2.930671
Alternative 2: 2.825410
Alternative 9: 2.755053
Alternative 8: 2.679458
Alternative 10: 2.597028
Alternative 7: 2.478609
Alternative 6: 2.411448

From the above values of the machine selection index, alternative 4 is the best choice among the alternatives considered for the flexible manufacturing cell under the given conditions. The ranking of machines based on the proposed methodology is 4-5-1-3-2-9-8-10-7-6; the ranking presented by Wang *et al.* (2000) was 4-5-3-1-2-8-9-10-7-6. The above results suggest the selection of alternative 4 for the FMC as the first right choice, alternative 5 as the second right choice, and alternative 6 as the last choice. These results are consistent with those presented by Wang *et al.* (2000). However, the ranking of certain alternatives obtained by using the proposed procedure is different from that proposed by Wang *et al.* (2000). For example, the third choice is alternative 1 as per the procedure proposed here, whereas it was alternative 3 in Wang *et al.* (2000). A closer look at the objective data of the four attributes PC, FS, MN, and P of these two alternatives reveals that there are significant differences between the two alternatives 1 and 3 in the case of PC ($581,818 *vs.* $586,060) and P (5,500 mm/min *vs.* 5,000 mm/min), that the difference is not high in the case of FS (54.49 m^2 *vs.* 51.24 m^2) and that there is no difference in the case of MN. Alternative 1 is best from the PC and P points of view, and alternative 3 is better from the FS point of view, and both alternatives are equal from the MN point of view. Thus, keeping in mind the relative importance of the attributes, proposing alternative 1 as the third choice based on the method proposed here seems to be more meaningful, compared to alternative 3 as proposed by Wang *et al.* (2000). Similarly the differences in the ranking of alternatives between the proposed procedure and the procedure suggested by Wang *et al.* (2000) can be explained. However, it may be added here that the weights of relative importance used by Wang *et al.* (2000) were different from those used in the present work. Further, it may be mentioned that the ranking depends upon the judgements of relative importance made by the decision maker.

The fuzzy method proposed by Wang *et al.* (2000) is cumbersome in terms of the representation of weights of relative importance, fuzzy distributions, rating and ranking models, computation time, *etc.* Further, the authors had converted the available objective values of the attributes, after normalization, into fuzzy values, which violates the basic rule of fuzzy logic, *i.e.*, the available objective values need not be fuzzified. Comparatively, the proposed GTMA provides a simple, straightforward and logical procedure for the machine selection problem in a flexible manufacturing cell.

Following graph theory and the matrix approach, the coefficients of similarity/dissimilarity can also be calculated for different machines using Equations 2.15 and 2.16.

It may be noted that GTMA offers a general methodology, and is applicable to any type of machine selection problem involving any number of machine selection attributes.

11.2.2 SAW Method

For start, the attributes are ranked in order of importance and 10 points are assigned to the least important attribute MN. The attribute FS is given 15 points to reflect its relative importance. P and PC are given 25 and 30 points, respectively. Thus, the weights of PC, FS, MN, and P are calculated as 0.375, 0.1875, 0.125, and 0.3125 respectively. Using these weights, and the normalized data of the attributes for different machines, the machine selection index values are calculated, and are arranged in descending order of the index.

Alternative 4:	0.9523
Alternative 9:	0.9024
Alternative 8:	0.8791
Alternative 5:	0.8734
Alternative 10:	0.8712
Alternative 1:	0.8710
Alternative 3:	0.8543
Alternative 7:	0.8471
Alternative 2:	0.8301
Alternative 6:	0.8254

From the above values of the machine selection index, it is clear that the alternative machine, designated as 4 is the best choice among the machines considered.

11.2.3 WPM

Using the same weights of the attributes as those selected for the SAW method, the following ranking of machines is obtained:

Alternative 4:	0.9515
Alternative 9:	0.8927
Alternative 5:	0.8716
Alternative 1:	0.8697
Alternative 10:	0.8609
Alternative 3:	0.8517
Alternative 7:	0.8356
Alternative 8:	0.8266
Alternative 2:	0.8240
Alternative 6:	0.8167

This method also suggests alternative 4 as the first choice and alternative 9 as the second choice.

11.2.4 AHP and its Versions

If the weights selected for the SAW method are used also in this method, then the ranking of machines obtained by using the relative as well as ideal mode AHP will be same. The multiplicative AHP method yields the same ranking as that given by WPM. However, let the decision maker prepare the following matrix:

	PC	FS	MN	P
PC	1	3	4	2
FS	1/3	1	2	1/2
MN	1/4	1/2	1	1/3
P	1/2	2	3	1

The normalized weights of each attribute are calculated following the procedure presented in Section 3.2.3, and these are: $W_{PC} = 0.467$, $W_{FS} = 0.16$, $W_{MN} = 0.095$, and $W_P = 0.278$.

The value of the machine selection index is now calculated using the above weights, and the normalized data of the attributes given in Table 11.2. The alternative machines are arranged in descending order of the machine selection index.

Alternative 4:	0.9509
Alternative 9:	0.9161
Alternative 8:	0.8983
Alternative 10:	0.8876
Alternative 5:	0.8729
Alternative 1:	0.8669
Alternative 7:	0.8642
Alternative 3:	0.8508
Alternative 6:	0.8400
Alternative 2:	0.8274

From the above values of the machine selection index, it is clear that the machine, designated as 4 is the best choice among the alternatives considered.

It may be observed that the above ranking is for the given preferences of the decision maker.

11.2.5 TOPSIS Method

The quantitative values of the machine selection attributes, which are given in Table 11.1, are normalized as explained in Section 3.2.6.

Relative importance of attributes (a_{ij}) is assigned using the AHP method as explained in Section 11.2.4. After performing the calculations, the alternative machines are arranged in descending order of their machine selection index. This can be arranged as 4-9-8-10-7-5-1-6-3-2.

11.2.6 Modified TOPSIS Method

Following the procedure of the modified TOPSIS method and using the same weights as those of the TOPSIS method, the following ranking of alternative machines is obtained:

Alternative 4:	0.7842
Alternative 9:	0.5755
Alternative 5:	0.5526
Alternative 1:	0.5475
Alternative 8:	0.5415
Alternative 3:	0.5038
Alternative 10:	0.4806
Alternative 2:	0.4557
Alternative 7:	0.4045
Alternative 6:	0.3471

It can be observed that all the above decision-making methods propose machine designated as 4 as the first right choice.

The example problem considered in this chapter is related to the selection of a group of machines required for a flexible manufacturing cell. However, the proposed decision-making methods are quite general, and can be applied also for the selection of a single machine tool for a given industrial application. For example, if a CNC machining center is required to be purchased by a firm, and a finite number of alternative CNC machining center configurations are available with objective as well as subjective information of the attributes, then the decision-making methods proposed in this chapter can be useful to the firm.

Ayag and Ozdemir (2006) considered the problem of selection of a CNC vertical turning center for general use by a company. Nineteen machine selection attributes were considered, and these were: productivity (spindle speed, power, cutting feed, traverse speed), flexibility (number of tools, rotary table), space (machine dimensions, adaptability, CNC type, taper number); precision (repeatability, thermal deformation), reliability (bearing failure rate, reliability of drive system, safety and environment, mist collector, safety door, fire extinguisher, training), and maintenance and service (repair service, regular maintenance). Three machine alternatives (Maho, Hass, and Seiki) were evaluated by Ayag and Ozdemir (2006) using the fuzzy AHP method. However, the fuzzy version of AHP proposed by Ayag and Ozdemir (2006) is a complicated one. Once the objective and/or subjective data of the above 19 attributes are available, then the decision-making methods proposed in this book can be used for machine selection. The pair-wise comparison of the 19 attributes can also be made easily.

It may be worth mentioning here that fuzzy set theory has serious difficulties in producing valid answers in decision-making based on fuzzifying judgements. No theorems are available dealing with its workability when applied indiscriminately as a number-crunching approach to numerical measurements that represent judgements. When numerical representation of judgements is allowed to vary in choice over the values of a fundamental scale, as in the analytic hierarchy process, these judgements are themselves already fuzzy. To make these even fuzzier can decrease the validity of the outcome, when the actual outcome is known. Also, improving the consistency of a judgement matrix does not necessarily increase the validity of the outcome. Validity is the goal in decision making, not consistency, which can be successively improved by manipulating the judgements as the answer becomes even farther removed from reality. Ayag and Ozdemir (2006) had not considered this fact in their work. Making fuzzy judgements fuzzier does not lead to a better outcome, and indeed often leads to a worse one. That is why this book proposes a logical method in Chapter 4 to take care of the above points, while assigning objective values to the subjective data of the attributes as well as deciding the relative importance of attributes.

References

Ayag Z, Ozdemir RG (2006) A fuzzy AHP approach to evaluating machine tool alternatives. Journal of Intelligent Manufacturing 17:179–190

Chan FTS, Swarnkar R (2005) Ant colony optimization approach to a fuzzy goal programming model for a machine tool selection and operation allocation problem in an FMS. Robotics & Computer Integrated Manufacturing doi:10.1016/j.rcim.2005.08.01

Chan FTS, Swarnkar R, Tiwari MK (2005) A fuzzy goal-programming model with an artificial immune system (AIS) approach for a machine tool selection and operation allocation problem in a flexible manufacturing system. International Journal of Production Research 43:4147–4163

Chtourou H, Masmoudi W, Maalej A (2005) An expert system for manufacturing systems machine group selection. Expert Systems with Applications 28:461–467

Djassemi M (2005) A simulation analysis of factors influencing the flexibility of cellular manufacturing. International Journal of Production Research 43:2101–2111

Karsak EE (2002) Distance-based fuzzy MCDM approach for evaluating flexible manufacturing system alternatives. International Journal of Production Research 40:3167–3181

Karsak EE, Kuzgunkaya O (2002) A fuzzy multiple objective programming approach for the selection of a flexible manufacturing system. International Journal of Production Economics 79:101–111

Karsak EE, Tolga E (2001) Fuzzy multi-criteria decision-making procedure for evaluating advanced manufacturing system investments. International Journal of Production Economics 69:49–64

Malek LA, Resare LJ (2000) Algorithm based decision support system for the concerted selection of equipment in machining/assembly cells. International Journal of Production Research 38:323–339

Mishra A, Prakash N, Tiwari MK, Lashkari RS (2006) A fuzzy goal-programming model of machine-tool selection and operation allocation problem in FMS: a quick converging simulated annealing-based approach. International Journal of Production Research 44:43–76

Rai R, Kameshwaran S, Tiwari MK (2002) Machine-tool selection and operation allocation in FMS: solving a fuzzy goal-programming model using a genetic algorithm. International Journal of Production Research 40:641–665

Sarkis J (1997) Evaluating flexible manufacturing systems using data envelopment analysis. The Engineering Economist 43:25–46

Talluri S, Whiteside MM, Seipel SJ (2000) A nonparametric stochastic procedure for FMS evaluation. European Journal of Operational Research 124:529–538

Tseng MC (2004) Strategic choice of flexible manufacturing technologies. International Journal of Production Economics 91:223–227

Wang T, Shaw CF, Chen YL (2000) Machine group selection in flexible manufacturing cell: a fuzzy multiple attribute decision-making approach. International Journal of Production Research 38:2079–2097

Yurdakul M (2004) AHP as a strategic decision-making tool to justify machine tool selection. Journal of Materials Processing Technology 146:365–376

12

Failure Cause Analysis of Machine Tools

12.1 Introduction

Machine tool reliability and maintainability significantly affect the three elements of competitiveness: quality, cost, and production time. Well-maintained machines produce tolerances better, help reduce scrap and reworking, and raise the consistency and quality of the part. Further, such machine tools increase uptime and yield good parts, thereby reducing total production cost. Machine tools form a complex system consisting of various subsystems/components, and failure of a machine tool may occur due to failure(s) in any of the subsystems/components. For example, a CNC machine tool will invariably incorporate some, if not all, of the following subsystems:

- Electronic subsystems: microprocessor- or mini-computer-based controllers, input/output devices (displays, keyboards, disk and tape drives, data ports), memory components, analog systems (A/D, D/A converters and power amplifiers).
- Electrical subsystems: motors, contactors, relays, limit switches, servo-feedback components, *etc.*
- Mechanical subsystems: gear boxes, slides and slide ways, drives, spindles, work holding devices, tool magazines and changers, swarf controllers, pallet systems, *etc.*
- Hydraulic subsystems: reservoirs, filters, pumps, valves, pressure relief valves, actuators, piston-cylinder arrangement, *etc.*
- Lubrication subsystems
- Coolant subsystems.

Failure of a machine tool may occur due to failure(s) in any of the elements of the subsystems. The failure may be attributed to specific failure causes. A failure cause is defined as a reason that makes the machine tool unable to perform its intended function. This may be attributed to failure events contributed by its subsystems, assemblies, or components, including the cutting process and the cutting conditions. In the present work, machine tool failures are examined by analyzing the contributing events for their failure cause.

There are numerous failure-causing elements in a machine tool. Some important common failure causes of a machine tool are given below:
Faulty components, wear between mating parts, fatigue of the components, casting and welding defects in the machine tool structure, thermal stresses, high cutting temperatures, excessive cutting forces, low rigidity, vibrations, noise, electrical and electronic troubles, geometrical inaccuracy, low hydraulic and pneumatic pressure (for clamping devices, rotating devices, and feed drives), failure of bearings, contamination of slideways (*e.g.*, due to swarf), loss of lubrication (slides, racks, ball-screw bearings, gears, and chains), malfunctions of valves, filter problems, cooling problems, pump cavitation, imbalance and disturbance in rives, chip conveying problems, clamping and indexing problems, tensioning problems in belts and chains, contactor troubles, motion control troubles, encoder problems, software troubles, servo adjustments, main process-related mechanisms, feed process-related mechanisms, auxiliary mechanisms, materials transportation system, environmental conditions, incorrect cutting conditions, nature of the machining process, incompatible cutting fluids, operator errors leading to poor operation, poor maintenance, *etc.*

A failure manifests itself as a deviation of the machine tool behavior from its specified behavior. Martin (1994) distinguished between two different types of machine tool failures. The soft or gradual failure develops gradually with time, and this is characteristic of many mechanical and hydraulic elements of the machine tool where wear takes place, causing a gradual degradation of the operation of the element. The hard or catastrophic failure takes place instantaneously, and the element is either on or off, this is characteristic of many electrical circuit elements, but does occur also in mechanical elements, *e.g.*, brittle fracture.

Interest in machine tool failure data was shown by an early study of machine tool reliability undertaken by the Machine Tool Industry Research Associates (Stewart, 1977) and reported that the average downtime due to breakdowns was of the order of 7.6%, and the failure rate was 1-2 breakdowns per NC machine per month. A study of 35 CNC machines (1981), based upon service engineers' records during one year warranty, quoted "an average availability of around 83%". Kilmartin and Hannan (1981) invoked the poor diagnostics of electronics in explaining much of the downtime, but over the following years evidence supporting the development of diagnostics in the domain was reported by Kegg (1984).

Continued interest in the reliability of machine tools gave rise to an initiative by the National Center for the System Reliability (NCSR) in the UK. This brought together a consortium of machine tool users and manufacturers, NCSR providing staff to collect the appropriate data. A report of NCSR (1988) provided data on the reliability of the CNC machine centers. The data were confidential to consortium members, but in general did highlight the more significant failure areas.

A machine tool is a complex system, and it is not possible to contemplate the condition monitoring of all parameters that describe the behavior of a machine tool. Consequently, a limited choice has to be made, and this should be based upon the information available on failures, their frequency, and the resulting downtime. Johansson (1981) proposed parameters for monitoring CNC lathes, such as: feed drive current, mains voltage, hydraulic oil pressure, acceleration time for spindle motor, interval for central lubrication, tool change time, temperature of spindle bearings, temperature of control box, temperature of spindle motor, temperature of hydraulic oil, oil filter degree of purity, and number of movements of X and Y slides.

Martin *et al.* (1990) applied the techniques of failure modes and effects analysis to analyze the catastrophic failures of machine tools. These are essentially logical decisions based upon system knowledge, and lend themselves to computerization and automatic decision making. Majstorovic and Milacic (1990) defined the basic architecture of expert systems for diagnosis and maintenance, and reviewed the current uses of expert systems. The authors reviewed 87 different expert systems; of 4.6% machine tool systems. Freyermuth (1991) described an expert system type analysis to define failures based on 'fault-symptom trees', which have similarities with fault trees. Angeli and Chatzinkikoraou (1985) developed an expert system to diagnose the faults in hydraulic systems. Marczewsky (1988) developed and implemented an expert system called 'Charley' to track machine tool conditions using vibration signatures from the machine tools. Puetz and Eichhorn (1987) proposed expert system shells for the failure diagnosis of CNC machine tools.

The main techniques used for the diagnosis of soft or gradual failures are: pattern recognition techniques (using artificial neural networks and fuzzy logic), expert systems, and mathematical model-based detection techniques. Williams (1990) had described different methods of automatic recognition of failure patterns. Pattern recognition techniques generally rely upon the use of failure dictionary-stored information upon the reaction of the system to certain failures. Marzi and Martin (1990) reported the design of a neural network that which analyzed the gradual failures represented by changes in the transient response at the outlet of the pump of a machine tool coolant system. Lee and Kramer (1993) proposed a methodology using neural networks to monitor machine tool behavior. A pattern discrimination model is used to measure the performance degradation quantitatively. Lee (1995) reviewed machine tool condition monitoring and fault diagnosis methods. Drake and Pan (1996) presented a method for diagnosing multiple failures and the levels of severity of individual faults in the flood coolant system of a CNC vertical milling machine tool. The method employed a neural network for pattern recognition with features extracted from the transient response of the coolant pressure on shut down.

Ye and Zhao (1996) developed a highly integrated system, integrating neural networks with a procedural decision-making algorithm, to implement hypothesis-test cycles in a manufacturing system diagnosis of tested failure events. Zeng and Wang (1991) described an experimental study to investigate the feasibility of employing fuzzy set theory in an integrated failure diagnostic system. The main monitored signal was assumed to be acceleration transformed into frequency spectra. Comparisons between the operating machine patterns and those in the failure dictionary were made to define the machine operating system by the use of the fuzzy fault assignment technique. Holloway and Krogh (1990) had proposed a behavioral model approach for failure detection and analysis in automated manufacturing systems. Their model provided the basis for on-line failure detection by generating expected system response signals that were compared with the actual sensor signals from the system. Failure analysis was accomplished by maintaining a current set of operational assumptions that identify the system components possibly causing deviations from the expected behavior.

Isermann (1991) described mathematical model based techniques for the detection of gradual failures in machine tools. The diagnosis techniques were described based upon the measurement of the variation of parameters, concluding with a description of a knowledge-based diagnosis system. Martin (1994) discussed model-based failure detection techniques covering the fields of modeling, parameter estimation, state estimation, use of observers, and parity space approaches. Alexander *et al.* (1993) developed a model for the diagnosis of CIM equipment. Poltavets (1994) presented fault diagnostic parameters (temperature, movement accuracy, vibration, and acoustics) and suggested wide use of highly effective computer equipment, mathematical modeling, and intensive development of sophisticated system investigation methods.

Rao (1997) reviewed key developments in the area of metal cutting machine tool design from a very practical perspective. Defining the drivers of machine tool design as higher productivity and higher accuracy, the author examined advances in design stemming from the needs of these two drivers. Kwon and Burdekin (1998) presented an adjustment technique for controller setting values in CNC machine tools by measurement of servo-induced feed drive errors. For measurement of the servo-induced errors, an experimental technique which incorporated two linear displacement sensors and a steel cube was developed, and servo feed drive errors were evaluated along a square corner test path. Based on evaluations of servo feed drive errors, different combinations of parameters in the machine control system and optimum setting parameters were found.

Hu *et al.* (1999) proposed a systematic approach for the failure diagnosis of flexible manufacturing systems that integrates condition monitoring, failure diagnosis and maintenance planning. Two diagnostic models for PLC-controlled flexible manufacturing systems were presented. In another work, Hu *et al.* (2000) designed an intelligent integrated fault-diagnosis system with a modular, and reconfigurable structure. The implementation of the integrated diagnosis was presented in detail. The system could monitor conditions, and diagnose the major failures of a flexible manufacturing system, and give corresponding maintenance planning as well. Huang and Liao (2000) developed a maintenance schedule and fault diagnosis system that integrates an artificial neural network and an expert system for a wire EDM setup. The faults considered were: wire breaking and unsatisfactory accuracy. Suggestions were made to eliminate/reduce these faults. Rao and Gandhi (2002) presented digraph and matrix methods for failure cause analysis of machine tools. Das *et al.* (2007) discussed reliability aspects of the design and analysis of cellular manufacturing systems. Luis *et al.* (2006) presented details of a sensor-less tool failure monitoring system for drilling machines.

Most of the techniques or approaches described above have drawbacks. In the case of systems such as pattern recognition techniques (using neural networks, fuzzy logic), and expert systems, a great deal of research is necessary in order to apply these types of systems to machine tool failure diagnosis. Significant amounts of data and time are required to define the normal healthy behavior of a machine tool. For example, pattern recognition techniques generally rely upon the use of failure dictionary stored information on the reaction of the system to certain failures. Because of the complexity of a machine tool, it will be possible to deal only with a limited number of failures and therefore, the necessity to define the most important failures. If a decision can be made regarding the significant failures, there still remains the decision as to which parameter will be sensitive to the failure. In certain cases, the parameter is easily definable, and the more complex cases will need research in themselves. In the laboratory, it is necessary to stimulate faults for which neither the time nor the resources are available to run the machine tools until failures occur. Even if this is done, most automatic data acquisition systems (DASs) generate excessive amounts of data, and the problem lies in data storage and analysis. In the case of mathematical modeling techniques, measurements have to be made on healthy systems to define the mathematical model, and to store the healthy response. These measurements will probably be different for what are normally the same machine tool type, and thus a separate machine needs its own measurement; it is not possible to take a measure of one machine and assume that this will apply to another.

As explained above, fault diagnosis and maintenance are knowledge-intensive, experimental tasks that may go beyond the capabilities of the practicing maintenance engineer. Moreover, *ad hoc* replacements further aggravate the problems at the operational stage, which not only culminates in loss of production and increase in machine downtime, but can also lead to human loss and injuries. This problem can be minimized to a large extent if failure cause analysis is considered. Its implementation at the design stage will lead to the design of failure-free reliable machine tool systems. It will also help in minimizing downtime, and avoiding *ad hoc* replacements.

The structure of a machine tool is highly important to understand and model its failure. The structure may be physical or abstract. In the system's structure, the components of the system/subsystem, the properties relevant to the problem are identified and their characteristic interdependence and interactions determined. The importance of a system's structure has been emphasized by many researchers (Czichos, 1978, 1980; Yoshikawa, 1982; Kokowa and Shingai, 1982; Kokowa *et al.*, 1983; Ishida *et al.,* 1985; Sethi and Agrawal, 1993; Gandhi and Agrawal, 1994, 1996; Clark and Paasch, 1997).

The subsystems/components of a machine tool are expected to perform appropriate function(s) to attain a desired output. The output of a given machine tool depends upon how well individual subsystems/components perform. The malfunctioning of a machine tool system is attributed to improper functional interaction between its components and subsystems. This means a function specifies intended behavior of an individual component or subsystem. Moreover, the structure of a machine tool system is important for understanding the connectivity of its components and subsystems. Therefore, both function and structure are key entities in failure consideration as a whole. However, to analyze the failure causes of machine tools, it is indispensable to consider the functional and structural interaction of the machine tool system especially at the design stage. This exercise, which aims to minimize the operational failures, can be implemented at the design stage, if appropriate procedures based on this approach are made available to the designer. Fault tree analysis (Fussell, 1975, 1976) has been extensively used in chemical and process industries for root cause analysis. However, this does not take into account the structure of the system explicitly. Hence there is a need for an appropriate procedure to analyze the failure causes of a machine tool. This aspect is considered in the present work using graph theory and the matrix approach. Graph theory is useful to represent the system structure and in conjunction with the matrix approach enables analysis of the problem in a more convenient way. Rao and Gandhi (2002) demonstrated this approach for machine tool failure cause analysis with the identification of failure contributing events and their interaction for a machine tool failure cause.

Failure cause of a machine tool is analyzed considering the contributing events and their interactions, and is demonstrated in the following sections.

12.2 Identifying Contributing Events of a Failure Cause

The contributing events of a failure cause of a machine tool are identified by examining various aspects such as affected system structure, mating components, cutting process, cutting conditions, and the tool and work piece. To illustrate this, an example of vibrations of machine tools is considered and the events of this failure cause (*i.e.*, vibrations) are considered by examining the following aspects:

- Machine tool: The machine tool structure deflects due to cutting forces and the weight of the moving subassemblies. The stiffness of the structure must be high with high damping characteristics to minimize the influence of dynamic loads. If this is not so, the frequency of vibration may coincide easily with the natural frequency of any mode of the machine tool, resulting in complete or partial destruction of the machine tool. Besides, vibration decreases the life of the machine tool.
- It is commonly experienced that, in any machine shop floor, if anywhere dynamic force through vibration is transmitted to the ground, then the machine shop floor will vibrate. This vibration may be transmitted to the machine tool through its foundation, and cause vibrations in the machine tool and damage the job surface.
- Incorrect machine tool leveling will also cause vibrations in the machine tool.
- The disturbances in the machine tool drives also lead to vibrations in the machine tool. These disturbances are generated due to many reasons. Some of the reasons are:
  - Rotating unbalanced masses: The effect of rotating unbalanced masses becomes more prominent when rotating bodies, or parts, are supported on the top of slender parts.
  - Faulty arrangement of drives: Faulty arrangement of drives also produces vibrations. If the driving gears have eccentricities, pitch errors, profiles errors, damaged portions, *etc.*, then they will produce non-uniform rotation which may contribute to machine tool vibrations. In the case of belt drives, if the section of the belt used is non-uniform then the effective pulley radius will change periodically causing a periodic variation of belt

tension. Belts that are too tight or too slack will also cause machine tool vibrations.

- Fault in the supporting bearings: If the bearings supporting the rotating members of the machine tool are faulty, the rotating members will not be fixed in position and will change position periodically depending on the nature of the fault. Moreover, if the frequency of the system is of the same order, then an appreciable vibration may be generated. Radial and axial play in the spindle may lead to vibrations.
- Reciprocating disturbances: The disturbance in the elements of machine tool executing rectilinear motion can also cause vibrations. This type of vibration may be due to reciprocating imbalance, or to stick-slip motion.

- Type of cutting and cutting conditions: Sometimes, when the cutting process itself is intermittent (*e.g.*, milling), or periodically discontinuous (*e.g.*, cutting with discontinuous chip formation), then cutting force fluctuates with a definite period. Due to this fluctuating or dynamic cutting force, which is transmitted to the machine tool via the cutting tool and the job, it is quite likely that a forced vibration will be generated due to the elastic nature of the system. If the frequency of force fluctuation falls in the range of natural frequency of the machine tool, then the vibrations will be severe. In addition, incorrect cutting conditions, *e.g.,* cutting speed, feed, and depth of cut, cause vibrations.
- The machine tool can vibrate due to the cutting process itself under particular conditions. In these cases, the active force is not from an outside element, but is due to the cutting process itself. These types of vibrations are self-induced, and commonly known as machine tool chatter. A slight disturbance in the cutting process caused by varying chip thickness, varying rate of penetration of the tool into the job, or variation in the angular speed of the job may cause such vibration.
- Work piece: If any inhomogeneity is present in the work material, an impulsive force will be generated due to a sudden increment in the hardness of the work material. As an effect of this impulse, a free vibration is set up in the cutting tool, and also in the machine tool body. Not-so-rigid work piece holding and its balancing, and slenderness of the work material also lead to vibrations.
- Cutting tool: Tool overhanging contributes to machine tool vibrations. Proper setting is required. A wrong geometry of the tool, or blunt tool lead to vibrations. Vibrations decrease the cutting tool life.
- Built-up edge on the cutting tool, formed due to the wrong cutting conditions, has an effect on vibrations similar to that of inhomogeneities in the work material. If the machine tool system is not dynamically stable, then the effect is considerable.
- Sudden impact load on the cutting tool sets up vibrations in the cutting tool and also in the machine tool body.

It may be added that the contributing events for other failure causes of a machine tool mentioned earlier can be identified in a similar way as that described above. The contributing events identified above are considered for modeling the machine tool failure cause using graph theory and the matrix approach, and this is discussed in the next section.

12.3 Machine Tool Failure Causality Digraph (MTFCD) and its Matrix Representation

The machine tool failure causality digraph (MTFCD) models a failure cause of a machine tool system, subsystem, or component, considering the failure contributing events and their interaction in terms of cause-effect relationship (*i.e.*, causality). A node V_i represents the i-th failure contributing event, and a directed edge e_{ij} from node i to node j represents the causality relation between i and j events. For example, if event i is the cause event, and j is the affected event, then a directed edge e_{ij} is drawn from node i to node j. If no causality relation exists between two events, then no edge is drawn between these nodes, *i.e.*, $e_{ij} = 0$. Sometimes it is possible that two events may be cause and effect events to one another. In such a case, two directed edges, e_{ij} and e_{ji}, are drawn, *i.e.*, one from node i to node j, and the other from node j to node i. If an event i is the cause and effect event to itself, then it shall be represented by a self-loop at that node. For a considered failure cause, all the contributing events are to be identified first, and also their causality relations. Therefore, every care needs to be taken in identifying all the failure contributing events for a failure cause. It is suggested that this exercise be carried out by a team consisting of designers, and operating and maintenance personnel.

To develop the machine tool failure causality digraph, a machine tool failure cause, *i.e.*, vibrations of the machine tool, is considered, and has been described in the previous section. However, for illustration six most important vibration contributing events are selected, and these events are listed below:

1. Machine tool leveling
2. Type of cutting and the cutting conditions
3. Inhomogeneities in the work material
4. Disturbance in machine tool drives
5. Cutting process
6. Tool setting and job holding

These six events are represented in the machine tool failure causality digraph shown in Figure 12.1 by six nodes. The directed edges are drawn keeping in mind the discussion presented earlier in this section. For example, machine tool leveling affects the disturbance in machine tool drives. So, a directed edge is drawn from node 1 to node 4 in the digraph. Similarly, other directed edges are drawn and the digraph is developed as shown in Figure 12.1. It is likely that there may be more causality relations between these events (*i.e.*, six events) and other events not shown (of other failure causes of machine tool), and these are represented as dashed directed edges. This is, however, for illustration only and is not considered further in the analysis.

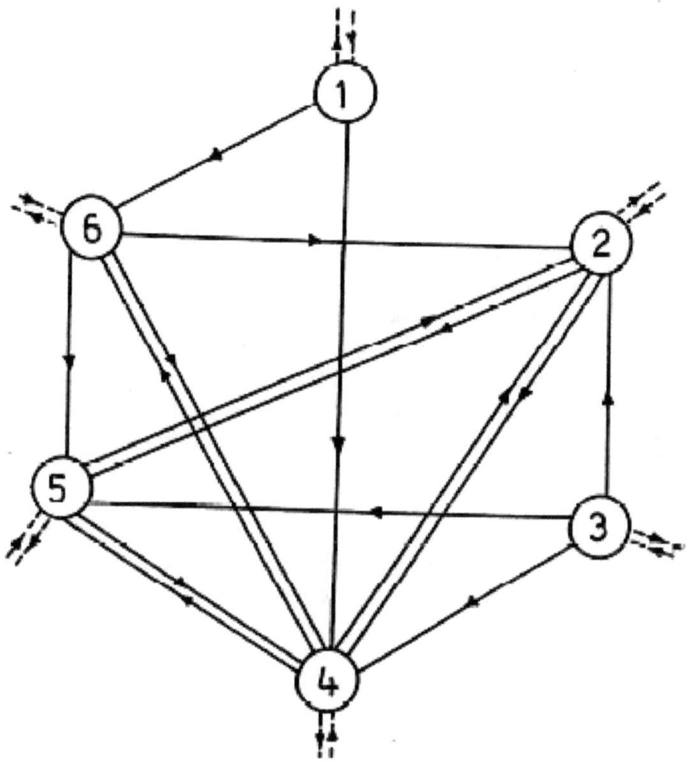

Figure 12.1. Machine tool failure causality digraph (from Rao and Gandhi 2002; reprinted with permission from Elsevier)

The machine tool failure causality digraph represents the graphical relationship among thee contributing events for a failure cause. If there are a large number of contributing events, then due to this large number of nodes and the directed edges, the digraph becomes complex. Visual appraisal also may not be easy. For instance, in the above example, if there were more than six contributing events, then obviously the related digraph would become complicated due to these events and their causality relations. So, to handle the machine tool failure causality digraph conveniently using a computer, a matrix approach is adapted. From this matrix form, an expression that becomes characteristic of the machine tool failure cause can be developed. This matrix is named the ‘machine tool failure severity and causality matrix’.

The machine tool failure severity and causality matrix for the failure cause ‘vibrations’ is written as matrix D.

$$D = \begin{array}{c|cccccc} \text{Events} & 1 & 2 & 3 & 4 & 5 & 6 \\ \hline 1 & S_1 & 0 & 0 & c_{14} & 0 & c_{16} \\ 2 & 0 & S_2 & 0 & c_{24} & c_{25} & 0 \\ 3 & 0 & c_{32} & S_3 & c_{34} & c_{35} & 0 \\ 4 & 0 & c_{42} & 0 & S_4 & c_{45} & c_{46} \\ 5 & 0 & c_{52} & 0 & c_{54} & S_5 & 0 \\ 6 & 0 & c_{62} & 0 & c_{64} & c_{65} & S_6 \end{array}$$

The diagonal element S_i represents a variable of severity of the i-th failure contributing event. Off-diagonal element c_{ij} represents the causality relation (of some degree) between the i-th and j-th events. It may be noted that this matrix considers both the severity of the failure contributing events and their causality relations for the considered machine tool failure cause 'vibrations'.

The permanent of the machine tool failure severity and causality matrix *i.e.*, per (D) is named 'machine tool failure causality function (MTFCF)'. For matrix D, MTFCF is written as:

$$\text{per}(D) = S_1 S_2 S_3 S_4 S_5 S_6 + (c_{24} c_{42} S_1 S_3 S_5 S_6 + c_{25} c_{52} S_1 S_3 S_4 S_6 + c_{45} c_{54} S_1 S_2 S_3 S_6 + c_{46} c_{64} S_1 S_2 S_3 S_5) + [(c_{24} c_{45} c_{52} S_1 S_3 S_6 + c_{25} c_{54} c_{42} S_1 S_3 S_6) + c_{24} c_{46} c_{62} S_1 S_3 S_5 + c_{46} c_{65} c_{54} S_1 S_2 S_3] + [((c_{25} c_{52})(c_{46} c_{64}) S_1 S_3) + (c_{24} c_{46} c_{65} c_{52} S_1 S_3 + c_{25} c_{54} c_{46} c_{62} S_1 S_3)] \quad (12.1)$$

MTFCF helps to analyze the failure cause from combinatorial consideration. This is desirable to give proper physical meaning to the events and their causality relations. Moreover, the permanent function does not contain negative sign, and thus no information is lost. *The reasons for adapting the permanent function, rather than characteristic and other such functions, are explained in Chapter 2.* Equation 12.1, *i.e.*, machine tool failure causality function, is the characteristic of the failure cause as it contains a number of terms that are its structure invariants. These are arranged in groupings. The first grouping represents the severity of six events (*i.e.*, $S_1 S_2 S_3 S_4 S_5 S_6$). The second group is absent, as an event can not become cause and effect to itself. The third grouping contains four terms. Each term represents a 2-event causality loop (*i.e.*, $c_{24} c_{42}$, $c_{25} c_{52}$, $c_{45} c_{54}$, $c_{46} c_{64}$), and the severity of four events (*i.e.*, $S_1 S_3 S_5 S_6$, $S_1 S_3 S_4 S_6$, $S_1 S_2 S_3 S_6$, $S_1 S_2 S_3 S_5$). Similarly, the other terms of MTFCF can be explained. It may be noted that Equation 12.1 is characteristic for the considered failure cause of the machine tools, *i.e.*, vibrations in this case.

12.4 General Machine Tool Failure Causality Function

The machine tool failure causality digraph (MTFCD) represents a machine tool failure cause, no matter how complicated it is. For a given machine tool failure cause, all the contributing events need to be identified first, and the causality relations of these identified events are to be determined for the failure cause. MTFCD is the key to the proposed machine tool failure causality analysis. The causality relations between the machine tool failure cause events must be thoroughly understood before assigning some value to these. If the causality relation between two failure cause events is wrongly understood as 0, then this 0 will cause many terms of the MTFCF to become 0, thereby leading to the loss of much information useful during the machine tool failure cause analysis. Hence it is desirable that one should interact with as many engineers as possible, preferably of different fields (design, operation, maintenance, *etc.*), to reproduce an exact machine tool failure cause representation. It must be emphasized that the process of constructing such a MTFCD would need the information and experience acquired to date, and if these aspects are taken care of in the digraph representation, this will substantially reduce the danger of failing to recognize the possible events and their causality relations. Keeping in mind these aspects, a general form of machine tool failure causality matrix is described in this section.

In general, if there are M number of failure contributing events and the causality relations exist among all the failure contributing events, then the failure severity and causality matrix, P, for the considered MTFCD is written as Equation 12.2, which is similar to Equation 2.10.

$$
P = \begin{array}{c} \text{Events} \\ 1 \\ 2 \\ 3 \\ - \\ - \\ M \end{array}
\begin{array}{c} \begin{array}{cccccc} 1 & 2 & 3 & - & - & M \end{array} \\
\left[\begin{array}{cccccc}
S_1 & c_{12} & c_{13} & - & - & c_{1M} \\
c_{21} & S_2 & c_{23} & - & - & c_{2M} \\
c_{31} & c_{32} & S_3 & - & - & c_{3M} \\
- & - & - & - & - & - \\
- & - & - & - & - & - \\
c_{M1} & c_{M2} & c_{M3} & - & - & S_M
\end{array}\right] \end{array}
\qquad (12.2)
$$

The MTFCF for this matrix P contains factorial M (M!) number of terms. In sigma form, this is written as Equation 12.3, which is similar to Equation 2.11.

$$
\begin{aligned}
\operatorname{per}(P) = {} & \prod_{i=1}^{M} S_i + \sum_{i=1}^{M-1} \sum_{j=i+1}^{M} \cdots\cdots\cdots \sum_{\substack{M=t+1 \\ \ldots,\, M \neq pus}}^{M} (c_{ij}\, c_{ji}) S_k\, S_l\, S_m\, S_n\, S_o \ldots\ldots S_t\, S_M \\
& + \sum_{i=1}^{M-2} \sum_{j=i+1}^{M-1} \sum_{k=j+1}^{M} \cdots\cdots\cdots \sum_{l=1} \sum_{\substack{M=t+1 \\ k,\, \ldots,\, M \neq pus}}^{M} (c_{ij}\, c_{jk}\, c_{ki} + c_{ik}\, c_{kj}\, c_{ji})\, S_l\, S_m\, S_n\, S_o \ldots\ldots S_t\, S_M \\
& + \Big[\sum_{i=1}^{M-3} \sum_{j=i+1}^{M} \sum_{k=i+1}^{M-1} \sum_{l=i+2}^{M} \cdots\cdots\cdots \sum_{\substack{M=t+1 \\ k,l,\, \ldots,\, M \neq pus}}^{M} (c_{ij}\, c_{ji})\, (c_{kl}\, c_{lk})\, S_m\, S_n\, S_o \ldots\ldots S_t\, S_M \\
& + \sum_{i=1}^{M-3} \sum_{j=i+1}^{M-1} \sum_{k=i+1}^{M} \sum_{l=j+1}^{M} \cdots\cdots\cdots \sum_{\substack{M=t+1 \\ k,l,\, \ldots,\, M \neq pus}}^{M} (c_{ij} c_{jk}\, c_{kl}\, c_{li} + c_{il}\, c_{lk}\, c_{kj}\, c_{ji})\, S_m\, S_n\, S_o \ldots\ldots S_t\, S_M \Big] \\
& + \Big[\sum_{i=1}^{M-2} \sum_{j=i+1}^{M-1} \sum_{k=j+1}^{M} \sum_{l=1}^{M-1} \sum_{m=l+1}^{M} \cdots\cdots\cdots \sum_{\substack{M=t+1 \\ k,l,m,\, \ldots,\, M \neq pus}}^{M} (c_{ij} c_{jk}\, c_{ki} + c_{ik}\, c_{kj}\, c_{ji})(\, c_{lm}\, c_{ml}) S_n\, S_o \ldots\ldots S_t\, S_M
\end{aligned}
$$

$$+\sum_{i=1}^{M-4}\ \sum_{j=i+1}^{M-1}\ \sum_{k=i+1}^{M}\ \sum_{l=i+1}^{M}\ \sum_{m=j+1}^{M} \cdots\cdots \sum_{\substack{M=t+1\\ k,l,m,\ldots,M \neq pus}}^{M} (c_{ij}c_{jk}c_{kl}\, c_{lm}\, c_{mi}+c_{im}\, c_{ml}c_{lk}\, c_{kj}\, c_{ji})S_n\, S_o \ldots S_t S_M]$$

$$+\,[(\sum_{i=1}^{M-3}\ \sum_{j=i+1}^{M-1}\ \sum_{k=i+1}^{M}\ \sum_{l=j+1}^{M}\ \sum_{m=1}^{M-1}\ \sum_{n=m+1}^{M} \cdots\cdots \sum_{\substack{M=t+1\\ k,l,m,n,\ldots,M \neq pus}}^{M} (c_{ij}c_{jk}c_{kl}c_{li}+c_{il}c_{lk}\, c_{kj}\, c_{ji})(c_{mn}\, c_{nm})S_o \ldots S_t\, S_M$$

$$+\sum_{i=1}^{M-5}\ \sum_{j=i+1}^{M-1}\ \sum_{k=j+1}^{M}\ \sum_{l=1}^{M-2}\ \sum_{m=l+1}^{M-1}\ \sum_{n=m+1}^{M} \cdots\cdots \sum_{\substack{M=t+1\\ k,l,m,n,\ldots,M \neq pus}}^{M} (c_{ij}c_{jk}c_{ki}+c_{ik}c_{kj}c_{ji})(c_{lm}c_{mn}c_{nl}+c_{ln}c_{nm}c_{ml})S_o \ldots S_t\, S_M$$

$$+\sum_{i=1}^{M-5}\ \sum_{j=i+1}^{M}\ \sum_{k=i+1}^{M-3}\ \sum_{l=i+2}^{M}\ \sum_{m=k+1}^{M-1}\ \sum_{n=k+2}^{M} \cdots\cdots \sum_{\substack{M=t+1\\ k,l,m,n,\ldots,M \neq pus}}^{M} (c_{ij}\, c_{ji})\,(c_{kl}\, c_{lk})\,(c_{mn}\, c_{nm})\, S_o \ldots S_t\, S_M$$

$$+\sum_{i=1}^{M-5}\ \sum_{j=i+1}^{M-1}\ \sum_{k=i+1}^{M}\ \sum_{l=i+1}^{M}\ \sum_{m=i+1}^{M}\ \sum_{n=j+1}^{M} \cdots\cdots \sum_{\substack{M=t+1\\ k,l,m,n,\ldots,M \neq pus}}^{M} (c_{ij}c_{jk}c_{kl}c_{lm}c_{mn}c_{ni}+c_{in}c_{nm}c_{ml}c_{lk}c_{kj}c_{ji})S_o \ldots S_t\, S_M]$$

$$+ \text{----------} \tag{12.3}$$

'pus' stands for 'previously used subscripts' *i.e.*, in Equation 12.3, k, l, m, n, …, and M take those subscripts that are other than previously used subscripts. The MTFCF contains terms arranged in (M +1) groups and these groups represent the severity measures of failure contributing events and the causality relation loops. The first group represents the measures of M events. The second group is absent, as there is no self-loop in the digraph. The third group contains 2-event causality relation loops and measures of (M-2) events. Each term of the fourth group represents a set of a 3-event causality relation loop, or its pair, and measures of (M-3) events. The fifth group contains two sub-groups. The terms of the first sub-group is a set of two 2-event causality relation loops and the measures of (M-4) events. Each term of the second sub-group is a set of a 4-event causality relation loop, or its pair, and the measures of (M-4) events. The sixth group contains two sub-groups. The terms of the first sub-group is a set of 3-event causality relation loop, or its pair, and a 2-event causality relation loop, and the measures of (M-5) events. Each term of the second sub-group is a set of a 5-event causality relation loop, or its pair, and the measures of (M-5) events. Similarly other terms of the equation are defined. Thus, the MTFCF fully characterizes the considered machine tool failure cause, as it contains all possible events and their causality relations.

12.5 Machine Tool Failure Cause Evaluation

It is desirable to evaluate the machine tool failure cause subjectively or objectively, and in terms of index/measure to ascertain the severity of the machine tool failure cause. The numerical value of the MTFCF is called the machine tool failure causality index (MTFCI). This index gives a measure of the severity of failure cause. To evaluate MTFCF, the values of S_i and c_{ij} are required. It is preferable to have these values based on shop-floor data or experience of the shop-floor personnel. If such objective value is not available, then a ranked value judgement on a fuzzy conversion scale may be adapted (*e.g.*, Tables 4.1 or 4.3). It is possible, for a failure cause, that some of the S_i values may be subjective, and the others objective. It is desirable to normalize the objective value of S_i on the same scale as the subjective value.

The causality relation, *i.e.*, c_{ij}, is also assigned on a scale. If the causality relation is strong between two events, then a value of 3 is assigned. If no causality relation exists between two events, then a value of 0 is assigned. This is suggested in Table 12.1.

Table 12.1. Quantification of causality relation between two events, c_{ij}

Causality relation between two events	Assigned value, c_{ij}
None	0
Weak	1
Medium	2
Strong	3

However, if one wishes a fuzzy assignment for the causality relation also, this can be done by following appropriate fuzzy conversion scale suggested by Chen and Hwang (1992), and modifying it suitably.

It may be mentioned that one can choose any scale for S_i or c_{ij}. However, a lower value for these is desirable to obtain manageable values of the MTFCI. It is possible that plant data pertaining to S_i and c_{ij} are not available. In such cases, these are assigned subjective values based on Tables 4.1 (or 4.3) and 12.1.

With the help of Tables 4.1 (or 4.3) and 12.1 and Equation 12.3, the numerical value of the MTFCF *i.e.*, MTFCI is calculated. This index gives a measure of the severity of failure cause. A higher value of MTFCI indicates that the considered failure cause is a serious one. A lower value of MTFCI indicates that the considered failure cause is not serious, and is therefore, desired.

Using MTFCI, two failure causes of a machine tool can be compared. The failure cause having higher value of MTFCI needs to be considered and efforts should be made to reduce the value of the index by taking appropriate failure minimization steps. Thus, different failure causes of a machine tool can be analyzed and arranged in decreasing order of the machine tool failure causality indices. The failure analyst can take suitable actions for their prevention in order of their severity as understood from the values of MTFCI.

12.6 Machine Tool Failure Cause Analysis

The machine tool failure causality function is a useful expression for the failure cause analysis of machine tools, as it represents the severity of the events and the causality relations. The analysis is carried out term by term.

(i) The first term represents the severity of M failure contributing events, and is given as:

$/ S_1 / S_2 / S_3 / \ldots\ldots\ldots /S_M /$

The slash represents a separation mark between the severity of two events. The analysis is to be carried out event-wise and turn by turn. A designer or practicing engineer needs to consider each and every event in detail. If the severity of an event is higher then more attention should be paid to this event, and to finding ways and means to minimize the severity of this event. For example, if the analysis is carried out for the failure cause 'vibrations' of a machine tool, the first term is $S_1 / S_2 / S_3 / S_4 / S_5 / S_6 /$. If the event 4 (*i.e.*, disturbances in machine tool drives) has more severity then in-depth study may reveal that this can be attributed to: rotating unbalanced masses, faulty arrangement of drives, faults in the supporting bearings, damaged gears, worn out belts, spindle play, manufacturing faults in the drive elements, *etc.* By the application of appropriate techniques, the severity of this event can be reduced. On the same lines, the severity of other events is considered.

(ii) When self-loops do not exist in the digraph, then this grouping will be absent.

(iii) When self-loops are absent, each term of the third grouping represents a set of 2-event causality loops and the severity of (M-2) events. This is given as:

$/ (c_{ij}\, c_{ji}) / S_k / S_l / \ldots\ldots\ldots /S_M /$

The entity to be analyzed first is $c_{ij}\, c_{ji}$. This is a 2-event causality loop and represents the resultant causality relation between i and j. If the analysis indicates that this value is comparatively high, then in-depth study is needed to reduce this entity to a lower value. For the present failure cause 'vibrations' of a machine tool, the third grouping is: $/ c_{24}\, c_{42} / S_1 / S_3 / S_5 / S_6 / + \ldots$

The first entity to be analyzed in the first term is $c_{24}\, c_{42}$. This means that the effect of type of cutting and cutting conditions *on* disturbances in machine tool drives, and the effect of disturbances in machine tool drives *on* type of cutting and cutting conditions are to be studied. These two events are cause and effect to each other, and $c_{24}\, c_{42}$ represents the resultant causality relation between these two events. Along with this resultant causality between the first two events, the severity of events 3, 5, and 6 is to be considered. Similarly, the other terms of the grouping can be analyzed.

(iv) When self-loops are absent, the fourth grouping contains the terms, each is a set of a 3-event causality loop, or its pair, and the severity of (M-3) events. This is given as:

$/ (c_{ij}\, c_{jk}\, c_{ki} + c_{ik}\, c_{kj}\, c_{ji}) / S_l / S_m / \ldots\ldots\ldots /S_M /$

The first entity to be analyzed is the 3-event causality loop $c_{ij}\, c_{jk}\, c_{ki}$, and its pair $c_{ik}\, c_{kj}\, c_{ji}$. If analysis indicates the entity's value comparatively higher, then

efforts should be made to reduce its value. For the present failure cause 'vibrations' of a machine tool, the fourth grouping is:
$/ (c_{24}\, c_{45}\, c_{52} + c_{25}\, c_{54}\, c_{42}) / S_1 / S_3 / S_6 / + \ldots\ldots$

The first entity in the first term to be analyzed is $c_{24}\, c_{45}\, c_{52}$. This is the resultant causality relation between events 2, 4, and 5. This means that in the considered failure cause 'vibrations', the causality relations between the type of cutting and the cutting conditions *and* disturbance in machine tool drives, and between disturbance in machine tool drives *and* the cutting process, and between the cutting process *and* type of cutting and cutting conditions are to be studied in detail, and this is expected to minimize this entity. Along with the resultant causality relation among these events, the severity of events 1, 3 and 6 is to be considered. In the same manner, $c_{25}\, c_{54}\, c_{42}$ is to be studied. Similarly, other entities of the other terms of this grouping can be critically analyzed. Ways and means to reduce the value of the entities can be suggested. Finally, this leads to the minimization of failure. Proceeding as described above, other groupings of the MTFCF can be assessed.

The above procedure analyzes the failure causes of the machine tools by identifying the failure-contributing events and their causality relations. Each and every entity in different groupings are analyzed, along with how they are contributing to the machine tool failure cause. The analysis, when carried out as described above, helps to identify the areas where improvements can be made, and leads to minimization of failures in machine tools. For the considered failure cause, 'vibrations', of a machine tool, these may be in terms of proper machine leveling along with proper vibration isolation arrangements, improving the stiffness and the damping characteristics of the machine tool, balancing of the rotating and non-rotating drives, correct arrangement of the drives, proper tool setting and job holding procedures, selection of right tool and work materials, right cutting conditions, maintaining proper belt tension, removing the spindle play, choosing right quality bearings, gears, *etc*. Thus, it would be possible to minimize vibrations, a common failure cause in machine tools.

The above procedure is applicable not only at the design stage of the machine tools but also at the operating stage. The designer or the practicing engineer needs to list the likely or observed failure causes in order of their probability of occurrence. Then the failure cause with the highest probability is attempted first in the above-described procedure. Similarly, other failure causes of the machine tool are analyzed.

Comparison of two failure causes can be done by calculating the value of coefficient of similarity/dissimilarity based on the numerical value of the terms of the MTFCF. The procedure is similar to that described in Section 2.5.2.

12.7 Methodology

The methodology for the proposed failure cause analysis of machine tools using graph theory and the matrix approach is given below:

1. Identify all failure causes attributed to a machine tool under consideration. This should be based on shop-floor data on machine tools, and the experience of persons involved in its operation, maintenance, and design.
2. Consider the first failure cause, and identify its contributing events and their interrelations. If any event is a cause and effect event to itself, consider that aspect also. Assign severity to the events (*i.e.*, S_i), and to the causality relations (*i.e.*, c_{ij}).
3. Develop the machine tool failure causality digraph considering the identified failure contributing events and their interrelations (*i.e.*, causality relations) in step 2. This digraph consists of nodes and directed edges. The number of nodes shall be equal to the number of failure-contributing events (*i.e.*, M). If an event is cause and effect to itself, then a self-loop is to be drawn at the node representing that event.
4. Develop the machine tool failure causality matrix for the machine tool failure causality digraph. This will be an M x M matrix with diagonal elements representing the severity of the failure contributing events (*plus* the causality relations for self-loops, if any, in the digraph) and off-diagonal elements representing the causality relations among the failure contributing events.
5. Obtain the machine tool failure causality function for the machine tool failure causality matrix, on the lines of Equation 12.3. Substitute the values of the severity of the failure contributing events and their causality relations obtained in step 2 into the machine tool failure causality function, and calculate the value of the machine tool failure causality index *i.e.*, MTFCI.
6. Carry out the failure cause analysis by critically examining each and every term of different groupings of the machine tool failure causality function. Suggest the ways and means to minimize the severity of the failure cause.
7. Repeat the steps 1 to 6 for all the other failure causes of the machine tool.
8. Arrange the MTFCI values for different failure causes in decreasing order. This gives an idea of the severity of the failure causes.
9. Evaluate the coefficients of similarity and dissimilarity for all the failure causes. List the values for all combinations. Compile the analysis, and document the failure causes for future analysis.

12.8 Summary

A methodology is presented in this chapter that is applied to a machine tool failure cause analysis, with the identification of failure-contributing events and their interaction for a machine tool failure cause. The procedure is useful for designers of reliable machine tools, and practicing engineers involved in failure minimization of the operating machine tool, leading to improved productivity and cost minimization. The proposed methodology helps in identifying areas of improvement, and minimizing the severity of failure causes, thereby leading to the development of a machine tool of increased reliability. The procedure is useful not only for the failure cause analysis of machine tools, but also for the failure cause analysis of any type of systems. Further, the procedure is useful for comparison and evaluation of failure causes.

References

Alexander SM, Vaidya CM, Graham JH (1993) Model for the diagnosis of CIM equipment. Computer-Electronics Engineering 19:175–183

Angeli C, Chatzinkikoraou A (1985) An expert system approach to fault diagnosis in hydraulic systems. Expert Systems 12:323–329

BREU-CT-91-0463 (1992) MIRAM-Machine management for increasing the reliability, availability, and maintainability. Heckler & Koch, University of Wales

Clark GE, Paasch RK (1997) Diagnostic modeling and diagnosability evaluation of mechanical systems. Journal of Mechanical Design 119:57–67.

Czichos H (1980) Systems Approach to the analysis of wear problems. In: International Conference on Friction, Lubrication, and Wear in Engineering, Melbourne, Australia

Czichos H (1978) Tribology-a systems approach to science and technology of friction, lubrication, and wear. Tribology Series-I, Elsevier, Amsterdam

Das K, Lashkari RS, Sengupta S (2007) Reliability consideration in the design and analysis of cellular manufacturing systems. International Journal of Production Economics 105:243–262

Drake PR, Pan D (1996) Multiple fault diagnosis for a machine tool's flood coolant system using a neural network. International Journal of Machine Tools & Manufacture 36:1247–1251

Freyermuth B (1991) Knowledge based incipient fault diagnosis of industrial robots. In: Safeprocess 91, Baden-Baden, Germany, pp 31–37

Fussell JB (1975) A review of fault tree analysis with emphasis on limitations. In: Proc. IFAC 6th World Congress, Boston

Fussel JB (1976) Fault tree analysis concepts and techniques. In: Henley EI, Lynn IW (eds) Generic techniques in systems and reliability assessment, Noordhoff International, Leyden

Gandhi OP, Agrawal VP (1994) A digraph approach to system wear evaluation and analysis. Journal of Tribology 116:268–274

Gandhi OP, Agrawal VP (1996) Failure cause analysis-a structural approach. Journal of Pressure Vessel Technology 118:434–440

Holloway LE, Krogh BH (1990) Fault detection and diagnosis in manufacturing system: a behavioral model approach. In: Proc. 2nd International Conference on CIM, Troy, New York

Hu W, Starr AG, Leung AYT (1991) Two diagnostic models for PLC controlled flexible manufacturing systems. International Journal of Machine Tools & Manufacture 39:1979–1991

Hu W, Starr AG, Zhou Z, Leung AYT (2000) A systematic approach to integrated fault diagnosis of flexible manufacturing systems. International Journal of Machine Tools & Manufacture 40:1587–1602

Huang JT, Liao YS (2000) A wire-EDM maintenance and fault diagnosis expert system integrated with an artificial neural network. International Journal of Production Research 38:2459–2470

Isermann R (1991) Fault diagnosis of machines via parameter estimation and knowledge processing. Safeprocess 91, Baden-Baden, Germany, pp 121–134

Ishida Y, Adachi N, Tokumaru H (1985) A topological approach to failure diagnosis of large scale system. IEEE Transactions on Systems, Man, and Cybernetics 20:129–141.

Johansson KE (1981) Field monitoring of NC machines - a systems approach. In: Innovation for Maintenance Technology Improvements, MFPG 3rd Meeting, National Bureau of Standards, New York

Kegg RL (1984) On line process diagnostics. CIRP Annals 33:469–473

Keller AZ, Perera VD (1981) Reliability analysis of CNC machine tools. In: Proc. 3rd National Conference on Reliability, Macclesfield

Kilmartin BR, Hannan RG (1981) An in-company study of NC machine utilization and its improvement by a systems approach. International Journal of Production Research 19:289–300

Kokowa M, Miyazaki S, Shingai S (1983) Fault location using digraph and inverse direction search with application. Automatica 19:729–735

Kokowa M, Shingai S (1982) Failure propagating simulation and non-failure path search in network systems. Automatica 18:335–341

Kwon HD, Burdekin M (1998) Adjustment of CNC machine tool controller setting values by an experimental method. International Journal of Machine Tools & Manufacture 38:1045–1065

Lee J (1995) Machine performance monitoring and proactive maintenance in computer integrated manufacturing: review and perspective. International Journal of Computer 1ntegratedManufacturing 8:370–380

Lee J, Kramer BM (1993) Analysis of machine degradation using a neural network approach. Journal of Manufacturing Systems 12:379–387

Luis AFG, Gilberto HR, Rocio PV, Rene DJRT, Wbaldo LT (2006) Sensorless tool failure monitoring system for drilling machines. International Journal of Machine Tools and Manufacture 46:381–386

Majstorovic V, Milacic V (1990) Expert system for diagnosis and maintenance. Maintenance 5:19–22

Marczewsky R (1988) An expert system for machine tool diagnosis. In: Proc. of 2nd Engineering Society of Detroit Conference on Expert Systems, Detroit

Martin KF (1994) A review by discussion of condition monitoring and fault diagnosis in machine tools. International Journal of Machine Tools & Manufacture 34:527–551

Martin KF, Hoh SM, Thorpe P (1990) Fault diagnosis of hard faults. In: International Conference on Manufacturing Systems and Environment, Japan Society of Mechanical Engineers, Tokyo

Marzi MH, Martin KF (1990) Artificial neural network condition monitoring and fault diagnosis. In: Proc. ASME International Conference on Neural Networks, San Diego, CA

NCSR (1988) CNC machining center. AMTA Reliability Publications, Risley

Poltavets OF (1994) Diagnostics in the machine tool: purpose, state, and prospects. Russian Engineering Research 14:6–8

Puetz RD, Eichhorn S (1987) Expert systems for fault diagnosis of CNC machines. In: Sviram D, Adey RD (eds) Knowledge-based expert systems for engineering. Computational Mechanics Publications, London

Rao SB (1997) Metal cutting machine tool design. Journal of Manufacturing Science and Engineering 119:713–716

Rao RV, Gandhi OP (2002) Failure cause analysis of machine tools using digraph and matrix methods. International Journal of Machine Tools & Manufacture 42:521–528

Sethi VK, Agrawal VP (1993) Hierarchical classification of kinematic chains: a multigraph approach. Mechanism and Machine Theory 28:601–614

Stewart E (1977) A survey of machine tool breakdowns. MTIRA Research Report No.63, In: Proc. of National Conference on Reliability, Macclesfield

Williams JH (1990) Transfer function techniques and fault location. Research Studies Press, Reading, MA

Ye N, Zhao B (1996) Hybrid intelligent system for fault diagnosis of advanced manufacturing systems. International Journal of Production Research 34:555–576

Yoshikawa H (1982) Multi-purpose modelling of mechanical systems-morphological model as a mesomode. In: Bjorke OP, Franksen OI (eds) System structures in engineering economic design and production. Tapir Publishers, Trondheim

Zeng L, Wang HP (1991) Machine fault classification: a fuzzy set approach. International Journal of Advanced Manufacturing Technology 6:83–94

13

Robot Selection for a Given Industrial Application

13.1 Introduction

The word **robot** was coined in 1920 by the Czech author K. Capek in his play *Rossum's Universal Robots*; it is derived from the Czech word *robota*, meaning 'worker'. An industrial robot is commonly defined as a reprogrammable multifunctional manipulator, designed to move materials, parts, tools, or other devices by means of variable programmed motions, and to perform a variety of other tasks. In a broader context, the term robot also includes manipulators that are activated directly by an operator.

Recent developments in information technology and engineering sciences have been the main reason for the increased utilization of robots in a variety of advanced manufacturing facilities. Robots with vastly different capabilities and specifications are available for a wide range of applications. The selection of robots to suit a particular application and production environment from among the large number available in the market has become a difficult task. Various aspects such as product design, production system, and economics, need to be considered before a suitable robot can be selected. The selection problem is particularly relevant in view of the likely lack of experience of prospective users in employing a robot. Indeed, robots are still a new concept in industry as a whole, and so it is not unusual for an industry to be a first-time robot purchaser. Many precision-based methods for robot selection have been developed to date.

Knott and Getto (1982) suggested a model to evaluate different robotic systems under uncertainty, and different alternatives were evaluated by computing the total net present values of cash flows of investment, labor components, and overheads. Offodile *et al.* (1987) developed a coding and classification system that was used to store robot characteristics in a database, and then selected a robot using economic modeling. While the attempt provides a valuable aid at the stage of final selection, such an exercise will be prohibitive at the initial selection stage when the number of potential robots is large, and many other considerations have to be taken into account. Imang and Schlesinger (1989) presented decision models for robot selection, and compared ordinary least squares and linear goal programming methods. Agrawal *et al.* (1991) employed the TOPSIS method for robot selection. However, the authors had not considered the subjective attributes.

Boubekri *et al.* (1991) developed an expert system for industrial robot selection considering functional, organizational, and economical attributes in the selection process. Wang *et al.* (1991) presented a decision support system that applies a fuzzy set method for robot selection. The objective attributes were evaluated via marginal value functions while the subjective attributes were evaluated via fuzzy set membership function. Data from both evaluations were finally processed such that a fuzzy set decision vector was obtained. However, the fuzzy method presented is a complicated one, and requires more computation.

Booth *et al.* (1992) proposed a decision model for the robot selection problem using both Mahalanobis distance analysis, *i.e.*, a multivariate distance measure, and principal-components analysis. Liang and Wang (1993) proposed a robot selection algorithm by combing the concepts of fuzzy set theory and hierarchical structure analysis. The algorithm was used to aggregate decision makers' fuzzy assessments about robot selection attributes weightings, and to obtain fuzzy suitability indices. The suitability ratings were then ranked to select the most suitable robot. Khouja and Offodile (1994) reviewed the literature on industrial robots selection problems and provided directions for future research. Khouja (1995) presented a two-phase robot selection model that involved the application of data envelopment analysis (DEA) in the first phase, and a multi-attribute decision-making model in the second phase.

Zhao and Yashuhiro (1996) introduced a genetic algorithm (GA) for an optimal selection and work station assignment problem for a computer-integrated manufacturing (CIM) system. Goh *et al.* (1996) proposed a revised weighted sum decision model that took into account both objective and subjective attributes of the robots under consideration. The model incorporated values assigned by a group of experts on different attributes in selecting the robots. Goh (1997) employed the analytic hierarchy process (AHP) method for robot selection. Parkan and Wu (1999) presented decision-making and performance measurement models with applications to robot selection. Particular emphasis was placed on a performance measurement procedure called operational competitiveness rating (OCRA) and a multiple attribute decision-making method, TOPSIS. The final selection was made on the basis of rankings obtained by averaging the results of OCRA, TOPSIS, and a utility model. However, the models had not considered the subjective attributes, and no explanation was given on how to assign the weightings to different robot selection attributes.

Khouja and Kumar (1999) used options theory and an investment evaluation procedure for selection of robots. Braglia and Petroni (1999) carried out investment evaluation using DEA for robot selection. Layek and Resare (2000) developed a decision support system (DSS) based on analytical algorithms to select machining centers and robots concurrently from the market milien. Chu and Lin (2003) pointed out the limitations of the Liang and Wang (1993) method, and proposed a fuzzy TOPSIS method for robot selection. However, the authors had converted the available objective values of the robot selection attributes into fuzzy values, which violates the basic rule of fuzzy logic, *i.e.*, the available objective values need not be fuzzified. Further, only a 5-point scale was adapted for the rating of robots under subjective attributes. Also, the fuzzy method was complicated, and requires more computation. Bhangale *et al.* (2004) listed a large number of robot selection attributes, and ranked the robots using TOPSIS and graphical methods, comparing the rankings given by these methods. However, the weights assigned by the authors to the attributes were not consistent. Karsak and Ahiska (2005) introduced a practical common weight MCDM methodology using the DEA method with an improved discriminating power for technology selection. Rao and Padmanabhan (2006) proposed a methodology based on digraph and matrix methods for evaluation of alternative industrial robots. A robot selection index was proposed that evaluates and ranks robots for a given industrial application. The index was obtained from a robot selection attributes function, in turn obtained from the robot selection attributes digraph. The digraph was developed based on robot selection attributes and their relative importance for the application considered. A step by step procedure for evaluation of a robot selection index was suggested.

The objective of a robot selection procedure is to identify the robot selection attributes, and obtain the most appropriate combination of the attributes in conjunction with the real requirements of the industrial application. A robot selection attribute is defined as a factor that influences the selection of a robot for a given industrial application. These attributes include: cost, configuration, load capacity, weight and size of the robot, type and number of end effectors, type of control, velocity of movements, type of programming, programming flexibility, reliability, repeatability, positioning accuracy, resolution, number of degrees of freedom, number of joints, their sequence and orientation, motion transformation characteristics, ease of operation, work volume, drive system, man-machine interface, vendor's service contract, training, delivery period, maintainability, ease of assembly, ease of disassembly, types and number of sensors used, availability or assured supply, management constraints, *etc.*

Efforts need to be extended to determine attributes that influence robot selection for a given industrial application, using a logical approach to eliminate unsuitable robots, and for selection of a proper robot to strengthen the existing robot selection procedure. Pertinent attributes and the alternative robots involved are to be identified. Values of the attributes and their relative importance are to be obtained. An objective or subjective value, or its range, may be assigned to each identified attribute as a limiting value, or threshold value, for its acceptance for the considered robot selection problem. An alternative robot with each of its selection attributes, meeting the acceptance value, may be short-listed. After short-listing the alternative robots, the main task to choose the alternative robot is to see how it serves the attributes considered.

The next section presents applications of the GTMA and fuzzy MADM methods for robot selection for a given industrial application.

13.2 Examples

Now, to demonstrate and validate the application of decision-making methods, two examples are considered.

13.2.1 Example 1

An example is considered to demonstrate the application of the GTMA and fuzzy MADM methods. This example problem considers five robot selection attributes, and three alternative robots. The objective and subjective information of the attributes is given in Table 13.1. Man–machine interface (MI) and programming flexibility (PF) are expressed subjectively in linguistic terms, and these attributes are assigned objective values with the help of Table 4.3. The objective data of the attributes are given in Table 13.2.

Table 13.1. Robot selection attributes information of example 13.2.1

Robot	PC ($1,000)	LC (kg)	RE (mm)	MI	PF
Robot 1	73	48	0.15	A	H
Robot 2	71	46	0.18	AA	VH
Robot 3	75	51	0.14	BA	H

PC: Purchasing cost LC: Load carrying capacity R: Repeatability error
MI: Man-machine interface PF: Programming flexibility
A: Average; AA: Above average; BA: Below average; H: High; VH: Very high

Table 13.2. Objective data of the robot selection attributes of example 13.2.1

Robot	PC ($1,000)	LC (kg)	RE (mm)	MI	PF
Robot 1	73	48	0.15	0.5	0.665
Robot 2	71	46	0.18	0.59	0.745
Robot 3	75	51	0.14	0.41	0.665

13.2.1.1 Application of GTMA

In the present work, the attributes considered are PC, LC, R, MI, and PF. The objective values of the robot selection attributes, which are given in Table 13.2, are to be normalized. LC, MI, and PF are beneficial attributes, and higher values are desirable. Values of these attributes are normalized, as explained in Section 2.4, and are given in Table 13.3 in the respective columns. PC and R are non-beneficial attributes, and lower values are desirable. The values of these attributes for different robots are normalized, and given in Table 13.3 in the respective columns.

Table 13.3. Normalized data of the robot selection attributes of example 13.2.1

Robot	PC	LC	RE	MI	PF
Robot 1	0.9726	0.9412	0.9333	0.8475	0.8926
Robot 2	1.0000	0.9020	0.7777	1.0000	1.0000
Robot 3	0.9467	1.0000	1.0000	0.6949	0.8926

Let the decision maker prepare the following relative importance assignments:

	PC	LC	RE	MI	PF
PC	---	0.745	0.5	0.865	0.745
LC	0.255	---	0.255	0.59	0.5
RE	0.5	0.745	---	0.865	0.745
MI	0.135	0.41	0.135	---	0.41
PF	0.255	0.5	0.255	0.59	---

The robot attributes digraph, robot attributes matrix of the digraph, and robot function for the matrix can be prepared. The value of the robot selection index is calculated using the values of A_i and a_{ij} for each robot.

The robot selection index values of different robots are given below in descending order:

Robot 2	6.1701
Robot 1	5.9386
Robot 3	5.7184

From the above values of the robot selection index, robot 2 is considered as the best choice among the robots considered for the given industrial application. The second choice is robot 1, and the third choice is robot 3.

13.2.1.2 SAW Method

Let the decision maker assign the following weights of importance to the attributes:

$W_{PC} = 0.40$, $W_{LC} = 0.08$, $W_R = 0.40$, $W_{MI} = 0.05$, and $W_{PF} = 0.08$

Using these weights, and the normalized data of the attributes for different robots, the robot selection index values are calculated, and are arranged in descending order of the index.

Robot 3	0.9579
Robot 1	0.9429
Robot 2	0.9032

As mentioned above, the ranking depends upon the weights of importance assigned to the attributes.

13.2.1.3 WPM

Application of WPM leads to the following ranking:

Robot 3	0.9554
Robot 1	0.9424
Robot 2	0.8969

13.2.1.4 AHP and its Versions

If the same weights as those selected for the SAW method are used in this method, then the ranking of robots obtained by using the relative as well as ideal mode AHP methods will be the same. The multiplicative AHP method also leads to the same ranking.

However, rather than the above, let the decision maker decide to use the AHP method to determine the weights (w_j) of the attributes, and prepare the following matrix:

$$\begin{array}{c} \\ PC \\ LC \\ R \\ MI \\ PF \end{array} \begin{array}{c} \begin{array}{ccccc} PC & LC & R & MI & PF \end{array} \\ \begin{bmatrix} 1 & 5 & 1 & 7 & 5 \\ 1/5 & 1 & 1/5 & 2 & 1 \\ 1 & 5 & 1 & 7 & 5 \\ 1/7 & 1/2 & 1/7 & 1 & 1/2 \\ 1/5 & 1 & 1/5 & 2 & 1 \end{bmatrix} \end{array}$$

Purchasing cost (PC) is considered as strongly more important than the load carrying capacity (LC) in this example. So, a relative importance value of 5 is assigned to PC over LC (*i.e.*, a_{12} = 5), and a relative importance value of 1/5 is assigned to LC over PC (*i.e.*, a_{21} = 1/5). PC and R are considered as equally important in this example. So, a relative importance value of 1 is assigned to PC over R, and a relative importance value of 1 is assigned to R over PC. Similarly, the relative importance among other attributes can be explained.

The normalized weights of each attribute are calculated and these are: W_{PC} = 0.3916, W_{LC} = 0.084, W_R = 0.3916, W_{MI} = 0.0485, and W_{PF} = 0.0841. The value of λ_{max} is 5.0204 and CR = 0.00455, which is much less than the allowed CR value of 0.1. Thus, there is good consistency in the judgements made.

The value of the robot selection index is now calculated, and the robots are arranged in descending order of the robot selection index.

Robot 3	0.9551
Robot 1	0.9416
Robot 2	0.9045

From the above values of the robot selection index, robot 3 is considered as the best choice among the robots considered for the given industrial application.

For the above weights of importance of attributes, multiplicative AHP also leads to the same ranking order of 3-1-2.

13.2.1.5 TOPSIS Method

The quantitative values of the robot selection attributes, which are given in Table 13.5, are normalized as explained in Section 3.2.6.

Relative importance of attributes (a_{ij}) is assigned using the AHP method as explained in Section 13.2.2.4, and these are W_{PC} = 0.3916, W_{LC} = 0.084, W_R = 0.3916, W_{MI} = 0.0485, and W_{PF} = 0.0841.

The weighted normalized matrix is calculated, and is shown below:

$$\begin{bmatrix} 0.2259 & 0.0482 & 0.2323 & 0.0277 & 0.0466 \\ 0.2323 & 0.0462 & 0.1936 & 0.0327 & 0.0522 \\ 0.2199 & 0.0512 & 0.2489 & 0.0227 & 0.0466 \end{bmatrix}$$

Ideal (best) and negative ideal (worst) solutions are calculated, and these are given as:

$V_{PC}^{+} = 0.2199$ $V_{PC}^{-} = 0.2323$
$V_{LC}^{+} = 0.0512$ $V_{LC}^{-} = 0.0462$
$V_{R}^{+} = 0.1936$ $V_{R}^{-} = 0.2489$
$V_{MI}^{+} = 0.0327$ $V_{MI}^{-} = 0.0227$
$V_{PF}^{+} = 0.0522$ $V_{PF}^{-} = 0.0446$

Separation measures are calculated, and these are:

$S_1^{+} = 0.0400$ $S_1^{-} = 0.0186$
$S_2^{+} = 0.0134$ $S_2^{-} = 0.0565$
$S_3^{+} = 0.0565$ $S_3^{-} = 0.0134$

The relative closeness of a particular alternative to the ideal solution is calculated, and these are:

$P_1 = 0.3169$, $P_2 = 0.8088$, and $P_3 = 0.1912$

This relative closeness to the ideal solution can be named as the 'robot selection index' in the present work.

The alternative robots are arranged in descending order of their robot selection index. This can be arranged as 2-1-3.

13.2.1.6 Modified TOPSIS Method

The positive ideal solution (R^{+}) and the negative ideal solution (R^{-}) are calculated, and are given below:

R_{PC}^{+} = 0.5614 R_{PC}^{-} = 0.5930
R_{LC}^{+} = 0.6087 R_{LC}^{-} = 0.5489
R_{R}^{+} = 0.5129 R_{R}^{-} = 0.6595
R_{MI}^{+} = 0.6740 R_{MI}^{-} = 0.4684
R_{PF}^{+} = 0.6209 R_{PF}^{-} = 0.5543

The weighted Euclidean distances are calculated as

D_1^{+} = 0.0402 D_1^{-} = 0.0734
D_2^{+} = 0.0933 D_2^{-} = 0.0531
D_3^{+} = 0.0531 D_3^{-} = 0.0933

The relative closeness of a particular alternative to the ideal solution is calculated (*i.e.*, robot selection index), and these are:

$P_{1\text{-mod}} = 0.6460$ $P_{2\text{-mod}} = 0.3625$ $P_{3\text{-mod}} = 0.6375$

The alternative robots are arranged in descending order of their robot selection index, as 1-3-2. From this, it appears that the ranking presented by using the modified TOPSIS method is not appropriate for the example problem considered.

13.2.2 Example 2

Bhangale *et al.* (2004) listed a large number of robot selection attributes, and ranked the robots using TOPSIS and graphical methods, comparing the rankings given by these methods. The example problem considering five attributes and seven alternative robots is shown in Table 13.4.

Table 13.4. Objective data of the robot selection attributes of example 13.2.2 (from Bhangale *et al.* 2004; reprinted with permission from Elsevier)

Robot	LC	RE	MS	MC	MR
Robot 1	60	0.4	2,540	500	990
Robot 2	6.35	0.15	1,016	3,000	1,041
Robot 3	6.8	0.1	1,727.2	1,500	1,676
Robot 4	10	0.2	1,000	2,000	965
Robot 5	2.5	0.1	560	500	915
Robot 6	4.5	0.08	1,016	350	508
Robot 7	3	0.1	1,778	1,000	920

LC: Load capacity (kg) RE: Repeatability error (mm)
MS: Maximum tip speed (mm/s) MC: Memory capacity in points or steps MR: Manipulator reach (mm)

13.2.2.1 Application of GTMA

Now, to demonstrate the proposed procedure of robot selection through GTMA, various steps of the methodology, given in Section 2.6, are carried out as described below:

In the present work, the attributes considered are LC, RE, MS, MC and MR. The objective values of the robot selection attributes, which are given in Table 13.4, are to be normalized. LC, MS, MC, and MR are beneficial attributes, and higher values are desirable. Values of these attributes are normalized, and are given in Table 13.5 in the respective columns. RE is a non-beneficial attribute, and lower values are desirable. The values of this attribute for different robots are normalized, and given in Table 13.5 in the respective columns.

Table 13.5. Normalized data of the robot selection attributes of example 13.2.2

Robot	LC	RE	MS	MC	MR
Robot 1	1	0.2	1	0.1667	0.5907
Robot 2	0.1058	0.53333	0.4	1	0.6211
Robot 3	0.1133	0.8	0.68	0.5	1
Robot 4	0.1667	0.4	0.3937	0.6667	0.5758
Robot 5	0.0417	0.8	0.2205	0.1667	0.5459
Robot 6	0.075	1	0.4	0.1167	0.3031
Robot 7	0.05	0.8	0.7	0.3333	0.5489

Relative importance of attributes (a_{ij}) is assigned values as explained in Section 2.4. Let the decision maker select the following assignments using the AHP procedure:

	LC	RE	MS	MC	MR
LC	-	1/6	1/7	1/7	1/5
RE	6	-	1/2	1/2	2
MS	7	2	-	1	3
MC	7	2	1	-	3
MR	5	1/2	1/3	1/3	-

The value of λ_{max} for this matrix is 5.0874 and CR = 0.0197, and, thus there is good consistency in the judgements made (of relative importance of attributes).

The value of the robot selection index is now calculated, and the robots are arranged in the descending order of the robot selection index.

Robot 3	92.004
Robot 1	88.074
Robot 2	84.929
Robot 7	81.391
Robot 4	77.954
Robot 6	72.986
Robot 5	71.296

From the above values of the robot selection index, robot 3 is considered as the best choice among the robots considered for the given industrial application. The second choice is Robot 1 and the last choice is robot 5. However, Bhangale *et al.* (2004) gave a ranking order of: robot 4 - robot 1 - robot 3 - robot 7 - robot 2 - robot 6 - robot 5. However, the relative importance matrix prepared by Bhangale *et al.* (2004) was completely inconsistent, and it is not possible to justify how the authors had calculated the weights of the relative importance of the attributes based on such a highly inconsistent judgement matrix. Thus, the ranking presented here for the proposed GTMA method is more genuine.

It may be mentioned that the ranking depends upon the judgements made by the user. The above ranking may change if the user assigns different relative importance values to the attributes.

13.2.2.2 AHP and its Versions

Let the decision maker prepares the following relative importance matrix:

	LC	RE	MS	MC	MR
LC	-	1/6	1/7	1/7	1/5
RE	6	-	1/2	1/2	2
MS	7	2	-	1	3
MC	7	2	1	-	3
MR	5	1/2	1/3	1/3	-

The normalized weights of each attribute are calculated following the procedure presented in Section 3.2.3, and these are: W_{LC} = 0.036, W_{RE} = 0.192,

W_{MS} = 0.326, W_{MC} = 0.326, and W_{MR} = 0.120. The value of λ_{max} for this matrix is 5.0874 and CR = 0.0195, and, thus there is good consistency in the judgements made.

The value of the robot selection index is now calculated using the above weights, and the normalized data of the attributes given in Table 13.2. The alternative robots are arranged in descending order of the robot selection index.

Robot 3	0.6623
Robot 2	0.6371
Robot 7	0.5581
Robot 1	0.5256
Robot 4	0.4976
Robot 6	0.4000
Robot 5	0.3468

From the above values of the robot selection index, it is clear that the robot, designated as 3 is the best choice among the robots considered for the given industrial application.

For the above weights of importance of attributes, multiplicative AHP leads to the same ranking order of 3-2-7-4-1-6-5.

References

Agrawal VP, Kohli V, Gupta S (1991) Computer aided robot selection: the 'multiple attribute decision making' approach. International Journal of Production Research 29:1629–1644

Bhangale PP, Agrawal VP, Saha SK (2004) Attribute based specification, comparison and selection of a robot. Mechanism & Machine Theory 39:1345–1366

Booth DE, Khouja M, Hu M (1992) A robust multivariate statistical procedure for evaluation and selection of industrial robots. International Journal of Operations & Production Management 12:15–24

Boubekri N, Sahoui M, Lakrib C (1991) Development of an expert system for industrial robot selection. Computers & Industrial Engineering 20:119–127

Bragilia M, Petroni A (1999) Evaluating and selecting investments in industrial robots. International Journal of Production Research 37:4157–4178

Chu TC, Lin YC (2003) A fuzzy TOPSIS method for robot selection. International Journal of Advanced Manufacturing Technology 21:284–290

Goh CH (1997) Analytic hierarchy process for robot selection. Journal of Manufacturing Systems 16:381–386.

Goh CH, Tung YCA, Cheng CH (1996) A revised weighted sum decision model for robot selection. Computers & Industrial Engineering 30:193–198

Imang MM, Schlesinger RJ (1989) Decision models for robot selection: a comparison of ordinary least squares and linear goal programming method. Decision Sciences 20:40–53

Karsak EE, Ahiska SS (2005) Practical common weight multi-criteria decision-making approach with an improved discriminating power for technology selection. International Journal of Production Research 43:1537–1554

Khouja M (1995) The use of data envelopment analysis for technology selection. Computers & Industrial Engineering 28:123–132

Khouja MJ, Kumar RL (1999) An options view of robot performance parameters in a dynamic environment. International Journal of Production Research 37:1243–1257

Khouja M, Offodile OF (1994) The industrial robots selection problem: a literature review and directions for future research. IIE Transactions 26:50–61

Knott K, Getto RD (1982) A model for evaluating alternative robot systems under uncertainty. International Journal of Production Research 20:155–165

Layek AM, Resare LJ (2000) Algorithm based decision support system for the concerted selection of equipment in machining/assembly cells. International Journal of Production Research 38:323–339

Liang GH, Wang MJ (1993) A fuzzy multi-criteria decision-making approach for robot selection. Robotics and Computer Aided Manufacturing 10:267–274

Offodile OF, Lambert PK, Dudek RA (1987) Development of a computer aided robot selection procedure (CARSP). International Journal of Production Research 25:1109–1112

Parkan C, Wu ML (1999) Decision-making and performance measurement models with applications to robot selection. Computers & Industrial Engineering 36:503–523

Rao RV, Padmanabhan KK (2006) Selection, identification and comparison of industrial robots using digraph and matrix methods Robotics and Computer Integrated Manufacturing 22:373–383

Wang MJ, Singh HP, Huang WV (1991) A decision support system for robot selection. Decision Support Systems 7:273–283

Zhao L, Yashuhiro T (1996) Genetic algorithm for robot selection and work station assignment problem. Computers & Industrial Engineering 31:599–602

14

Selection of an Automated Inspection System

14.1 Introduction

As automation increases in all aspects of manufacturing processes and operations, the need for automated inspection has become obvious. Flexible manufacturing systems and manufacturing cells have led to the adoption of advanced measuring techniques and systems. In fact, installation and utilization of these systems is now necessary and essential in manufacturing.

In the past, a batch of parts was manufactured and sent to be measured in a separate quality control room; if this batch passed measurement inspection, it was put into inventory. Automated inspection, however, is based on various on-line sensor systems that monitor the dimensions of the parts while they are being made, and if necessary use these measurements as input to correct the process (Kalpakjian and Schmid, 2000).

Automated inspection techniques can be divided into two broad categories: (1) contact inspection and (2) non-contact inspection. In contact inspection, physical contact is made between the object and the measuring or gaging instrument, whereas in non-contact inspection no physical contact is made. The principal contact inspection technologies are:

- Conventional measuring and gaging instruments, manual and automated
- Coordinate measuring machines (CMMs) and related techniques
- Stylus type surface texture measuring machines

Conventional measuring and gaging techniques and CMMs measure dimensions and related specifications. Surface texture measuring machines measure surface characteristics such as roughness and waviness.

Non-contact inspection methods utilize a sensor located at a certain distance from the object to measure or gage the desired features (Groover, 2001). The non-contact inspection technologies can be classified into two categories: (1) optical and (2) non-optical. Optical inspection technologies make use of light to accomplish the measurement or gaging cycle. The most important optical technology is machine vision; however, other optical techniques are important in certain industries. Non-optical inspection technologies utilize energy forms other than light to perform the inspection; these other energies include various electrical fields, radiation and ultrasonics.

The characteristics and quality of measuring instruments or gages are generally described by various specific attributes such as accuracy, repeatability, sensitivity, amplification, calibration, stability, linearity, drift, precision, resolution, speed of response, volumetric performance, maintainability, reliability, initial cost, operation cost, throughput rate, environmental factor requirement (temperature, humidity, dust and so on), flexibility in software interface, size and type of parts to be measured, operator skills required, *etc.*

The selection of an automated inspection system requires consideration of various attributes as mentioned above. Very limited research work was done on this selection aspect. Elshennaway (1989) presented a methodology for the performance evaluation of CMMs. Golomski (1990) had discussed the selection of automated inspection device from accounting point of view. The author had shown that using automated inspection equipment can reduce the indirect cost of inspection but increase the depreciation cost as well as the maintenance cost. Pandey and Kengpol (1995) presented a methodology for selecting the best possible automated inspection device for use in FMSs. The problem had been modeled as that of multicriterion decision making and solved using Preference Ranking Organization METHod for Enrichment Evaluations (PROMETHEE). The study had demonstrated the effectiveness of multicriterion decision making approach.

Now, to demonstrate and validate the application of proposed decision making methods, an example is considered. First GTMA is applied and subsequently few MADM methods are applied to rank and select the automated inspection systems.

14.2 Example

Pandey and Kengpol (1995) presented a methodology for selecting the best possible automated inspection device for use in FMSs. The authors had surveyed the Thailand industries and considered 11 attributes and 4 alternative automated inspection systems. The eleven attributes considered were accuracy, volumetric performance, repeatability, resolution, maintainability, reliability, initial cost, operation cost, throughput rate, environmental factor requirement and flexibility in software interface. The four alternative automated inspection systems considered were CMM1(USA), CMM2(Japan), AVI(USA), LASER SCAN (Japan).The corresponding data is presented in Table 14.1.

14.2.1 Application of Graph Theory and Matrix Approach (GTMA)

Various steps of the methodology, proposed in Section 2.6, are carried out as described below:

Step 1: In the present work, the attributes considered are the same as of those Pandey and Kengpol (1995), and these are: accuracy (A), volumetric performance (V), repeatability (R), resolution (S), maintainability (M), reliability (L), initial cost

(I), operation cost (O), throughput rate (T), environmental factor requirement (E), and flexibility in software interface (F).

Table 14.1. Data of the automated inspection system selection attributes (from Pandey and Kengpol 1995; reprinted with permission from Elsevier)

Attributes	A	B	C	D
Accuracy	90	80	60	75
Volumetric performance	80	70	50	70
Repeatability	80	80	50	70
Resolution	70	70	80	60
Maintainability	60	60	80	70
Reliability	85	80	70	70
Initial cost	40	30	20	25
Operation cost	2	7	1	4
Throughput rate	70	70	80	80
Environmental factor requirement	80	80	60	70
Flexibility in software interface	80	60	60	70

A: CMM1 (USA); B: CMM2 (Japan); C: AVI (USA); D: LASER SCAN (Japan)

The quantitative values of the automated inspection system selection attributes, which are given in Table 14.1, are to be normalized. A, V, R, S, M, L, T, and F are beneficial attributes, and higher values are desirable. Values of these attributes are normalized, and are given in Table 14.2 in the respective columns. I, O and E are non-beneficial attributes, and lower values are desirable. The values of these attributes for different alternative automated inspection systems are normalized, and given in Table 14.2 in the respective columns.

Table 14.2. Normalized data of the automated inspection system selection attributes

Attributes	A	B	C	D
Accuracy	1	0.8889	0.6667	0.8333
Volumetric performance	1	0.875	0.625	0.875
Repeatability	1	1	0.625	0.875
Resolution	0.875	0.875	1	0.75
Maintainability	0.75	0.75	1	0.875
Reliability	1	0.9412	0.8235	0.8235
Initial cost	0.5	0.6667	1	0.8
Operation cost	0.5	0.1428	1	0.25
Throughput rate	0.875	0.875	1	1
Environmental factor requirement	0.75	0.75	1	0.8571
Flexibility in software interface	1	0.75	0.75	0.875

Let the decision maker (*i.e.*, user organization) prepare the following relative importance assignments:

	A	V	R	S	M	L	I	O	T	E	F
A	-	0.590	0.500	0.665	0.665	0.665	0.590	0.665	0.5	0.745	0.59
V	0.41	-	0.410	0.590	0.590	0.590	0.500	0.590	0.41	0.665	0.50
R	0.50	0.59	-	0.665	0.665	0.665	0.590	0.665	0.5	0.745	0.59
S	0.335	0.41	0.335	-	0.5	0.5	0.41	0.5	0.335	0.665	0.41
M	0.335	0.41	0.335	0.5	-	0.5	0.41	0.5	0.335	0.500	0.41
L	0.335	0.41	0.335	0.5	0.5	-	0.5	0.5	0.41	0.59	0.5
I	0.41	0.5	0.41	0.59	0.59	0.59	-	0.59	0.41	0.665	0.5
O	0.335	0.41	0.335	0.5	0.5	0.5	0.41	-	0.335	0.5	0.41
T	0.5	0.59	0.5	0.665	0.665	0.59	0.5	0.665	-	0.745	0.59
E	0.255	0.335	0.255	0.335	0.5	0.41	0.335	0.5	0.255	-	0.335
F	0.41	0.5	0.41	0.59	0.59	0.59	0.5	0.59	0.41	0.665	-

Step 2:

1. The automated inspection system selection attributes digraph, showing the presence as well as relative importance of the above attributes, is similar to Figure 2.2, but 11 attributes is drawn. This is not shown here due to obvious reasons.
2. The automated inspection system selection attributes matrix of this digraph is written based on Equation 2.10. However, it is not shown here.
3. The automated inspection system selection attributes function is written. However, it may be added that as a computer program is developed for calculating the permanent function value of a matrix, this step can be skipped.

4 & 5. The automated inspection system selection index (AIS-SI) is calculated using the values of A_i and a_{ij} for each alternative automated inspection system. The AIS-SI values of different automated inspection systems are given below in descending order:

AVI (USA):	31158.7734
CMM (USA):	29780.7563
LASER SCAN (Japan):	27462.2604
CMM (Japan):	25897.6459

From the above values of AVS-SI, it is understood that the automated inspection system AVI (USA) is the right choice for the given inspection application under the given conditions. The next choice is CMM (USA), and the last choice is CMM (Japan). However, Pandey and Kengpol (1995) suggested CMM (USA) as the first choice, LASER SCAN (Japan) as the second choice, AVI (USA) as the third choice, and CMM (Japan) as the last choice.

14.2.2 AHP and its Versions

Let the decision maker prepare the following matrix:

	A	V	R	S	M	L	I	O	T	E	F
A	1	3	1	4	5	4	3	5	2	6	3
V	1/3	1	1/3	2	3	2	1	3	1/2	4	1
R	1	3	1	4	5	4	3	5	2	6	3
S	1/4	1/2	1/4	1	2	1	1/2	2	1/3	3	1/2
M	1/5	1/3	1/5	1/2	1	1/2	1/3	1	1/4	2	1/3
L	1/4	1/2	1/4	1	2	1	1/2	2	1/3	3	1/2
I	1/3	1	1/3	2	3	2	1	3	1/2	4	1
O	1/5	1/3	1/5	1/2	1	1/2	1/3	1	1/4	2	1/3
T	1/2	2	1/2	3	4	3	2	4	1	5	2
E	1/6	1/4	1/6	1/3	1/2	1/3	1/4	1/2	1/5	1	1/4
F	1/3	1	1/3	2	3	2	1	3	1/2	4	1

In the above matrix, accuracy (A) and repeatability (R) are considered more important than the remaining attributes.

The normalized weights of each attribute are calculated following the procedure presented in Section 3.2.3, and these are: $W_A = 0.2071$, $W_V = 0.0858$, $W_R = 0.2071$, $W_S = 0.0518$, $W_M = 0.0325$, $W_L = 0.0518$, $W_I = 0.0858$, $W_O = 0.0325$, $W_T = 0.1376$, $W_E = 0.0219$, and $W_F = 0.0858$. The value of λ_{max} is 11.1958 and CR = 0.01332, which is much less than the allowed CR value of 0.1. Thus, there is good consistency in the judgements made.

The value of AIS-SI is now calculated using the above weights and the normalized data of the attributes given in Table 14.2. This leads to the ranking given by the revised AHP or ideal mode of AHP method. The alternative automated inspection systems are arranged in descending order of the AIS-SI:

CMM (USA):	0.9050
CMM (Japan):	0.8493
LASER SCAN (Japan):	0.8487
AVI (USA):	0.7925

It may be noted that the ranking depends upon the judgements of relative importance of attributes made by the decision maker.

For the above weights of importance of attributes, multiplicative AHP leads to the following ranking order:

CMM (USA):	0.8837
LASER SCAN (Japan):	0.8301
CMM (Japan):	0.8154
AVI (USA):	0.7738

14.2.3 TOPSIS Method

Following the steps of the methodology given in Section 3.2.6, the TOPSIS method gives the ranking order shown below:

CMM (USA):	0.6811
CMM (Japan):	0.6074
LASER SCAN (Japan):	0.5977
AVI (USA):	0.3813

It may be observed that the ranking given by the TOPSIS method is same as that given by the ideal AHP method.

14.2.4 Modified TOPSIS Method

This methods leads to the following ranking order:

CMM (USA):	0.6308
LASER SCAN (Japan):	0.5724
AVI (USA):	0.5280
CMM (Japan):	0.4572

The ranking suggested by this method is same as that proposed by Pandey and Kengpol (1995) using the PROMETHEE method.

In this particular example of automated inspection system selection, proposing CMM (USA) as the first right choice seems to be more logical and objective. AVI (USA) is better than the other alternative inspection systems with respect to six of 11 attributes. However, the weights of importance assigned to the attributes play an important role in the selection process.

References

Elshennaway AK (1989) The role of inspection in automated manufacturing. Computers & Industrial Engineering 17:327–333

Golomski WA (1990) Justification Methods for Computer Integrated Manufacturing Systems. In: Accounting Aspects of Automated Inspection Systems, Elsevier, Amsterdam

Groover MP (2001) Automation, production systems and computer integrated manufacturing. Prentice Hall of India, New Delhi

Kalpakjian S, Schmid SR (2000) Manufacturing engineering and technology. Addison Wesley Longman (Singapore), Indian branch, Delhi

Pandey PC, Kengpol A (1995) Selection of an automated inspection system using multiattribute decision analysis. International Journal of Production Economics 39:289–298

15

Selection of Material Handling Equipment

15.1 Introduction

Material handling equipment selection is an important function in the design of a material handling system, and thus a crucial step for facilities planning. Using proper material handling equipment can enhance the production process, provide effective utilization of manpower, increase production, and improve system flexibility. The importance of material handling equipment selection cannot be overlooked. However, with the wide range of material handling equipment available today, determination of the best equipment alternative for a given production scenario is not an easy task (Chan *et al.*, 2001).

Material handling accounts for 30-75% of the total cost of a product, and efficient material handling can be responsible for reducing the manufacturing system operations cost by 15-30% (Sule, 1994). These values underscore the importance of material handling costs as an element in improving the cost structure of a product. The determination of a material handling system involves both the selection of suitable material handling equipment, and the assignment of material handling operations to each individual piece of equipment. Hence, material handling system selection can be defined as the selection of material handling equipment to perform material handling operations within a given working area considering all aspects of the products to be handled (Sujono and Lashkari, 2007).

The material handling system (MHS) plays a crucial role in flexible manufacturing systems. When inadequately designed, the MHS can indeed interfere severely with the overall performance of the system, and lead to substantial losses in productivity and competitiveness, and to unacceptably long lead times. Thus, to avoid such pitfalls, MHS selection is considered as an important issue in manufacturing industries.

Material handling equipment has been classified into the following main groups of industrial trucks, conveyors, automated guided vehicles (AGVs), cranes, storage/retrieval systems and industrial robots (Sule, 1994; Kulak, 2005). This module includes examples of 40 move equipment types, and six storage equipment with their performance attributes. Table 15.1 presents the material handling equipment types, and Table 15.2 presents the material handling equipment selection attributes for manufacturing systems.

Table 15.1. Types of material handling equipment (from Kulak 2005; reprinted with permission from Elsevier)

(1) Industrial trucks:	
	Handcart, tier platform truck, handlift truck, power-driven handtruck, power-driven platform truck, forklift truck, narrow-aisle trucks, material lift, tractortrailer train, drum truck, drum lifter
(2) Conveyors:	
	Belt conveyor, roller conveyor, chute conveyor, slat conveyor, screw conveyor, chain conveyor, plain chain conveyor, trolley conveyor, wheel conveyor, tow conveyor, bucket conveyor, cart-on-track conveyor, pneumatic tube conveyor, overhead monorail conveyor
(3) Automated guided vehicles (AGV):	
	Manual load/unload AGV, low-lift AGV, high-lift AGV, tugged AGV, roller deck AGV, stationary deck AGV, lift deck AGV
(4) Cranes:	
	Stacker crane, tower crane, gantry crane, jib crane
(5) Storage/retrieval systems:	
	Unit load AS/RS, man-on-board AS/RS, shelf storage system, pallet rack system, block stocking on floor, block stocking in rack
(6) Robots:	
	Pneumatic robot, electric robot, hydraulic robot, mechanized manipulator

In the literature, there are various studies focusing on the solution of the complicated problem of material handling equipment selection. Malmborg *et al.* (1987) developed a prototype expert system considering 17 equipment attributes and 47 devices for industrial truck type selection. Velury and Kennedy (1992) studied the selection of relevant factors that need to be considered in the design of a bulk material handling system, and the selection of equipment once these factors had been considered. A model was presented that took into account economics, characteristics of the equipment, environmental characteristics, and compatibilities between equipment types.

Swaminathan *et al.* (1992) developed EXCITE, the expert consultant for in-plant transportation equipment, addressing 35 equipment types, and 28 material, move, and method attributes. Chu *et al.* (1995) developed a computer-aided material handling equipment selection system called ADVISOR. Park (1996) developed an intelligent consultant system for material handling equipment selection, including 50 equipment types and 29 attributes, *i.e.*, move attributes, material characteristics, operation requirements, and area constraints. Kim and Eom (1997) introduced a material handling selection expert system. Fisher *et al.* (1998) introduced MATHES, the 'material handling equipment selection expert systems', for the selection of material handling equipment from 16 possible choices. MATHES including 172 rules dealing with path, volume of flow, sizes of unit, and distance between departments as parameters. MATHES II had been provided with the same procedure as MATHES. However, MATHES II had a larger working scope, and greater consultation functions.

Table 15.2. Material handling equipment selection attributes for manufacturing systems (from Kulak 2005; reprinted with permission from Elsevier)

Material:
- Material type: individual unit, pallet unit, loose, bulk, packed, bar-stock, *etc.*
- Material weight: light, medium, heavy
- Bottom surface: flat, non-flat
- Material nature: fragile, sturdy
- Material size: small, medium, large
- Annual demands of the material: <X, X, or above

Operation:
- Function: move, storage/retrieval
- Operation control: controllable, uncontrollable
- Automation: required, not required
- Method of transportation: carry, tow
- Transfer frequency (per shift): <X, X, or above
- Storage/retrieval order: first-in-first-out (FIFO), first-in-last-out (FILO)

Move:
- Type: transportation, conveying, loading/unloading, storage/retrieval
- Direction: decline, horizontal, vertical
- Level: on floor, above floor
- Area and path: fixed, variable
- Distance: <X, X, or above
- Height: <X, X, or above

Area constraints:
- Floor space: available, not available
- Aisle width: <X, X, or above
- Truss height: <X, X, or above
- Rack deep: single, double X: predefined value

Chan *et al.* (2001) described the development of an intelligent material handling equipment selection system called material handling equipment selection advisor (MHESA). The MHESA was composed of three modules: (1) a database to store equipment types with their specifications; (2) a knowledge-based expert system for assisting material handling equipment selection; and (3) an analytic hierarchy process (AHP) model to choose the most favorable equipment type. The concept proposed by the authors could automate the design of a material handling equipment selection system. Fonseca *et al.* (2004) developed a knowledge-based system for conveyor equipment selection.

Lashkari *et al.* (2004) presented an integrated approach to operation allocation (OA) and material handling systems selection (MHSS) in cellular manufacturing systems. The OA model assigns the operations of a set of part types to a group of machines, and provides this information as input to the MHSS model. The MHSS model allocates equipment for handling the parts between machines, as well as at the single machine level. This information was then fed back as an input to the OA model. An iterative algorithm was developed to solve the two models sequentially, and a numerical example was provided to demonstrate the applicability of the models.

Intelligent computer systems have been developed, such as experts systems, and decision support systems for the selection of material handling equipments. One of the successful applications of experts systems was SEMH, selection of equipment for material handling. SEMH searches its knowledge base to recommend the degree of mechanization, and the type of material handling equipment to be used, based on various characteristics, *i.e.*, type, weight, size, *etc.* (Fonseca *et al.*, 2004).

Kulak (2005) developed a decision support system called FUMAHES-fuzzy multi-attribute material handling equipment selection. FUMAHES consists of a database, a rule-based system, and multi-attribute decision-making modules. The database includes detailed data about equipment types and their properties. The rule-based system module provides rules that are utilized by inference engine for determining the most appropriate material handling equipment type. Ultimately, a final decision was made for the most proper equipment among the alternatives of the same type, using the information axiom of axiomatic design principles.

Sujono and Lashkari (2007) proposed a method for simultaneously determining operation allocation and material handling system selection in an FMS environment with multiple performance objectives. The 0–1 integer programming model was developed to select machines, assign operations of part types to the selected machines, and assign material handling equipment to transport the parts from machine to machine, as well as to handle the part in a given machine. The selection was based on the compatibility between the material handling equipment and the parts. The objective was to minimize the costs of operations, material handling, and machine setups, and to maximize the part–equipment compatibility.

Chakraborthy and Banik (2006) focused on the application of the AHP technique in selecting the optimal material handling equipment for a specific material handling equipment type. The relative importance of each criterion, subcriterion and sub-subcriteria was measured using pair-wise comparison matrices, and the overall ranking of each alternative equipment was then determined.

A better approach in material handling equipment selection can be the identification of pertinent attributes, and of the potential alternative equipment by a team consisting of experts at different levels. The values of the attributes (A_i) with the equipment specifications and requirements can be obtained, and their relative importance (a_{ij}) can be determined and stored in a database. An objective or subjective value, or its range, may be assigned to each identified attribute as a limiting value or threshold value for its acceptance for the material handling equipment selection problem considered. Alternative material handling equipment with each of its selection attributes, meeting the acceptance value, may be short-listed. After short-listing the alternative material handling equipment, the main task to choose the alternative material handling equipment is to assess how it serves the considered attributes. The multiple attribute decision-making methods proposed in this book can be used for this purpose.

Now, an example is considered to demonstrate the application of GTMA and other decision-making methods for material handling equipment selection for an industrial application.

15.2 Example

A case of material handling equipment selection, similar to the one presented by Kulak (2005), is considered. Here, the problem under consideration is to determine the most appropriate conveyor among the alternatives of the same type. The related objective and subjective data of the attributes are given in Table 15.3. The flexibility attribute is defined subjectively. Hence, appropriately using Table 4.3, the objective values are assigned, and shown in Table 15.4.

Table 15.3. Objective and subjective data of the conveyor selection attributes

Alternative conveyor	FC	VC	SC	IW	W	F
1	2	0.45	12	15	10	Very good
2	2.3	0.44	13	20	10	Excellent
3	2.25	0.45	11	30	20	Excellent
4	2.4	0.46	10	25	15	Very good

FC: Fixed costs per hour VC: Variable costs per hour SC: Speed of conveyor (m/min)
IW: Item width (cm) W: Item weight (kg) F: Flexibility

Table 15.4. Objective data of the conveyor selection attributes

Alternative conveyor	FC	VC	SC	IW	W	F
1	2	0.45	12	15	10	0.745
2	2.3	0.44	13	20	10	0.955
3	2.25	0.45	11	30	20	0.955
4	2.4	0.46	10	25	15	0.745

15.2.1 Application of Graph Theory and Matrix Approach (GTMA)

Various steps of the methodology, proposed in Section 2.6, are carried out as described below:

Step 1: In the present work, the attributes considered are the same as those considered by Kulak (2005) and these are: fixed costs per hour (FC), variable costs per hour (VC), speed of conveyor (SC), item width (IW), item weight (W), and flexibility (F). The objective values of the attributes, which are given in Table 15.4, are to be normalized. FC and VC are non-beneficial attributes, and the remaining four attributes are considered as beneficial attributes. The conveyor should have low fixed and variables costs, higher speed, ability to handle large item widths, and weights, and have higher flexibility. Values of these attributes are normalized, and are given in Table 15.5 in the respective columns.

Table 15.5. Normalized data of the conveyor selection attributes

Alternative conveyor	FC	VC	SC	IW	W	F
1	1	0.9778	0.9231	0.5	0.5	0.7801
2	0.8696	1	1	0.6667	0.5	1
3	0.8889	0.9778	0.8461	1	1	1
4	0.8333	0.9565	0.7692	0.8333	0.75	0.7801

Let the decision maker assign equal importance to the attributes as shown below:

	FC	VC	SC	IW	W	F
FC	---	0.50	0.50	0.50	0.50	0.50
VC	0.50	---	0.50	0.50	0.50	0.50
SC	0.50	0.50	---	0.50	0.50	0.50
IW	0.50	0.50	0.50	--	0.50	0.50
W	0.50	0.50	0.50	0.50	--	0.50
F	0.50	0.50	0.50	0.50	0.50	---

Step 2:

Conveyor selection attributes digraph, conveyor selection attributes matrix of the digraph, and conveyor selection attributes function for the matrix can be prepared. The value of the conveyor selection index is calculated using the values of A_i and a_{ij} for each alternative conveyor. The conveyor selection index values of different conveyors are given below in descending order:

Conveyor 3 : 27.7449
Conveyor 2 : 21.8354
Conveyor 4 : 21.3136
Conveyor 1 : 19.3483

Thus, GTMA suggests conveyor 3 as the correct choice for the considered material handling application and conveyor 1 as the last choice. The difference between the index values of conveyors 2 and 4 is small, and these may be considered as equal to each other.

15.2.2 SAW Method

Considering equal weights of importance of the six conveyor selection attributes and using the normalized data of the attributes given in Table 15.5, the SAW method leads to the following values of conveyor selection index:

Conveyor 3 : 0.9521
Conveyor 2 : 0.8394
Conveyor 4 : 0.8204
Conveyor 1 : 0.7802

Thus, the SAW method also suggests conveyor 3 as the correct choice for the material handling application considered, and conveyor 1 as the last choice.

15.2.3 WPM

Considering equal weights of importance of the six conveyor selection attributes, the conveyor selection index value for each conveyor is calculated, and the values are arranged as given below:

Conveyor 3	:	0.9501
Conveyor 4	:	0.8177
Conveyor 2	:	0.8135
Conveyor 1	:	0.7486

WPM also suggests conveyor 3 as the correct choice for the considered material handling application and conveyor 1 as the last choice. The difference between the index values of conveyors 2 and 4 is very small, and these may be considered as equal to each other.

15.2.4 AHP and its Versions

The AHP method gives the same results as those of the SAW method. The multiplicative AHP method gives the same results as those of WPM.

15.2.5 TOPSIS Method

Following the steps of the methodology given in Section 3.2.6, the TOPSIS method gives the following weighted normalized matrix:

$$\begin{bmatrix} 0.0743 & 0.0833 & 0.0866 & 0.0539 & 0.0580 & 0.0724 \\ 0.0855 & 0.0815 & 0.0940 & 0.0719 & 0.0580 & 0.0928 \\ 0.0836 & 0.0833 & 0.0794 & 0.1079 & 0.1161 & 0.0928 \\ 0.0892 & 0.0852 & 0.0721 & 0.0899 & 0.0871 & 0.0724 \end{bmatrix}$$

The ideal (best) and negative ideal (worst) solutions are obtained, and these are given as:

$V_{FC}^{+} = 0.0743$ $V_{FC}^{-} = 0.0892$
$V_{VC}^{+} = 0.0815$ $V_{VC}^{-} = 0.0852$
$V_{SC}^{+} = 0.0940$ $V_{SC}^{-} = 0.0721$
$V_{IW}^{+} = 0.1079$ $V_{IW}^{-} = 0.0539$
$V_{W}^{+} = 0.1161$ $V_{W}^{-} = 0.0580$
$V_{F}^{+} = 0.0928$ $V_{F}^{-} = 0.0724$

The separation measures are:

$S_1^{+} = 0.0822$ $S_1^{-} = 0.0208$
$S_2^{+} = 0.0692$ $S_2^{-} = 0.0352$
$S_3^{+} = 0.0173$ $S_3^{-} = 0.0823$
$S_4^{+} = 0.0478$ $S_4^{-} = 0.0462$

The relative closeness of a particular alternative to the ideal solution is calculated; this is named the 'conveyor selection index (CSI)' in the present example, and these values are arranged in descending order as:

Conveyor 3 : 0.8267
Conveyor 4 : 0.4915
Conveyor 2 : 0.3369
Conveyor 1 : 0.2020

TOPSIS method also suggests conveyor 3 as the correct choice for the material handling application considered, and conveyor 1 as the last choice.

15.2.6 Modified TOPSIS Method

For equal relative importance weights of six conveyor selection attributes, the modified TOPSIS method gives the following results:

The weighted Euclidean distances are:

$D_1^+ = 0.2012$ $D_1^- = 0.0509$
$D_2^+ = 0.1694$ $D_2^- = 0.0861$
$D_3^+ = 0.0423$ $D_3^- = 0.2017$
$D_4^+ = 0.1171$ $D_4^- = 0.1132$

The conveyor selection index values are calculated and these are arranged in descending order as:

Conveyor 3 : 0.8267
Conveyor 4 : 0.4915
Conveyor 2 : 0.3369
Conveyor 1 : 0.2020

The results presented by the modified TOPSIS method are exactly the same as those given by simple TOPSIS method.

All decision-making methods presented in this chapter suggest conveyor 3 as the correct choice for the application considered, and conveyor 1 as the last choice. In this example, the conveyor selection attributes are considered to have equal relative importance. However, the methods covered in this chapter can deal with unequal relative importance values of the attributes also. Further, in the present example, the most appropriate material handling equipment selection from among the alternatives of the same type (*i.e.*, conveyor) is considered. The proposed methods can also be applied with equal ease for the selection problem of choosing the most appropriate material handling equipment from among the alternatives of different type.

References

Chakraborthy S, Banik D (2006) Design of a material handling equipment selection model using analytic hierarchy process. International Journal of Advanced Manufacturing Technology 28:1237–1245

Chan FTS, Ip RWL, Lau H (2001) Integration of expert system with analytic hierarchy process for the design of material handling equipment selection system. Journal of Materials Processing Technology 116:137–145

Chu HK, Egbelu PJ, Wu CT (1995) ADVISOR: a computer-aided material handling equipment selection system. International Journal of Production Research 33:3311–3329

Fisher EL, Farber JB, Kay MG (1998) MATHES: an expert system for material handling equipment selection. Engineering Costs and Production Economics 14:297–310

Fonseca DJ, Uppal G, Greene TJ (2004) A knowledge-based system for conveyor equipment selection. Expert Systems with Applications 26:615–623

Kim KS, Eom JK (1997) Expert system for selection of material handling and storage systems. International Journal of Industrial Engineering 4:81–89

Kulak O (2005) A decision support system for fuzzy multi-attribute selection of material handling equipments. Expert Systems with Applications 29:310–319

Lashkari RS, Boparai R, Paulo J (2004) Towards an integrated model of operation allocation and material handling selection in cellular manufacturing systems. International Journal of Production Economics 87:115–139

Malmborg CJ, Simons MH, Choudhary JV (1987) A prototype expert system for industrial truck type selection. Industrial Engineering 19:58–65

Park YB (1996) ICMESE: intelligent consultant system for material handling equipment selection and evaluation. Journal of Manufacturing Systems 15:325–336

Sujono S, Lashkari RS (2007) A multi-objective model of operation allocation and material handling system selection in FMS design. International Journal of Production Economics 105:116–133

Sule DR (1994) Manufacturing facilities: location, planning and design. PWS Publishing Company, Boston

Swaminathan SR, Matson JO, Mellichamp, JM (1992) EXCİTE: expert consultant for in-plant transportation equipment. International Journal of Production Research 30:1969–1983

Velury J, Kennedy J (1992) A systematic procedure for the selection of bulk material handling equipment. International Journal of Production Economics 27:233–240

16

Selection of Rapid Prototyping Process in Rapid Product Development

16.1 Introduction

In the development of a new product, there is invariably a need to produce a single example, or prototype, of a designed part or system, before the allocation of large amounts of capital funds to new production facilities or assembly lines. The main reason for this need is that the capital cost is extremely high, and production tooling takes much time to prepare; consequently, a working prototype is needed for 'troubleshooting' and for design evaluation, before a complicated system is ready to be produced and marketed (Kalpakjian and Schmid, 2000). In a competitive market, the speed with which a product flows from concept to marketable product plays a crucial role. It is well known that products that are introduced before their competitors are generally more profitable and enjoy a larger share of the market. At the same time, there are important concerns regarding the production of high-quality products. For these reasons, there is a concerted effort to bring high-quality products to market quickly.

A new technology that considerably speeds the iterative product development process is the concept and practice of rapid prototyping (RP). The advantages of rapid prototyping include:

- Cost reduction up to 50%.
- Processing time reduction up to 75%. Physical models from CAD data files can be manufactured in a matter of hours to allow rapid evaluation of manufacturability and design effectiveness.
- Better visualization and concept verification.
- High design flexibility to enable short-term component modifications.
- Usage of prototype in subsequent manufacturing operations to obtain the final parts.
- Cost-effective component production for demonstration purposes, and functional test samples.
- Use of rapid prototyping operations for production of rapid tooling for manufacturing operations.

Rapid prototyping processes can be classified into three major groups: subtractive, additive, and virtual. As the names imply, subtractive processes involve material removal from a work piece larger than the final part; additive processes build up a part by adding material incrementally; and virtual processes use advanced computer-based visualization technologies.

Rapid prototyping systems have been used mainly in manufacturing industries such as automobiles, electric home appliances and aerospace. Generally, RP processes begin with a stereolithography (STL) file that describes a model created by a CAD surface or a solid modeler. The RP models can be used to visualize or verify designs, to check for form, fit and function, or to produce a tooling (or master) pattern for casting or molding (Williams *et al.*, 1996). Due partly to the rapid growth of RP technology, the selection of the most appropriate RP process to meet users' requirements from among a number of RP systems has become increasingly important. However, it is difficult for users with RP experience as well as those who employ a service bureau, to select a suitable system because there are so many RP systems worldwide, and the best selection depends on many attributes. Furthermore, each system has its own strengths, defects, applications, utilities, and limitations. This is a complex problem that cannot be solved readily using conventional statistical techniques alone. Selection of an appropriate process requires a sound understanding of the interactions between the part quality, part properties, cost, build envelope, build time (speed), and other concerns (Byun and Lee, 2004).

Several studies have focused on developing RP systems selection procedures. Schmidt (1994) made a benchmark comparison of rapid prototyping technologies. Bauer *et al.* (1996) developed the rapid prototyping system selector, a software tool that helps find the best RP system to manufacture a physical prototype. It aimed to help RP users, designers, or RP job shops choose the best combination of materials and RP machines to fabricate a prototype rather than to select the most suitable RP process based on specific selection attributes. Phillipson (1997) developed RP Advisor for choosing an appropriate RP system that uses build time, cost and quality, as selection attributes. It considered six commercially available RP systems and also ran on MS Access. The best RP system was chosen using multi-attributes optimization theory; however, the system did not consider various attributes such as material properties and had limitation in calculations and ease of use. Pham and Gault (1998) presented an overview of RP technologies, and comments on their strengths and weaknesses. A taxonomy was also suggested, along with a preliminary guide to process selection based on the end use of the prototype. Bibb (1999) developed a computer-based RP design advice system to help small manufacturing enterprises solve problems encountered when attempting to apply RP and tooling technologies in product development

Kengpol and O'Brien (2001) outlined a decision support tool to assess the value of investing in time compression technologies (TCTs) to achieve rapid product development. The authors presented a data structure to monitor the effectiveness of a decision and a decision model that consolidated quantitative and qualitative variables through the use of the analytic hierarchy process (AHP), cost/benefit and statistical analyses. Masood and Soo (2002) presented an expert system-based RP system selection program incorporating 39 RP systems commercially available from 21 RP manufacturers worldwide. The program allowed the user to choose one of four options, namely, quick selection, detailed selection, build technology, or machine style for system selection, with each option considers a systematic selection attributes. The program is a rule based expert system, and recommends the RP system along with its full specifications on the basis of interactive question–answer session, with the user. The system is believed to be the first expert system-based RP selection program, and has the potential for future expansion into a full-fledged RP selector system.

Masood *et al.* (2003) presented a generic mathematical algorithm to determine the best part orientation for building a part in a layer-by-layer rapid prototyping system. Byun and Lee (2004) presented an effective methodology for selecting the RP system most appropriate for the end use of the part when multi-attributes included both uncertain and crisp data. The major factors used for RP process selection included accuracy, surface roughness, strength, elongation, the cost of the part and build time. Crisp data, such as accuracy and roughness, were obtained by the new test part, which was utilized for the benchmarking of the capabilities of the various RP systems. It was designed with conjoint analysis to reflect the users' knowledge and experience. With the part cost and build time classified as imprecise data, linguistic variables, which were treated as fuzzy numbers with triangular membership functions, were used. A modified TOPSIS approach was also proposed to analyze both quantitative and qualitative data. This method determined ranks between the RP systems and effectively reflected the information produced for the decision using the multiple attributes. The ranks were then altered using the weights assigned by a pair-wise comparison matrix to provide the final ranking. However, the fuzzy approach used by the authors for assigning the values to the cost and build time attributes makes them even fuzzier, and requires more computation. In another work, Byun and Lee (2006) attempted to determine the optimal build-up direction of a part for different RP systems. The best orientation was selected using the simple additive weighing method. The validity of the algorithm was demonstrated by a few examples.

Some studies mentioned above were based on a rule-based knowledge representation with IF–THEN rules. It is difficult to order the ranking of the most suitable RP systems using conditional expressions such as IF, THEN, ELSE, or CASE statements, although various authors attempted to do so. In addition, because some did not consider various selection attributes, the selection of an RP system could not provide the desirable solution (Byun and Lee, 2004). Most of all, when the selection attributes include vague values such as ease of use, and environmental affinity for an RP system or an attribute for which it is difficult to obtain an exact numerical value (as with the build time and part cost), decision makers experience several difficulties in selecting an appropriate RP system by conventional decision methods using deterministic values (Byun and Lee, 2004). There is a need for a simple, systematic and logical scientific method, or mathematical tool, to guide user organizations in taking a proper RP system selection decision that can solve the problems mentioned by Byun and Lee (2004). The objective of an RP system selection procedure is to identify the RP system selection attributes, and obtain the most appropriate combination of RP system selection attributes in conjunction with the real requirement. An RP system selection attribute is defined as a factor that influences the selection of RP system for making the prototype of a given product. Thus, efforts need to be extended to determine attributes that influence RP system selection, using a simple logical approach, to eliminate unsuitable RP systems and for selection of a proper RP system to strengthen the existing RP system selection procedure. This is considered in this chapter using the graph theory and matrix approach (GTMA) and fuzzy MADM methods.

Now, to demonstrate and validate the application of decision-making methods, an example is considered. First GTMA is applied, and subsequently a few MADM methods are applied to rank and select the RP systems.

16.2 Example

Byun and Lee (2004) developed a decision support system for the selection of a rapid prototyping process using the modified TOPSIS method. On the basis of the data obtained by the questionnaires from different user groups such as the service bureau, governmental institutes and industry users, the authors argued that attributes such as dimensional accuracy, surface roughness, part cost, build time and material properties (tensile strength and elongation) were the major ones in assessing RP parts, as these can provide sufficient information for the selection of an appropriate RP process.

A case study of a designed test part comparing six RP systems was conducted. Six attributes, accuracy (A), surface roughness (R), tensile strength (S), elongation (E), cost of the part (C) and build time (B), were identified as evaluation attributes for the selection of the RP system. The build time included the pre-processing time, building time and post-processing time. The part cost included both the material and the labor costs. Attributes C and B were expressed in linguistic terms. The test part design and dimensions are shown in Figure 16.1 and the objective and subjective data of the attributes are given in Table 16.1.

Table 16.1. Data of the RP system selection attributes (from Byun and Lee 2004; with kind permission from Springer Science and Business Media)

RP system	A	B	S	E	C	B
SLA3500	120	6.5	6.5	5	VH	M
SLS2500	150	12.5	40	8.5	VH	M
FDM8000	125	21	30	10	H	VH
LOM1015	185	20	25	10	SH	SL
Quadra	95	3.5	30	6	VH	SL
Z402	600	15.5	5	1	VVL	VL

A: Accuracy R: Surface roughness S: Tensile strength
E: Elongation C: Cost of the part B: Build time
VVL: Very very low VL: Very low SL: Slightly low
M: Medium SH: Slightly high H: High
VH: Very high

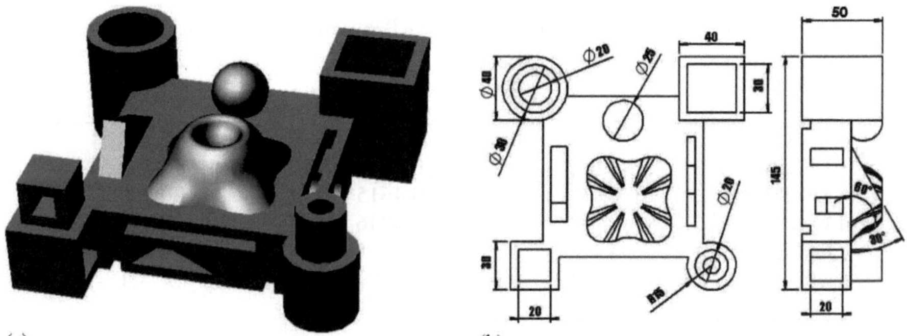

(a) (b)

Figure 16.1. Test part design and its dimensions (from Byun and Lee 2004; with kind permission from Springer Science and Business Media)

16.2.1 Application of Graph Theory and Matrix Approach (GTMA)

Various steps of the methodology, proposed in Section 2.6, are carried out as described below:

Step 1: In the present work, the attributes considered are the same as those of Byun and Lee (2004) and these are: accuracy (A), surface roughness (R), tensile strength (S), elongation (E), cost of the part (C) and build time (B). Attributes C and B, which are expressed in a fuzzy manner are quantified using Table 4.3. The objective data of all six attributes are given in Table 16.2.

Table 16.2. Objective data of the RP system selection attributes

RP system	A	B	S	E	C	B
SLA3500	120	6.5	6.5	5	0.745	0.5
SLS2500	150	12.5	40	8.5	0.745	0.5
FDM8000	125	21	30	10	0.665	0.745
LOM1015	185	20	25	10	0.59	0.41
Quadra	95	3.5	30	6	0.745	0.41
Z402	600	15.5	5	1	0.135	0.255

The quantitative values of the RP system selection attributes, which are given in Table 16.2, are to be normalized. S and E are beneficial attributes and higher values are desirable. Values of these attributes are normalized, as explained in Section 2.4, and are given in Table 16.3 in the respective columns. The values given for A and R in Table 16.1 are in fact related with dimensional inaccuracy and surface roughness. Hence A, R, C, and B are considered as non-beneficial attributes and lower values are desirable. The values of these attributes for different RP systems are normalized, and are given in Table 16.3 in the respective columns.

Table 16.3. Normalized data of the RP system selection attributes

RP system	A	B	S	E	C	B
SLA3500	0.7917	0.5385	1	0.5	0.1812	0.51
SLS2500	0.6333	0.28	0.6154	0.85	0.1812	0.51
FDM8000	0.76	0.1667	0.4615	1	0.2030	0.3423
LOM1015	0.5135	0.175	0.3846	1	0.2288	0.6219
Quadra	1	1	0.4615	0.6	0.1812	0.6219
Z402	0.1583	0.2258	0.0769	0.1	1	1

Relative importance of attributes (a_{ij}) is also assigned values, as explained in Section 2.4. Let the decision maker (*i.e.*, user organization) select the following assignments:

$$
\begin{array}{c} \\ A \\ R \\ S \\ E \\ C \\ B \end{array}
\begin{array}{cccccc}
A & R & S & E & C & B \\
\end{array}
$$

$$
\begin{array}{c|cccccc|}
 & A & R & S & E & C & B \\
A & - & 0.5 & 0.665 & 0.665 & 0.745 & 0.745 \\
R & 0.5 & - & 0.665 & 0.665 & 0.745 & 0.745 \\
S & 0.335 & 0.335 & - & 0.5 & 0.665 & 0.665 \\
E & 0.335 & 0.335 & 0.5 & - & 0.665 & 0.665 \\
C & 0.255 & 0.255 & 0.335 & 0.335 & - & 0.5 \\
B & 0.255 & 0.255 & 0.335 & 0.335 & 0.5 & -
\end{array}
$$

In this case, A and R are considered more important than the remaining four attributes. S and E are considered more important than C and B. However, in actual practice, the user organization depending on the requirements can judiciously decide these values of relative importance. The assigned values are for demonstration purposes only.

Step 2:

1. The RP system selection attributes digraph, showing the presence as well as relative importance of the above attributes is similar to Figure 2.2, but with six attributes drawn. This is not shown here due to obvious reasons.
2. The RP system selection attributes matrix of this digraph is written based on Equation 2.10. However, it is not shown here.
3. The RP system selection attributes function is written but not shown here. However, it may be added that as a computer program is developed for calculating the permanent function value of a matrix, this step can be skipped.

4 & 5. RP system selection index (RPSI) is calculated using the values of A_i and a_{ij} for each alternative RP system. The RPSI values of the different RP systems are given below in descending order:

Quadra	10.7085
SLA3500	9.4606
SLS2500	8.0812
LOM1015	7.6081
FDM8000	7.5736
Z402	6.6198

From the above values of RPSI, it is understood that the RP system Quadra is the correct choice for the given application under the given conditions. The next choice is SLA3500 and the last choice is Z402. These results match those obtained by Byun and Lee (2004) using the TOPSIS approach. In their work also, Byun and Lee (2004) proposed Quadra as the first choice, SLA3500 as the second choice, and Z402 as the last choice.

Suppose, in the relative importance matrix, that the assignments are made as shown below:

	A	R	S	E	C	B
A	-	0.5	0.665	0.665	0.335	0.335
R	0.5	-	0.665	0.665	0.335	0.335
S	0.335	0.335	-	0.5	0.255	0.255
E	0.335	0.335	0.5	-	0.255	0.255
C	0.665	0.665	0.745	0.745	-	0.5
B	0.665	0.665	0.745	0.745	0.5	-

In this example, C and B are considered more important than the other four attributes. A and R are considered more important than S and E. For these assignments, the RPSI values for different RP systems are calculated, and are given below in descending order:

Quadra	10.5126
SLA3500	9.4849
SLS2500	8.1792
LOM1015	7.7134
FDM8000	7.6659
Z402	6.5914

From the above values of RPSI, it is understood that the RP system Quadra is the correct choice for the given application. The next choice is SLA3500, and the last choice is Z402. Byun and Lee (2004) assessed this relative importance matrix case also and proposed Z402 as the first choice, LOM1015 as the second choice and FDM8000 as the last choice. A closer look at the values of the alternatives reveals that Z402 has the least values compared to the other alternatives for the four attributes A, R, S and E, but better for the two attributes C and B. Thus, proposing Z402 as the first choice may not be appropriate. Further, the fuzzy calculations made by Byun and Lee (2004) are not necessary, and also require more computation. Moreover, the relative importance matrix prepared by Byun and Lee (2004) shows inconsistency in judgements, with the calculated value of CR almost equaling the maximum allowed value of 0.1. For example, if A is 3 times as important as S, and 1/5 asf important as C, then S can not be 1/3 as important as C. Such errors were present in the relative importance matrix prepared by Byun and Lee (2004).

16.2.2 SAW Method

The procedure suggested by Edwards *et al.* (1982) to assess weights for each of the attributes to reflect relative importance to the RP system selection decision is followed here. First, the attributes are ranked in order of importance and 10 points each are assigned to the least important attributes C and B. Attributes S and E are considered as equally important in the present example, and given 20 points each to reflect their relative importance. A and R are considered as equally important, and given 50 points each. The final weights are obtained by normalizing the sum of the points to one. Thus, the weights of A, R, S, E, C, and B are calculated as 0.3125, 0.3125, 0.125, 0.125, 0.0625, and 0.0625, respectively. Using these weights, and the normalized data of the attributes for different RP systems, the RPSI values are calculated, and are arranged in descending order of the index.

Quadra	0.8079
SLA3500	0.6464
SLS2500	0.5118
FDM8000	0.5064
LOM1015	0.4414
Z402	0.2671

The SAW method also suggests the RP system Quadra as the correct choice for the given problem of RP system selection, and Z402 as the last choice.

16.2.3 WPM

Using the same weights of attributes as those selected for the SAW method, RPSI for each RP system is calculated, and the values are arranged as given below:

Quadra	0.7431
SLA3500	0.6054
SLS2500	0.4628
FDM8000	0.4029
LOM1015	0.3700
Z402	0.1921

WPM also suggests the RP system Quadra as the correct choice for the given problem of RP system selection, and Z402 as the last choice.

16.2.4 AHP and its Versions

The AHP method may use the same weights as those selected for the SAW method. In that case, the ranking of the RP systems will be the same. However, let the decision maker prepare the following matrix:

	A	R	S	E	C	B
A	1	1	3	3	5	5
R	1	1	3	3	5	5
S	1/3	1/3	1	1	3	3
E	1/3	1/3	1	1	3	3
C	1/5	1/5	1/3	1/3	1	1
B	1/5	1/5	1/3	1/3	1	1

In the above matrix, accuracy (A) and surface roughness (R) are considered more important than the remaining four attributes. S and E are considered more important than C and B.

The normalized weights of each attribute are calculated following the procedure presented in Section 3.2.3, and these are: W_A = 0.3185, W_R = 0.3185, W_S = 0.1291, W_E = 0.1291, W_C = 0.0524, and W_B = 0.0524. The value of λ_{max} is 6.077 and CR = 0.0124, which is much less than the allowed CR value of 0.1. Thus, there is good consistency in the judgements made.

The value of the RP system selection index is now calculated using the above weights, and the normalized data of the attributes given in Table 16.3. This leads to the ranking given by the revised AHP or ideal mode of AHP. The alternative RP systems are arranged in descending order of the RPSI:

Quadra	0.8182
SLA3500	0.6552
SLS2500	0.5173
FDM8000	0.5138
LOM1015	0.4429
Z402	0.2459

For the above weights of importance of attributes, the multiplicative AHP method leads to the same ranking order.

It may be observed that the ranking depends upon the judgements of relative importance of attributes made by the decision maker.

16.2.5 TOPSIS Method

Following the steps of the methodology given in Section 3.2.6, the TOPSIS method gives the following ranking order:

SLA3500	0.8434
Quadra	0.8256
SLS2500	0.7149
FDM8000	0.5870
LOM1015	0.5608
Z402	0.1681

It may be observed that SLA3500 is proposed as the first choice, and Quadra as the second choice.

16.2.6 Modified TOPSIS Method

This methods leads to the following ranking order:

SLA3500	0.7808
Quadra	0.7415
SLS2500	0.6863
FDM8000	0.5745
LOM1015	0.5570
Z402	0.2170

Like TOPSIS, this method also suggests SLA3500 as the first choice, and Quadra as the second choice.

16.2.7 Compromise Ranking Method (VIKOR)

Step 1: The objective is to evaluate the six rapid prototyping systems, and the attributes are: accuracy (A), surface roughness (R), tensile strength (S), elongation (E), cost of the part (C), and build time (B). The best, *i.e.*, $(m_{ij})_{max}$, and the worst, *i.e.*, $(m_{ij})_{min}$, values of all attributes are also determined.

Step 2: The values of E_i and F_i are calculated using Equations 3.26 and 3.27 and are given below. The same weights as used in the SAW method are considered and these are: $W_A = 0.3125$, $W_R = 0.3125$, $W_S = 0.125$, $W_E = 0.125$, $W_C = 0.0625$, and $W_B = 0.0625$.

$E_1 = 0.0155+0.0536+0.1196+0.0695+0.0625+0.03125 = 0.35195$

$E_2 = 0.034+0.1607+0+0.0208+0.0625+0.03125 = 0.30925$

$E_3 = 0.0186+0.3125+0.0357+0+0.0543+0.0625 = 0.4836$

$E_4 = 0.0557+0.2947+0.0536+0+0.0466+0.0198 = 0.4704$

$E_5 = 0+0+0.0357+0.0556+0.0625+0.0198 = 0.1736$

$E_6 = 0.3125+0.2143+0.125+0.125+0+0 = 0.7768$

$E_{i\text{-}min} = 0.1736$ $E_{i\text{-}max} = 0.7768$

$R_1 = 0.1196$ $R_2 = 0.1607$ $R_3 = 0.3125$ $R_4 = 0.2947$ $R_5 = 0.0625$

$R_6 = 0.2392$ $R_7 = 0.3125$

$F_{i\text{-}min} = 0.0625$ $F_{i\text{-}max} = 0.3125$

Step 3: The values of P_i are calculated using Equation 3.28 and for v = 0.5.

$P_1 = 0.26205$ $P_2 = 0.30884$ $P_3 = 0.75695$ $P_4 = 0.7104$

$P_5 = 0$ $P_6 = 1$

Step 4: The alternatives are arranged in ascending order, according to the values of P_i. Similarly, the alternatives are arranged according to the values of E_i and F_i separately. Thus, three ranking lists are obtained. The best alternative, ranked by P_i, is the one with the minimum value of P_i.

$P_5 = 0$	$E_5 = 0.1736$	$F_5 = 0.0625$
$P_1 = 0.26205$	$E_2 = 0.30925$	$F_1 = 0.1196$
$P_2 = 0.30884$	$E_1 = 0.35195$	$F_2 = 0.1607$
$P_4 = 0.7104$	$E_4 = 0.4704$	$F_4 = 0.2947$
$P_3 = 0.75695$	$E_3 = 0.4836$	$F_3 = 0.3125$
$P_6 = 1$	$E_6 = 0.7768$	$F_6 = 0.3125$

Step 5: For the given attribute weights, the compromise solution alternative rapid prototyping system 5 (*i.e.*, Quadra), which is best ranked by the measure P, is suggested, as it satisfies both the conditions given in Section 3.2.7.

References

Bauer J, Klingenberg HH, Müller H (1996) Computer based rapid prototyping system selection and support. In: Proc. Time Compression Technologies Conference, The Heritage Motor Center, Gaydon, UK

Bibb R (1999) The development of a rapid prototyping selection system for small companies. Thesis, School of Product Design and Engineering, University of Wales; Cardiff

Byun HS, Lee KS (2004) A decision support system for the selection of a rapid prototyping process using the modified TOPSIS method. International Journal of Advanced Manufacturing Technology 26:1338–1347

Byun HS, Lee KH (2006) Determination of the optimal build direction for different rapid prototyping processes using multi-criterion decision making. Robotics and Computer-Integrated Manufacturing 22:69–80

Edwards W, Newman JR, Snapper K, Seaver D (1982) Multiattribute Evaluation. SAGE Publications, Newbury Park, California

http://www.biba.uni-bremen.de/groups/rp/rptsel_e.html

Kalpakjian S, Schmid SR (2000) Manufacturing engineering and technology. Addison Wesley Longman (Singapore), Indian branch, Delhi

Kengpol A, O'Brien C (2001) The development of a decision support tool for the selection of advanced technology to achieve rapid product development. International Journal of Production Economics 69:177–191

Masood SH, Soo A (2002) A rule based expert system for rapid prototyping system selection. Robotics & Computer Integrated Manufacturing 18:267–274

Masood SH, Rattanawong W, Iovenitti P (2003) A generic algorithm for a best part orientation system for complex parts in rapid prototyping. Journal of Materials Processing Technology 139:110–116

Pham DT, Gault RS (1998) A comparison of rapid prototyping technologies. International Journal of Machine Tools and Manufacture 38:1257–1287

Phillipson DK (1997) Rapid prototyping machine selection program. In: Proc. 6th European Conference Rapid Prototyping and Manufacturing, Nottingham, UK

Schmidt LD (1994) A benchmarking comparison of commercial techniques in rapid prototyping. Rapid Prototyping and Manufacturing Conference, Dearborn, MI

Williams RE, Komaragiri SN, Melton VL, Bishu RR (1996) Investigation of the effect of various build methods on the performance of rapid prototyping (stereolithography). Journal of Materials Processing Technology 61:173–178

17

Selection of Software in Manufacturing Industries

17.1 Introduction

Application of software in design and manufacturing processes is one of the resolutions many industries have resorted to in the 21st century. This has been a result of increased complexity of products, globalization, rapid changes in technology, and so on. The idea was that the application of software would increase the competitive advantage of an industry. Various types of software are used by manufacturing industries, such as product development process (PDP) software, product data management (PDM) software, product life-cycle management (PLM) software, enterprise resource planning (ERP) software, computer-aided design (CAD) software, computer-aided manufacturing (CAM) software, computer aided engineering (CAE) software, *etc.*

The software used in various industries can be either COTS or in-house developed. COTS is acronym for commercial off-the-shelf, an adjective that describes software or hardware products that are ready-made, and available for sale to the general public. For example, Microsoft Office is a COTS product that is a packaged software solution for businesses. COTS products are designed to be implemented easily into existing systems without the need for customization. Given the high interest in motivation to the use of commercially available software in manufacturing industries, the evaluation and selection of COTS products is an important activity in software development projects. Selecting an appropriate COTS product is often a non-trivial task in which multiple attributes need to be carefully considered (Shyur, 2006). Many decision makers select COTS products according to their experience and intuition. However, this approach is obviously subjective, and its weakness has been addressed by Kontio (1996), Leung and Leung (2002), Mikhailov and Singh (2003).

The analytic hierarchy process (AHP) has been widely used by both researchers and practitioners in COTS selection peocessess (Hong and Nigam, 1981; Min, 1992; Finnie *et al.*, 1995; Kontio, 1996, Lai *et al.*, 1999, 2002; Wei *et al.*, 2005). Santhanam and Kyparisis (1995, 1996) proposed a nonlinear programming model to optimize resource allocation, and their model considered interdependencies between projects in the information system selection process.

Carney and Wallnau (1998) observed that there are almost as many perspectives on the topic of software evaluation as there are evaluation techniques. The authors developed some basic principles applicable for evauation of commercial off-the-shelf software. Lee and Kim (2000) used the analytic network process (ANP) and goal programming for interdependent information system for project selection. Sarkis and Sundarraj (2000) discussed various factors for strategic evaluation of enterprise information technologies. Teltumbde (2000) presented a framework for evaluating ERP projects.

Badri *et al.* (2001) presented a goal programming model to select an information system project considering multiple criteria including benefits, hardware, software and other costs, risk factors, preferences of decision makers and users, completion time, and training time constraints. Lai *et al.* (2002) reported the results of a case study where the analytic hierarchy process (AHP) technique was employed to support the selection of a multimedia authorizing system (MAS) in a group decision environment. Three MAS products were identified and ultimately ranked using the AHP. Six software engineers, who were technically competent and experienced, participated in the study.

Morisio *et al.* (2002) investigated COTS-based software development within a particular NASA environment, with an emphasis on the processes used. Fifteen projects using a COTS-based approach were studied, and their actual process was documented. This process was evaluated to identify essential differences in comparison to traditional software development. The authors concluded that the main differences, and the activities for which projects require more guidance, were requirements definition and COTS selection, high-level design integration, and testing. Starting from these empirical observations, a new process and set of guidelines for COTS-based development were developed and presented.

Sarkis and Talluri (2004) presented a decision framework that could aid members of the supply chain and a supply chain director in deciding which electronic commerce technology media and software would be most suitable for the whole supply chain. The techniques used in this approach included both qualitative and quantitative measurements for the evaluation or justification of these systems. The framework used an integrative set of models based on the analytical hierarchy process and goal programming.

Wei *et al.* (2005) presented a comprehensive framework for selecting a suitable enterprise resource planning (ERP) system using an AHP-based approach. A real-world example was presented to demonstrate the feasibility of the framework. Mulebeke and Zheng (2006) carried out a case study to introduce analytic network process (ANP) as a multiple attribute strategic decision making approach to help in the selection of appropriate software to suit the product development process of a particular product. Shyur (2006) modeled the COTS evaluation problem, and proposed a five-phase COTS selection model combining the techniques of the analytic network process (ANP) and modified TOPSIS. ANP was used to determine the relative weights of multiple attributes. The modified TOPSIS approach was used to rank the competing COTS products in terms of their overall performance.

For an enterprise to gain a competitive advantage, managers need to have outlined a set of objectives. Usually, these objectives are a reflection of market and business drivers. In acquisition of software, this software has to be able to satisfy a basic evaluation criterion based on its performance attributes, and also needs to meet the market and business drivers of the industrial enterprise, thereby satisfying the overall objectives. Wei *et al.* (2005) suggested to consider factors (or attributes) such as total costs, implementation time, functionality, user friendliness, flexibility, reliability, vendor's reputation, technical capability, and service while selecting ERP software.

Shyur (2006) proposed different evaluation criteria and related attributes. The criteria (and the attributes) are: cost (license fee, modular pricing, maintenance, documentation, consultant fee, resource utilization, conversion cost, *etc.*), supplier's support (vendor responsiveness, consulting, hotline, training, technical support personnel, continuing enhancement, time-sharing access, warranty, documentation, financial stability, local branch office, third vendor support, growth of customer base, active R&D, *etc.*), technological risk (non-robust and incomplete packages, complex and undefined, COTS-to-legacy-system interfaces, middleware technology bugs, poor custom code, and poor system performance, software maturity, hardware maturity, *etc.*), closeness of fit to the company's business (main target, included functionality, *etc.*), ease of implementation (shorter implementation time, user friendliness, multisite implementation, *etc.*), flexibility to easy change as the company's business changes (adaptability, openness for customer development, openness for working with other systems, *etc.*), and system integration (internal connectivity, external connectivity, *etc.*).

The initial approach in software selection can be the identification of pertinent criteria (and the attributes), and the alternative software involved in the selection problem, *i.e.*, to obtain the values of the attributes (A_i) with the system specifications and requirements, and their relative importance (a_{ij}). An objective or subjective value, or its range, may be assigned to each identified attribute as a limiting value, or threshold value, for its acceptance for the software selection problem considered. Alternative software with each of its selection attribute, meeting the acceptance value, may be short-listed. After short-listing the alternative software, the main task in choosing the alternative is to see how it serves the attributes considered.

Now, an example of software selection is considered to demonstrate the applicability of the decision methods proposed in this book.

17.2 Example

Shyur (2006) modeled a COTS evaluation problem, and proposed a five-phase COTS selection model combining the techniques of analytic network process (ANP) and modified TOPSIS. ANP was used to determine the relative weights of multiple attributes. The modified TOPSIS approach was used to rank the competing COTS products in terms of their overall performance. To illustrate how the approach was used for the COTS evaluation problem, an empirical study of a real case was conducted. Using the AHP method, four alternative softwares and seven criteria were considered. The criteria considered were cost (CO), supplier's support (SS), ease of implementation (EI), closeness of fit to the company's business (FB), flexibility to easy change as the company's business changes (FC), technological risk (TR), and system integration (SI). Here, each criterion is a broader one, and includes many attributes) as mentioned in Section 17.1. The weights of these criteria were obtained using ANP method and these were 0.242, 0.360, 0.042, 0.102, 0.030, 0.157, and 0.067 respectively. All seven criteria were considered as beneficial, and the normalized values were calculated, and are given in Table 17.1.

Table 17.1. Normalized values of software selection criteria (from Shyur 2006; reprinted with permission from Elsevier)

Software	CO	SS	EI	FB	FC	TR	SI
A_1	0.55	0.70	0.39	0.64	0.61	0.30	0.55
A_2	0.46	0.35	0.55	0.40	0.41	0.69	0.39
A_3	0.28	0.35	0.63	0.32	0.30	0.59	0.39
A_4	0.64	0.52	0.39	0.56	0.61	0.30	0.63

The attribute values of TR determined by Shyur (2006) showed that TR was a beneficial criterion. Using the modified TOPSIS procedure, Shyur (2006) obtained the following ranking for the softwares considered:

A_4 0.652
A_1 0.645
A_2 0.433
A_3 0.236

Rather than using the modified TOPSIS method, if AHP is selected with the same criteria weights for the above software selection problem, then the following ranking results:

A_1 0.5690
A_4 0.5231
A_2 0.4480
A_3 0.3806

Thus, application of the AHP method leads to A_1 as the first choice, and A_4 as second choice.

Rather than using the modified TOPSIS method, if the simple TOPSIS method is selected, then the following ranking is obtained, which suggests A_1 as the first choice:

A_1 0.6908
A_4 0.5520
A_2 0.3556
A_3 0.2261

Independently applying the modified TOPSIS method (using the same weights of importance of the criteria) leads to the ranking order of A_1-A_4-A_2-A_3. It seems that Shyur (2006) had some errors in computing the closeness coefficient values of the alternative softwares.

Now, let the graph theory and matrix approach be used with the following assignments of relative importance:

	CO	SS	EI	FB	FC	TR	SI
CO	-	0.41	0.865	0.745	0.865	0.665	0.745
SS	0.59	-	0.955	0.865	0.955	0.745	0.865
EI	0.135	0.045	-	0.335	0.5	0.255	0.41
FB	0.255	0.135	0.665	-	0.745	0.41	0.59
FC	0.135	0.045	0.5	0.245	-	0.255	0.41
TR	0.335	0.255	0.745	0.59	0.745	-	0.59
SI	0.255	0.135	0.59	0.41	0.59	0.41	-

The value of the software selection index is calculated using the values of A_i and a_{ij} for each alternative software. The software selection index values of different softwares are given below in descending order:

A_1	15.4045
A_4	14.5715
A_2	12.3263
A_3	10.8532

From the above values of the software selection index, software A_1 is interpreted as the best choice among the software alternatives considered for the given software selection problem.

17.3 General Remarks

This chapter has presented the details of selection of software in manufacturing industries. The selection of suitable software would increase the competitive advantage of an industry. Once all possible software alternatives are identified and appropriately screened, the best alternative software can be chosen by using any of the fuzzy decision-making methods proposed in this book.

References

Badri MA, Davis D, Davis D (2001) A comprehensive 0–1 goal programming model for project selection. International Journal of Project Management 19:243–252

Carney DJ, Wallnau KC (1998) A basis for evaluation of commercial software. Information and Software Technology 40:851–860

Finnie GR, Wittig GE, Petkov DI (1995) Prioritizing software development productivity factors using the analytic hierarchy process. Journal of Systems and Software 22:129–139

García F, Piattini M, Ruiz F, Canfora G, Visaggio CA (2006) FMESP: Framework for the modeling and evaluation of software processes. Journal of Systems Architecture 52:627–639

Hong S, Nigam R (1981) Analytic hierarchy process applied to evaluation of financial modeling software. In: Proc. of the 1st International Conference on Decision Support Systems, Atlanta

Kontio J (1996) A case study in applying a systematic method for COTS selection. In: Proc. IEEE, ICSE-18, pp 201–209

Lai VS, Trueblood RP, Wong BK (1999) Software selection: a case study of the application of the analytical hierarchical process to the selection of a multimedia authoring system. Information & Management 36:221–232

Lai VS, Bo KW, Cheung W (2002) Group decision making in a multiple criteria environment: A case using the AHP in software selection. European Journal of Operational Research 137:134–144

Lee JW, Kim SH (2000) Using analytic network process and goal programming for interdependent information system project selection. Computers & Operations Research 27:367–382

Leung KPRH, Leung HKN (2002) On the efficiency of domain-based COTS product selection method. Information and Software Technology 44:703–715

Mikhailov L, Singh MG (2003) Fuzzy analytic network process and its application to the development of decision support systems. IEEE Transactions on Systems, Man, and Cybernetics Part C. Applications and Reviews 33:33–41

Min H (1992) Selection of software: the analytic hierarchy process. International Journal of Physical Distribution and Logistics Management 122:42–52

Morisio M, Seaman CB, Basili VR, Parra AT, Kraft SE, Condon SE (2002) COTS-based software development: Processes and open issues. Journal of Systems and Software 61:189–199

Mulebeke JAW, Zheng L (2006) Analytical network process for software selection in product development: Journal of Engineering and Technology Management 23:337–352

Santhanam R, Kyparisis GJ (1995) A multiple criteria decision model for information system project selection. Computers & Operations Research 22:807–818

Santhanam R, Kyparisis GJ (1996) A decision model for interdependent information system project selection. European Journal of Operational Research 89:380–399

Sarkis J, Sundarraj RP (2000) Factors for strategic evaluation of enterprise information technologies. International Journal of Physical Distribution & Logistics Management 30:196–220

Sarkis J, Talluri S (2004) Evaluating and selecting e-commerce software and communication systems for a supply chain. European Journal of Operational Research 159:318–329

Shyur HJ (2006) COTS evaluation using modified TOPSIS and ANP. Applied Mathematics and Computation 177:251–259

Teltumbde A (2000) A framework of evaluating ERP projects. International Journal of Production Research 38:4507–4520

Wei CC, Chien CF, Wang MJJ (2005) An AHP-based approach to ERP system selection. International Journal of Production Economics 96:47–62

18

Welding Process Selection for a Given Application

18.1 Introduction

Welding is a process of joining two or more pieces of the same or dissimilar materials to achieve complete coalescence. Almost all materials can be welded but not always by the same welding process. Some welding processes are known to be associated with specific jobs and industries (Parmar, 1995). For example, submerged arc welding is the sole process used in joining thick plates in long linear seams in ships, pressure vessels, bridges, structural work and nuclear reactors. Gas tungsten arc welding is used extensively by the aircraft industry, rocket and missile fabricators, and chemical and nuclear fabricators. Resistance spot welding is used mainly for lap welding of thin sheets, particularly in the welding of automobile and refrigerator bodies. In all these cases, the selection of the said processes can be attributed to the fact that the desired quality weld joint is accomplished at the least cost. Thus cost is the main selection criterion. In such specific cases, there may not be any other choice, and the exercise for selection of a welding process would be redundant. However, there are many instances where a number of processes can be nearly equally effective in producing the end product (Muralidharan *et al.*, 1999). Each process will have its merits and demerits, and hence from a group of welding processes, a particular process has to be chosen based on its overall benefits. The process selection is aimed for such situations, and the need for a systematic procedure for efficient and effective selection of a welding process is overdue. Efforts need to be extended to determine the factors that influence welding process selection for a given application (*i.e.*, end product), using a logical approach, to eliminate unsuitable welding processes, and for selection of a welding process to strengthen the existing welding process selection procedure.

Darwish *et al.* (1997) developed a knowledge-based system for identifying the most appropriate welding processes to suit specific circumstances. Thirty welding processes of industrial importance were incorporated into the system. Only the product type and some of the process capabilities, namely, material type, material thickness, method of use, quality level, joint type, and welding position, were used to determine the best selection among competitive welding processes. Yeo and Neo (1998) emphasized the need for inclusion of environmental performance for decision making of welding processes. Muralidharan *et al.* (1999) discussed the application of AHP for the selection of a welding process considering subjective attributes.

A welding process is defined by independent and dependent process variables among which exist different relations. Independent process variables are the process input variables, and include: properties/characteristics of the work material (such as mechanical, metallurgical, chemical, and thermal properties, *etc.*), welding conditions (such as welding current, voltage, welding speed, *etc.*), electrode size, welding machine capacity, operator's inherent skill, *etc.* Dependent process variables are the process output variables, and include: welding joint properties (such as mechanical properties, metallurgical properties, *etc.*), processed weld quality, welded joint distortion, cleaning required after welding, post-heat treatment, consumption of consumables, cost involved, *etc.* Output process variables are functions of input process variables. A welding process selection attribute is defined as a welding process variable (*i.e.*, dependent or independent). Welding process selection can be carried out based upon these attributes.

Now, an example is considered to demonstrate and validate the proposed GTMA and fuzzy MADM methods for welding process selection.

18.2 Example

The following three arc welding processes are most commonly used to join mild steel (0.2% C) of 6 mm thickness, known to be the best weldable metal in arc welding processes: (i) shielded metal arc welding (SMAW), (ii) gas tungsten arc welding (GTAW), and (iii) gas metal arc welding (GMAW).

The selection of a welding process for mild steel is usually based on economic considerations and welded joint properties. The economic analysis of the welding can be broadly divided into four components: equipment cost, consumables cost, labor cost, and overhead cost. Apart from this, mechanical analysis and the metallurgical analysis of the joint are also considered before selecting a particular process. In general, process selection in the above case will consider only the above attributes, *i.e.*, objective attributes. Indeed, subjective attributes of each welding process seem to have been neglected so far in the selection process.

18.2.1 Graph Theory and Matrix Approach (GTMA)

Various steps of the GTMA methodology proposed in Section 2.6 are carried out as given below:

Step 1:The attributes considered are: weld quality (WQ), operator fatigue (OF), skill required (SR), cleaning required after welding (CR), availability of consumables (AC) and initial preparation required (IP). These attributes are subjective, and are to be assigned values. As the data regarding the objective attributes are not available, only subjective attributes are considered. When the objective data of the attributes are also available, then such data are to be normalized as described in Section 2.4. The welding process selection attributes OF, SR, CR, and IP are non-beneficial, and lower values are desirable. A welding process is preferred if it offers lower values of these attributes. Attributes WQ and AC are considered as beneficial attributes. A welding process is preferred if it offers higher values of these attributes. Values of all six attributes are assigned, and are given in Table 18.1 in the respective columns. Table 18.1 shows the values of A_i for different welding processes. The normalized values of the attributes are given in Table 18.2.

Table 18.1. Welding process attribute values for the example considered

Welding process	WQ	OF	SR	CR	AC	IP
SMAW	0.5	0.5	0.5	0.665	0.745	0.5
GTAW	0.745	0.665	0.745	0.5	0.5	0.665
GMAW	0.59	0.745	0.665	0.59	0.665	0.745

Table 18.2. Normalized data of the attributes of the example considered

Welding process	WQ	OF	SR	CR	AC	IP
SMAW	0.6711	1	1	0.7519	1	1
GTAW	1	0.7519	0.6711	1	0.6711	0.7519
GMAW	0.7919	0.6711	0.7519	0.8475	0.8926	0.6711

Relative importance of attributes (a_{ij}) is assigned values, using Table 2.4. Let the decision maker prepare the following assignments:

	WQ	OF	SR	CR	AC	IP
WQ	-	0.59	0.665	0.665	0.745	0.865
OF	0.41	-	0.59	0.665	0.745	0.745
SR	0.335	0.41	-	0.59	0.665	0.745
CR	0.335	0.335	0.41	-	0.59	0.59
AC	0.255	0.255	0.335	0.41	-	0.59
IP	0.135	0.255	0.255	0.41	0.41	-

Step 2:

1. The welding process selection attributes graph, showing the presence as well as relative importance of the above attributes, is similar to Figure 2.2 but with six attributes drawn. However, this is not shown here for obvious reasons.
2. The welding process selection attributes matrix of this graph can be written based on Equation 2.10.
3. The welding process selection attributes function is written. However, as a computer program is developed for calculating the permanent function value of a matrix, this step can be skipped.

4 & 5. The welding process selection index is calculated using the values of A_i and a_{ij} for each alternative welding process. The welding process selection index values of different welding processes are given below in descending order:

SMAW	18.1927
GTAW	14.7962
GMAW	13.6832

From the above values of the welding process selection index, it is interpreted that the SMAW process is the preferred process, followed by GTAW and GMAW for the considered example of welding 6-mm mild steel.

18.2.2 SAW method

Let the attributes be ranked in order of importance and 10 points be assigned to the least important attribute, IP. Attribute AC is given 15 points to reflect its relative importance. CR, SR, OF, and WQ are given 20, 30, 40, and 50 points, respectively. Thus, the weights of WQ, OF, SR, CR, AC, and IP are calculated as 0.303, 0.242, 0.182, 0.121, 0.091, and 0.061, respectively. Using these weights, and the normalized data of the attributes for different welding processes, the welding process selection index values are calculated, and are arranged in descending order of the index.

SMAW	0.8703
GTAW	0.8350
GMAW	0.7639

From the above values of the welding process selection index, it is clear that the SMAW process is the best choice among the welding processes considered.

18.2.3 WPM

For the same weights of relative importance of the attributes as those used in the SAW method, WPM leads to the following ranking of the welding processes:

SMAW	0.8561
GTAW	0.8226
GMAW	0.7605

This method also suggests SMAW as the correct choice for the given welding application.

18.2.4 AHP and its Versions

If the same weights as those used in the SAW method are selected for this method, then the ranking of welding processes obtained by using the relative as well as ideal mode AHP will be the same. The multiplicative AHP method also leads to the same ranking as that given by WPM. However, let the decision maker prepare the following matrix of relative importance:

$$\begin{array}{c|cccccc|} & WQ & OF & SR & CR & AC & IP \\ WQ & 1 & 2 & 3 & 3 & 4 & 5 \\ OF & 1/2 & 1 & 2 & 3 & 4 & 4 \\ SR & 1/3 & 1/2 & 1 & 2 & 3 & 4 \\ CR & 1/3 & 1/3 & 1/2 & 1 & 2 & 3 \\ AC & 1/4 & 1/4 & 1/3 & 1/2 & 1 & 2 \\ IP & 1/5 & 1/4 & 1/4 & 1/3 & 1/2 & 1 \end{array}$$

The normalized weights of each attribute are calculated following the procedure presented in Section 3.2.3, and these are W_{WQ} = 0.3534, W_{OF} = 0.2526, W_{SR} = 0.1669, W_{CR} = 0.1103, W_{AC} = 0.0695, and W_{IP} = 0.0473. The consistency ratio is 0.029, and thus there exists good consistency.

The value of the welding process selection index is now calculated using the above weights, and the normalized data of the attributes given in Table 18.2. The alternative welding processes are arranged in descending order of the welding process selection index.

SMAW	0.8564
GTAW	0.8478
GMAW	0.6908

Thus, the SMAW process is suggested as the best for the given application.

It may be observed that the above ranking is for the given preferences of the decision maker.

18.2.5 TOPSIS Method

The quantitative values of the welding process selection attributes, which are given in Table 18.1, are normalized as explained in Section 3.2.6.

Relative importance of attributes (a_{ij}) is assigned using the AHP method as explained in Section 18.2.4. After performing the calculations, the alternative welding processes are arranged in descending order of their welding process selection index. This can be arranged as GTAW-SMAW-GMAW. This ranking is somewhat surprising. For the same weights of attributes as those used in the AHP method in Section 18.2.4, TOPSIS gives a different ranking. The reason may be that the TOPSIS method is biased toward the alternative in which the highest value of the attribute with maximum weight of relative importance occurs.

References

Darwish SM, Tamimi AA, Habdan SA (1997) A knowledge base for metal welding process selection. International Journal of Machine Tools and Manufacture 37:1007–1023

Muralidharan C, Anantharaman N, Balasubramanian V, Deshmukh SG (1999) Selection of welding process using analytic hierarchy process. Institution of Engineers (India), Mechanical Engineering Division 80:151–154

Parmar RS (1995) Welding technology. Khanna Publishers, New Delhi

Yeo SH, Neo KG (1998) Inclusion of environmental performance for decision making of welding processes. Journal of Materials Processing Technology 82:78–88

19

Geometric Moldability Analysis of Parts

19.1 Introduction

In molding or casting manufacturing processes, material is reshaped in a hollow mold. A simple reusable mold consists of two rigid halves that are removed in opposite directions; the orientation of the removal directions is called the parting direction. In order for part geometry to be de-moldable, it must be oriented relative to the parting direction so that the two mold halves can be removed from the part via translation along the parting direction, without colliding with the part. Surfaces where collisions occur, preventing extraction of the part, are called undercuts. They occur where the mold extends into the area between the part and the parting surface, relative to the parting direction. Forming undercuts requires additional mold inserts that increase the cost of the mold; it is desirable to avoid these if possible. Finding a feasible two part molding orientation (one without undercuts) for an arbitrary geometry is subject to geometric accessibility constraints; not all geometries admit such an orientation (McMains and Chen, 2004).

A major cost for a part to be molded lies in the mold design and its operation, which are greatly influenced by certain design decisions made in the part design process. Thus, it is important to establish a link between part design and mold design that allows part designers to make design decisions that do not adversely affect mold design. This leads to the research of design for moldability. The moldability of a part refers to its manufacturability in the injection molding process and reflects, its ability to be extracted from the mold core and cavity when the mold opens. A part with good moldability requires simple mold structure and short operation time. Generally, the moldability of a part depends on its geometry and other factors, such as the material used for molding. Geometric moldability analysis aims to discern and represent the constraints on moldability that emerge strictly from the geometry and topology of the part to be molded (Yin *et al.*, 2004). On the other hand, parting direction (PD), parting line (PL), and undercut features (UF) are the three main parameters in mold design that greatly influence the moldability of parts. A combination of these parameters determines a cavity design scheme (CDS), by dictating the number and shape of cores, draw depth (DD), draft angles, and the machining complexity of the parting surface, and affects all subsequent steps in mold design, such as designs of the feed, rejection, and cooling systems. Hence, to perform geometric moldability analysis, it is necessary to identify these parameters from the part's geometry early in the part design cycle.

Ravi and Srinivasan (1990) presented a scientific approach developed for design of parting surfaces of patterns, molds and dies used in the manufacture of cast, forged, injection-molded and die-cast components. This had enabled computer-aided generation of parting surfaces, and the determination of projected area, flatness and draw for a parting surface, identification of surfaces to which draft is provided, recognition of component segments causing undercuts, testing for dimensional stability, and location of flash, machined surfaces and feeders. Influencing criteria for parting-surface design were formulated, and developed into algorithms implemented on a personal computer. However, the authors did not address the problem of how to generate PL or parting surfaces according to the considered decision criteria.

Chen *et al.* (1993) defined two levels of visibility, complete and partial visibility, on the basis of conditions for demoldability. The viewing directions from which a surface was completely visible were represented as a convex region on the unit sphere, called the visibility map of the surface. Algorithms were given for subdividing a given object into pockets, for which visibility and demoldability could be determined independently, for constructing visibility maps, and for selecting an optimal pair of parting directions for a mold that minimizes the number of cores. An example illustrated the algorithms. Woo (1994) described a way in which a three-dimensional work piece was mapped onto the unit sphere, and its visibility was determined. For applications, manufacturing machines were classified by their degrees of freedom into point, line, and surface visible processes. Algorithms for optimal work piece orientation were then formulated as simple intersections on the sphere.

Wuerger and Gadh (1997a, 1997b) applied a similar approach to select part orientation and die-open direction by detecting features that obstruct possible opening directions, and then by computing their restrictions with respect to the whole part. Weinstein and Manoochehri (1996, 1997) divided the part surfaces into concave and convex regions, which influence the PD and PL location of the part, respectively. Majhi *et al.* (1999) presented algorithms to compute a flattest undercut free parting line for a convex polyhedron based on different flatness criteria. However, the algorithms fail to determine PD or PL when there exist internal undercuts, or interacting undercuts to be formed by two or more cores. Hui (1997) developed a subdivision technique based on the concept of blockage in a given direction, to evaluate the geometry of an undercut. Based on the notion of internal and external undercut, the moldability of a component was studied. A local and a global blockage tests were introduced to detect any interference between the molded component and a side core, or a split core. A search strategy was also developed for selecting a suitable combination of main parting, side core, and split core directions.

Kurth and Gadh (1997) dealt with two special kinds of interacting undercut feature, by transiting these into two-dimensional space. However, the approach was simply a rule of thumb approximation and is limited to special cases of interacting undercut features. Yin *et al.* (2000) proposed a procedure to solve a specific accessibility and setup problem in mold parting, NC-machining and CMMs inspection path planning, to perform visibility analysis with respect to the geometry of the part, the shape of the effector, and degrees of freedom between part/effector. A new method for computing visibility cones was formulated by identifying C-obstacles in configuration space (C-Space), in which a general and efficient algorithm was presented and implemented using visibility culling. The algorithm was efficient even in very complex cases.

Chen *et al.* (2000) presented a method for automatic mold parting direction selection in computer-aided design of molds and dies. When given a three-dimensional CAD model, the minimum volume bounding box of the model was found first. Three pairs of possible parting directions were defined based on the bounding box surface normal vectors. These were in the length direction, the width direction, and the height direction of the bounding box. Then, a dexel model was constructed along each parting direction. Parting lines corresponding to the three possible parting directions were estimated using a slice method. Finally, criteria such as undercut, draw, projected area and flatness in all parting directions were quantified using the dexel model of the part. A designer's preferences, coded as fuzzy weighting factors, were setup for the evaluation of the most promising parting direction.

Ye *et al.* (2001) described a hybrid method to recognize undercut features from molded parts with planar, quadric, and free-form surfaces. The hybrid method took advantage of graph-based and hint-based approaches. Various undercut features, including interacting undercut features, were defined by extended attributed face-edge graphs (EAFEG). Unlike conventional graph-based methods, which recognize features by graph matching, the new approach recognized the undercut features by searching the cut-sets of subgraphs. Face properties and parting lines were used as hints to guide the search of cut-sets. To recognize undercut features from parts with free-form surfaces, a convex-hull algorithm was used to determine the face properties (positive, negative, and horizontal). The case study showed that the method can recognize various undercut features successfully. However, this method is time consuming. In addition, it can not guarantee correct results.

Yin *et al.* (2001) presented a virtual prototyping (VP) approach for geometric moldability analysis of near-net-shape manufactured parts. A virtual prototype (VP) of a mold, which was a realistic digital product model, was generated by combining automated and interactive approaches to evaluate the moldability of a part in the early stages of the product development cycle. The automated approaches for generating a VP were proposed to construct the parting surface, cores, and cavity of the mold based on the recognized undercut features. Interaction with the VP in the virtual reality environment allowed the designers to evaluate the moldability of a part in an intuitive way. A new volume-based feature recognition method and data structure using a non-directional blocking graph (NDBG) was developed to recognize both isolated and interacting undercut features in a uniform way. A system configuration for the VP was developed and implemented using virtual reality technologies.

McMains and Chen (2004) considered the problem of whether a given geometry can be molded in a two-part, rigid, reusable mold with opposite removal directions. An algorithm was developed for solving the opposite direction moldability problem for a two-dimensional polygon bounded by edges that may be either straight or curved. A structure called the normal graph of the polygon was defined that represented the range of normals of the polygon's edges, along with their connectivity.

Yin *et al.* (2004) found that most existing methods focus on PD selection, PL determination, or UF recognition, with more or less limitations. Since the mold design parameters or attributes influence each other, it is necessary to develop a methodology to consider these in an integrated manner. Hence, the authors presented a methodology for moldability analysis by finding the optimal cavity design scheme (CDS) based on manufacturing and cost considerations using part geometry, where a CDS refers to a combination of the parting direction, parting line (PL), and undercut features (UF). The methodology took advantage of geometric reasoning and fuzzy evaluation, and consisted of two main stages: (1) generating all possible design alternatives, and (2) choosing the best alternative. In the first stage, after recognizing the potential UF from the given part, a spherical arrangement was constructed by partitioning the unit direction sphere using outward normals of the part's surfaces, with the property that each cell in this arrangement had a unique combination of PL and UF set. Thus, all design alternatives were identified. In the second stage, the fuzzy multiple attribute decision-making model was employed to choose the optimal scheme from the set of design alternatives with respect to a set of criteria related to the number and volume of undercuts, flatness of the PL, draw depth, and draft angle.

Now, the example problem presented by Yin *et al.* (2004) is considered to demonstrate the applicability of GTMA and fuzzy MADM methods to the geometric moldability analysis problem.

19.2 Example

Yin *et al.* (2004) generated 14 CDSs by geometric reasoning, considered five attributes and used fuzzy weighted average and ranking methods to evaluate possible CDSs. These attributes include: number of external undercut features (NEU), volume of undercut features (VU), flatness of the PL (FPL), draw depth (DD), and draft factor (DF). The values of these attributes are desired to be relatively low. The authors presented the normalized attribute values of plausible design alternatives for a mouse part. The normalized values are shown in Table 19.1.

Table 19.1. Normalized data of the possible cavity design alternatives (from Yin *et al.*, 2004; reprinted with permission from Elsevier)

Design scheme	NEU	VU	FPL	DD	DF
1	0.2	0.001	0.6672	0.363	1
2	0.2	0.001	0.8862	0.616	1
3	0.2	0.001	0.6852	0.290	1
4	0.2	0.001	0.4013	0.254	1
5	0.2	0.001	0.5663	0.346	0.8
7	0.2	0.001	0.6562	0.323	0.8
8	0.2	0.001	0.2157	0.282	1
9	1	1	0.9937	0.9964	0.8
10	0.25	0.043	1	1	1
11	0.2	0.001	0.9981	0.290	0.2
12	0.2	0.001	0.3042	0.283	0.8
13	0.2	0.001	0.5501	0.270	0.6
14	0.25	0.001	0.9994	0.561	0

19.2.1 Graph Theory and Matrix Approach (GTMA)

In the present work, the attributes considered are the same as of those of Yin *et al.* (2004) and these are: number of external undercut features (NEU), volume of undercut features (VU), flatness of the PL (FPL), draw depth (DD) and draft factor (DF). The quantitative values of these attributes are already normalized as given in Table 19.1. Let the decision maker prepare the following assignments of relative importance:

	NEU	VU	FPL	DD	DF
NEU	-	0.745	0.59	0.665	0.745
VU	0.255	-	0.335	0.41	0.335
FPL	0.41	0.665	-	0.59	0.665
DD	0.335	0.59	0.41	-	0.59
DF	0.255	0.665	0.335	0.41	-

The CDS selection attributes digraph, CDS selection attributes matrix of the digraph, and CDS selection function for the matrix can be prepared. The value of the CDS selection index is calculated using the values of A_i and a_{ij} for each CDS. The CDS selection index values of different CDSs are given below in descending order:

9	8.0153
10	3.7619
2	3.0146
1	2.5383
3	2.4800
7	2.3298
6	2.2769

4	2.2018
14	2.1124
8	2.0668
5	2.0506
11	2.0437
13	2.0477
12	2.0100

From the above values of the CDS selection index, CDS 9 is understood as the best choice among the alternatives considered, CDS 10 as the second best choice, and CDS 12 as the last choice. The ranking order proposed by Yin *et al.* (2004) was 9-10-2-14-1-3-11-7-6-4-13-5-8-12. This also suggests CDS 9 as the first choice, CDS 10 as the second choice, and CDS 12 as the last choice.

19.2.2 SAW Method

For a start, the attributes are ranked in order of importance and 10 points each are assigned to the least important attributes VU and DF. Attribute DD is given 20 points to reflect its relative importance. FPL is considered more important, and given 30 points. NEU is considered as very important compared to all other attributes, and is given 40 points. The final weights are obtained by normalizing the sum of the points to one. Thus, the weights of NEU, VU, FPL, DD, and DF are calculated as 0.364, 0.091, 0.273, 0.182, and 0.091, respectively. Using these weights, and the normalized data of the attributes for different CDSs, the CDS selection index values are calculated and are arranged in descending order of the index.

9	0.9804
10	0.6409
2	0.5179
14	0.4649
11	0.4164
1	0.4121
3	0.4037
7	0.3836
6	0.3633
13	0.3268
5	0.3255
4	0.3197
12	0.2802
8	0.2741

The SAW method also suggests CDS 9 as the correct choice, and CDS 10 as the second choice for the given CDS selection problem.

19.2.3 AHP Method

The AHP method may use the same weights as those selected for the SAW method. In that case, the ranking of the CDSs will be the same. However, let the decision maker prepare the following matrix:

	NEU	VU	FPL	DD	DF
NEU	1	5	3	4	5
VU	1/5	1	1/4	1/3	1
FPL	1/3	4	1	3	4
DD	1/4	3	1/3	1	3
DF	1/5	1	1/4	1/3	1

The normalized weights of each attribute are calculated following the procedure presented in Section 3.2.3, and these are: $W_{NEU} = 0.4673$, $W_{ML} = W_{DF} = 0.0658$, $W_{FPL} = 0.260$, and $W_{DD} = 0.1409$. The value of λ_{max} is 5.2046 and CR = 0.0461, which is much less than the allowed CR value of 0.1.

The value of CDS selection index is now calculated using the above weights, and the normalized data of the attributes given in Table 19.1. The alternative CDSs are arranged in descending order of the CDS selection index:

9	0.9849
10	0.5853
2	0.4757
14	0.4559
11	0.4072
1	0.3834
3	0.3778
7	0.3619
6	0.3417
13	0.3139
5	0.3119
4	0.2989
12	0.2648
8	0.2546

The AHP method also suggests CDS 9 as the correct choice and CDS 10 as the second choice for the given CDS selection problem. The ranking is the same as that obtained by using the SAW method.

19.2.4 TOPSIS Method

Using the same weights as those for the AHP method, and following the steps of the methodology given in Section 3.2.6, The TOPSIS method gives the following ranking order of CDSs:

9	0.9866
10	0.2542
14	0.2179
11	0.1951
2	0.1921
3	0.1383
1	0.1363
7	0.1285
6	0.1096
13	0.0995

5	0.0955
4	0.0813
8	0.0635
12	0.0574

The TOPSIS method also suggests CDS 9 as the correct choice and CDS 10 as the second choice for the given CDS selection problem. CDS 12 is proposed as the last choice.

19.2.5 Modified TOPSIS Method

For the same weights as those used in the AHP method, the modified TOPSIS method gives the following ranking order:

9	0.9705
10	0.3217
2	0.2461
14	0.2454
11	0.2189
1	0.1872
3	0.1689
7	0.1530
6	0.1436
4	0.1328
8	0.1328
13	0.1309
5	0.1276
12	0.1130

The modified TOPSIS method also suggests CDS 9 as the correct choice, CDS 10 as the second choice, and CDS 12 as the last choice.

19.3 General Remarks

This chapter presented a methodology for moldability analysis by determining the optimal CDS using part geometry. Once all possible design alternatives are generated using geometric reasoning, the best alternative can be chosen by using any of the fuzzy decision-making methods proposed in this book.

References

Chen LL, Chou SY, Woo TC (1993) Parting directions for mold and die design. Computer Aided Design 25:762–768

Chen YH, Wang YZ, Leung TM (2000) An investigation of parting direction based on dexel model and fuzzy decision making. International Journal of Production Research 38:1357–1375

Hui KC (1997) Geometric aspects of the moldability of parts. Computer Aided Design 29:197–208

Kurth GR, Gadh R (1997) Virtual prototyping of die-design: determination of die-open directions for near-net shape manufactured parts with extruded or rotational features. Computer Integrated Manufacturing Systems 10:69–81

Majhi J, Gupta P, Janardan R (1999) Computing a flattest, undercut-free parting line for a convex polyhedron, with application to mold design. Computational Geometry 13:229–252

McMains S, Chen S (2004) Determining moldability and parting directions for polygons with curved edges. In: ASME International Mechanical Engineering Congress and Exposition, Anaheim, CA

Ravi B, Srinivasan MN (1990) Decision criteria for computer-aided parting surface design. Computer Aided Design 22:11–18

Weinstein M, Manoochehri S (1996) Geometric influence of a mold part on the draw direction range and parting line locations. Journal of Mechanical Design 118:29–39.

Weinstein M, Manoochehri S (1997) Optimum parting line design of molded and cast parts for manufacturability. Journal of Manufacturing Systems 16:1–12.

Woo TC (1994) Visibility maps and spherical algorithms. Computer Aided Design 26:6–16

Wuerger D, Gadh R (1997a) Virtual prototyping of die design part one: theory and formulation. Journal of Concurrent Engineering: Research Applications 5:307–315

Wuerger D, Gadh R (1997b) Virtual prototyping of die design part two: algorithmic, computational, and practical considerations. Journal of Concurrent Engineering: Research Applications 5:317–326

Ye XG, Fuh JYH, Lee KS (2001) A hybrid method for recognition of undercut features from molded parts. Computer Aided Design 33:1023–1034

Yin ZP, Ding H, Xiong YL (2000) Visibility theory and algorithms with application to manufacturing processes. International Journal of Production Research 38:2891–2909

Yin ZP, Ding H, Xiong YL (2001) Virtual prototyping of mold design: geometric moldability analysis for near-net-shape manufactured parts by feature recognition and geometric reasoning. Computer Aided Design 33:137–154

Yin ZP, Ding H, Li HX, Xiong YL (2004) Geometric moldability analysis by geometric reasoning and fizzy decision making. Computer Aided Design 36:37–50

20

Evaluation of Metal Stamping Layouts

20.1 Introduction

In stamping, sheet metal parts of various levels of complexity are produced rapidly, often in very large volumes, using hard tooling. The production process operates efficiently, and material costs can typically represent 75% or more of total costs in stamping facilities (Industry Canada, 1998). Due to the high volume of parts produced, even small inefficiencies in material utilization per part can lead to very large amounts of wasted material over a die's life. Hence, the choice of an efficient strip-layout is an important step during die design, because only the optimum layout can reduce wastage of the strip material, and reduce the overall cost of production.

Originally, strip-layout problems were solved manually, for example, by cutting blanks from cardboard and manipulating these to obtain a good layout. In recent years, however, the trend is two-dimensional layout solutions based on computer-aided design. Fogg and Jamieson (1975) described the attributes influencing the optimization of press tool die layouts, and presented a solution using computer aids. Adamowicz and Albano (1976) presented the nesting aspects of two-dimensional shapes in rectangular modules. Chow (1979) presented nesting of a single shape on a strip. Albano and Saruppo (1980) used heuristic search methods for optimum allocation of two-dimensional irregular shapes. Nee and Venkatesh (1984) presented a heuristic algorithm for optimum layout of metal stamping blanks. Nee (1984) reported a presented computer-aided layout of metal stamping blanks. Qu and Sanders (1987) presented a nesting algorithm for irregular parts, and attributes affecting trim losses. Martin and Stephenson (1988) discussed about the emplacement of objects into boxes.

Nee (1989) described PC-based computer aids in sheet metal working. Karoupi and Loftus (1991) discussed accommodating diverse shapes within hexagonal pavers. Prasad and Somasundaram (1991) presented a heuristic algorithm named CASNS for the nesting of irregularly shaped sheet metal blanks. In another work, Prasad and Somasundaram (1992) presented an automated die design system, CADDS, for sheet metal blanking. Jain *et al.* (1992) used a non-traditional optimization technique called simulated annealing for optimal blank nesting. Ismail and Hon (1992) presented the nesting aspects of two-dimensional shapes for press tool design. Joshi and Sudit (1994) reviewed the procedures for solving single pass strip-layout problems. Theodoracates and Grimsley (1995) used simulated annealing and polynomial time cooling schedules for the optimal packing of arbitrarily shaped polygons.

Prasad *et al.* (1995) presented a sliding algorithm for optimal nesting of arbitrarily shaped sheet metal blanks. Cheok *et al.* (1996) reported an intelligent planning tool for the design of progressive dies. Huang *et al.* (1996) discussed automated design of progressive dies. Ismail *et al.* (1996) presented a feature based design concept of progressive dies. Cheok and Nee (1998) described various configurations of progressive dies. Choi *et al.* (1998) presented a compact and practical CAD system for blanking or piercing of irregularly shaped metal products and stator and rotor parts. Singh and Sekhon (1996) used digraph and matrix methods for evaluation of metal stamping layouts. In another work, Singh and Sekhon (1998) presented a low-cost modeler for two-dimensional metal stamping layouts. Nye (2000) used computer-aided design methods for stamping strip-layouts for optimum raw material utilization. In another work, Nye (2001) described a new algorithm for optimizing the layout of an irregular convex polygonal blank in a strip. This algorithm could orient a single blank such that the utilization of the strip material was maximized.

Rao (2004) applied the AHP method for evaluation of metal stamping layouts. A strip-layout selection index was proposed that could be used for evaluation and ranking of strip-layouts. Ciurana *et al.* (2006) presented an activity model for defining sheet metal process planning. Gomes and Oliveira (2006) used simulated annealing and linear programming methods together for solving irregular strip packing problems. Kamalapurkar and Date (2006) viewed wastage in totality and attempted to minimize the total wastage arising from layout as well as rejections. Highly strained regions in a sheet metal blank were identified. Based on the permissible window of variation in the material properties, a 'defect map' was generated on the sheet. The blanks were laid, out and the possible number of rejections was predicted probabilistically, leading to the prediction of actual utilization of the material. Bortfeldt (2006) suggested a genetic algorithm for two-dimensional strip packing problems that functions without any encoding of solutions. Rather fully defined layouts were manipulated as such by means of specific genetic operators. GA performed best, compared to eleven competing methods from the literature.

Most of the above approaches were aimed mainly at achieving better material utilization. However, the strip-layout with maximum material saving may not be the best strip-layout. Indeed, the die construction may become more complex which could offset the savings due to material economy unless a large number of parts are to be produced. Whatever the chosen procedure for obtaining alternative strip-layouts, it is desirable to make an optimum choice among the available layouts. The decision of selecting a particular strip-layout depends on attributes (*i.e.*, attributes) such as material utilization, die cost, stamping operational cost, required production rate, job accuracy, nature of stock (strip or coil), slitting allowance, balance of blanking pressure, space available in the die set, press specifications, bridge width, shearing strength and thickness of the blank material, facilities available on the shop floor, *etc.* (Fogg and Jamieson, 1975; Donaldson *et al.*, 1976; Paquin and Crowley, 1986). Ideally, it is essential ideally to choose the attributes that are relevant to the particular problem at hand. The actual identification of evaluating attributes may involve discussions with experts working in the fields of production, die making, tool design, and product design. Further more, the relative importance of one attribute over the other is also required. Very few published studies are available on the aspect of die design problem.

The objective of a strip-layout selection procedure is to identify the strip-layout attributes and obtain the most appropriate combination of the attributes in conjunction with the real requirement of the stamping operation. Thus, efforts need to be extended to determine attributes that influence strip-layout selection for a given stamping operation, using a logical approach, to eliminate unsuitable strip-layouts, and for selection of a proper strip-layout to strengthen the existing strip-layout selection procedure. This is considered in this chapter using the graph theory and matrix approach (GTMA) and fuzzy MADM methods.

20.2 Example

Singh and Sekhon (1996) presented a strip-layout selection methodology using the digraph and matrix approach. The authors considered an annual production of 400,000 numbers of a blank, shown in Figure 20.1. The dimensions shown are in centimeters. Six alternative strip-layouts, shown in Figure 20.2, were synthesized.

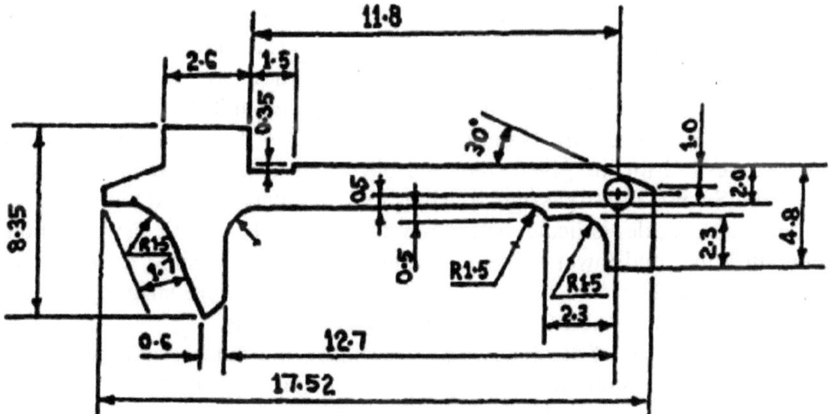

Figure 20.1. Blank profile (from Singh and Sekhon 1996; reprinted with permission from Elsevier)

Five strip-layout selection attributes were identified relevant to the case, and these were: economical material utilization (U_r), die cost (D_c), stamping operational cost (O_c), required production rate (P_r) and job accuracy (J_a). Table 20.1 presents the estimated quantitative values of U_r, D_c, O_c, P_r, and assigned qualitative values of J_a.

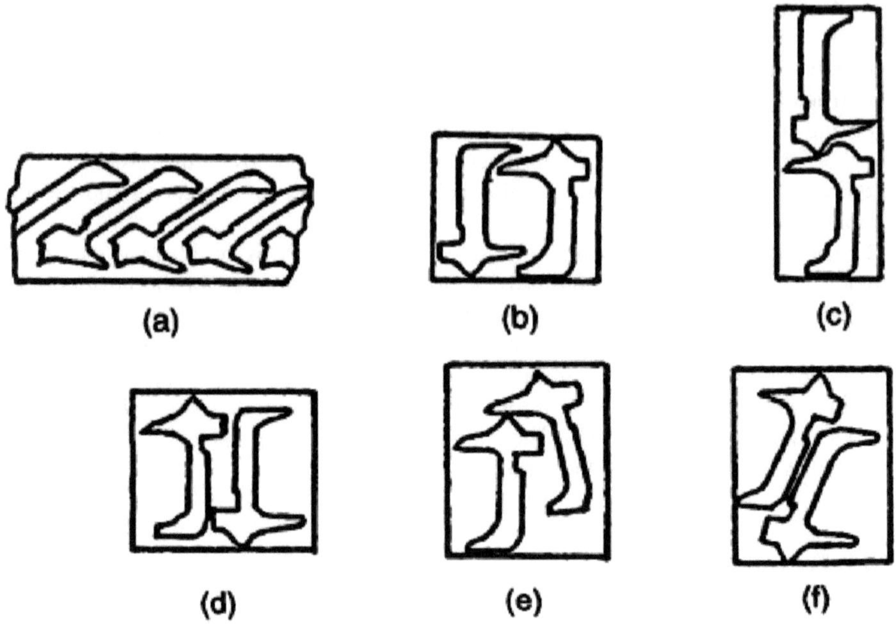

Figure 20.2. Alternative strip-layouts (from Singh and Sekhon 1996; reprinted with permission from Elsevier)

20. 2.1 Application of Graph Theory and Matrix Approach (GTMA)

In the present work, the strip-layout selection attributes (*i.e.*, attributes) considered are the same as those of Singh and Sekhon (1996), and these are: Ur, D_c, O_c, P_r and J_a. The objective values of the attributes, which are given in Table 20.1, are to be normalized. U_r, P_r, and J_a are beneficial attributes, and higher values of these attributes are desired for the given stamping operation. It may be mentioned here that Singh and Sekhon (1996) assigned values to J_a qualitatively, and higher values on the qualitative scale indicate better job accuracy. D_c and O_c are the non-beneficial attributes, and lower values of these attributes are desired for the given stamping operation. The values of these attributes for different strip-layouts are normalized, and given in Table 20.2 in the respective columns.

Relative importance of attributes (a_{ij}) is assigned the values, as explained in Section 2.4. Let the decision maker prepare the following assignments:

	U_r	D_c	O_c	P_r	J_a
U_r	-	0.665	0.745	0.5	0.5
D_c	0.335	-	0.665	0.335	0.335
O_c	0.255	0.335	-	0.255	0.255
P_r	0.5	0.665	0.745	-	0.5
J_a	0.5	0.665	0.745	0.5	-

Table 20.1. Alternative strip-layout data (from Singh and Sekhon 1996; reprinted with permission from Elsevier)

Layout	U_r	D_c	O_c	P_r	J_a
(a)	0.26	25,000	130	80	4
(b)	0.40	28,560	138	120	3
(c)	0.33	31,109	90	150	3
(d)	0.32	31,702	150	125	2
(e)	0.31	32,390	160	110	2
(f)	0.31	32,663	116	108	2

U_r is expressed in %; D_c, in Rupees; O_C in Rupees/1,000 pieces; P_r in pieces/minute.

Table 20.2. Normalized data of the alternative strip-layouts

Layout	U_r	D_c	O_c	P_r	J_a
(a)	0.65	1	0.6923	0.5333	1
(b)	1	0.8754	0.6522	0.8	0.75
(c)	0.825	0.8036	1	1	0.75
(d)	0.8	0.7886	0.6	0.8333	0.5
(e)	0.775	0.7718	0.5625	0.7333	0.5
(f)	0.775	0.7654	0.7759	0.72	0.5

The strip-layout selection attributes digraph, strip-layout attributes matrix of the digraph, and strip-layout selection function for the matrix can be prepared. The value of the strip-layout selection index is calculated, using the values of A_i and a_{ij} for each strip-layout.

The strip-layout selection index values of the different strip-layouts are given below in descending order:

Strip-layout (c): 6.5096
Strip-layout (b): 5.6506
Strip-layout (a): 5.1466
Strip-layout (f): 4.5414
Strip-layout (d): 4.4613
Strip-layout (e): 4.1421

From the values of the strip-layout selection index, it is clear that the strip-layout, designated as (c) is the best choice among the strip-layouts considered for the given stamping operation. The next choice is strip-layout (b), and strip-layout (e) is the last choice. This ranking matches with that suggested by Singh and Sekhon (1996). However, Singh and Sekhon (1996) had not applied the digraph and matrix methods correctly. For example, Singh and Sekhon (1996) assumed a range for each attribute, and the attribute values were normalized within that range. A different range might lead to different normalized values. When actual values of the attribute are available, there is no need to assume a range, and the values of the attribute can be normalized within the available values (as the comparison is between the available alternatives). Further, the relative importance values assigned by Singh and Sekhon (1996) were not logical, and sometimes contradictory. By contrast, the procedure explained in this chapter is systematic and logical.

Following the graph theory and matrix approach, the coefficients of similarity/dissimilarity are also calculated for different strip-layouts using Equations 2.15 and 2.16. The coefficient of similarity values are given in Table 20.3. These are useful for strip-layout documentation, and for easy storage and retrieval of the strip-layout data for the given stamping operation.

Table 20.3. Values of coefficient of similarity for the strip-layouts of the example considered

Strip-layout	(b)	(c)	(d)	(e)	(f)
(a)	0.9108	0.7906	0.8668	0.8048	0.8824
(b)		0.8680	0.7895	0.7330	0.8037
(c)			0.6853	0.6363	0.6976
(d)				0.9285	0.9824
(e)					0.9121

20.2.2 SAW Method

The procedure suggested by Edwards and Newman (1982) to assess weights for each of the attribute to reflect its relative importance to the strip-layout selection decision is followed here. The attributes are ranked in order of importance, and 10 points are assigned to the least important attribute, O_c. Attribute D_c is given 20 points to reflect its relative importance. U_r, P_r, and J_a are considered as equally important and given 30 points each. The final weights are obtained by normalizing the sum of the points to one. Thus, the weights of U_r, P_r, and J_a are calculated as 0.25 each, and the weights of D_c and O_c as 0.1667, and 0.0833 respectively. Using these weights, and the normalized data of the attributes for different strip-layouts, the strip-layout selection index values are calculated, and are arranged in descending order of the index.

Strip-layout (c): 0.8610
Strip-layout (b): 0.8377
Strip-layout (a): 0.7702
Strip-layout (d): 0.7148
Strip-layout (f): 0.6909
Strip-layout (e): 0.6776

From the above values of the strip-layout selection index, it is clear that the strip-layout, designated as (c) is the best choice among the strip-layouts considered.

20.2.3 WPM

Using the same weights of the attributes as those selected for the SAW method, the following ranking of strip-layouts is obtained:

Strip-layout (c): 0.8552
Strip-layout (b): 0.8307
Strip-layout (a): 0.7442
Strip-layout (d): 0.7000
Strip-layout (f): 0.6806
Strip-layout (e): 0.6665

The ranking is the same as that obtained by using the SAW method in this example.

20.2.4 AHP and its Versions

If the same weights as those for the SAW method are used in this method, then the ranking of strip-layouts obtained by using the relative as well as ideal mode AHP will be the same. The multiplicative AHP method yields the same ranking as that given by WPM.

Let the decision maker prepare the following matrix:

$$\begin{array}{c} \\ U_r \\ D_c \\ O_c \\ P_r \\ J_a \end{array} \begin{array}{c} \begin{array}{ccccc} U_r & D_c & O_c & P_r & J_a \end{array} \\ \begin{bmatrix} 1 & 3 & 5 & 1 & 1 \\ 1/3 & 1 & 3 & 1/3 & 1/3 \\ 1/5 & 1/3 & 1 & 1/5 & 1/5 \\ 1 & 3 & 5 & 1 & 1 \\ 1 & 3 & 5 & 1 & 1 \end{bmatrix} \end{array}$$

U_r, P_r, and J_a are considered as equally important. These three attributes are considered as moderately more important than D_c, and strongly more important than O_c, and the relative importance values are assigned accordingly in the above matrix. The assigned values in this paper are for demonstration purposes only.

The normalized weights of each attribute are calculated following the procedure presented in Section 3.2.3, and these are W_{Ur} = 0.2815, W_{Dc} = 0.1054, W_{Oc} = 0.0501, W_{Pr} = 0.2815, and W_{Ja} = 0.2815. The value of λ_{max} is 5.0417 and CR = 0.0093, which is much less than the allowed CR value of 0.1. Thus, there is good consistency in the judgements made.

The value of the strip-layout selection index is now calculated using the above weights and the normalized data of the attributes given in Table 20.2. This leads to the ranking given by the revised AHP or ideal mode of AHP. The alternative strip-layouts are arranged in descending order of the strip-layout selection index.

Strip-layout (c): 0.8597
Strip-layout (b): 0.8428
Strip-layout (a): 0.7546
Strip-layout (d): 0.7137
Strip-layout (f): 0.6811
Strip-layout (e): 0.6749

From the above values of the strip-layout selection index, it is clear that the strip-layout designated as (c) is the best choice among the strip-layouts considered.

For the above weights of importance of attributes, the multiplicative AHP method also leads to the same ranking order.

It may be observed that the above ranking is for the given preferences of the decision maker. The ranking depends upon the judgements of relative importance of attributes made by the decision maker.

20.2.5 TOPSIS Method

The quantitative values of the strip-layout selection attributes, which are given in Table 20.1, are normalized as explained in Section 3.2.6.

Relative importance of attributes (a_{ij}) is assigned using the AHP method, as explained in Section 20.2.4.

The weighted normalized matrix, V_{6x5}, is calculated and is shown below:

$$\begin{bmatrix} 0.0921 & 0.0354 & 0.0200 & 0.0783 & 0.1666 \\ 0.1417 & 0.0405 & 0.0213 & 0.1175 & 0.1245 \\ 0.1169 & 0.0441 & 0.0139 & 0.1468 & 0.1245 \\ 0.1134 & 0.0450 & 0.0231 & 0.1224 & 0.0830 \\ 0.1098 & 0.0459 & 0.0246 & 0.1077 & 0.0830 \\ 0.1098 & 0.0464 & 0.0178 & 0.1057 & 0.0830 \end{bmatrix}$$

Ideal (best) and negative ideal (worst) solutions are calculated, and these are given as:

$V_{Ur}^{+} = 0.1417$ $V_{Ur}^{-} = 0.0921$
$V_{Dc}^{+} = 0.0355$ $V_{Dc}^{-} = 0.0464$
$V_{Oc}^{+} = 0.0139$ $V_{Oc}^{-} = 0.0246$
$V_{Pr}^{+} = 0.1468$ $V_{Pr}^{-} = 0.0783$
$V_{Ja}^{+} = 0.1660$ $V_{Ja}^{-} = 0.0830$

Separation measures are calculated, and these are:

$S_{(a)}^{+} = 0.0848$ $S_{(a)}^{-} = 0.0838$
$S_{(b)}^{+} = 0.0516$ $S_{(b)}^{-} = 0.0759$
$S_{(c)}^{+} = 0.0491$ $S_{(c)}^{-} = 0.0846$
$S_{(d)}^{+} = 0.0920$ $S_{(d)}^{-} = 0.0489$
$S_{(e)}^{+} = 0.0983$ $S_{(e)}^{-} = 0.0343$
$S_{(f)}^{+} = 0.0986$ $S_{(f)}^{-} = 0.0333$

The relative closeness of a particular alternative to the ideal solution is calculated, and these are:

$P_{(a)} = 0.4971$; $P_{(b)} = 0.5952$; $P_{(c)} = 0.6326$;
$P_{(d)} = 0.3473$; $P_{(e)} = 0.2586$; $P_{(f)} = 0.2526$

This relative closeness to ideal solution can be named the 'strip-layout selection index' in the present work.

The alternative strip-layouts are arranged in descending order of their selection index. This can be arranged as (c)-(b)-(a)-(d)-(e)-(f).

20.2.6 Modified TOPSIS Method

The positive ideal solution (R^{+}) and the negative ideal solution (R^{-}) are calculated.

$R_{Ur}^{+} = 0.5035$ $R_{Ur}^{-} = 0.3273$
$R_{Dc}^{+} = 0.3362$ $R_{Dc}^{-} = 0.4393$

$R_{Oc}^{+} = 0.2769$ $R_{Oc}^{-} = 0.4924$
$R_{Pr}^{+} = 0.5216$ $R_{Pr}^{-} = 0.2782$
$R_{Ja}^{+} = 0.5898$ $R_{Ja}^{-} = 0.2949$

The weighted Euclidean distances are calculated as

$D_{(a)}^{+} = 0.1618$ $D_{(a)}^{-} = 0.1613$
$D_{(b)}^{+} = 0.1026$ $D_{(b)}^{-} = 0.1444$
$D_{(c)}^{+} = 0.0949$ $D_{(c)}^{-} = 0.1654$
$D_{(d)}^{+} = 0.1789$ $D_{(d)}^{-} = 0.0925$
$D_{(e)}^{+} = 0.1921$ $D_{(e)}^{-} = 0.0647$
$D_{(f)}^{+} = 0.1885$ $D_{(f)}^{-} = 0.0686$

The relative closeness of a particular alternative to the ideal solution is calculated (*i.e.*, strip-layout selection index), and these are:

$P_{(a)\text{-mod}} = 0.4992$; $P_{(b)\text{-mod}} = 0.5848$; $P_{(c)\text{-mod}} = 0.6353$
$P_{(d)\text{-mod}} = 0.3409$; $P_{(e)\text{-mod}} = 0.2518$; $P_{(f)\text{-mod}} = 0.2667$

The alternative strip-layouts are arranged in descending order of their strip-layout selection index. This can be arranged as (c)-(b)-(a)-(d)-(f)-(e). This ranking is more logical than the ranking presented by the simple TOPSIS method.

It can be observed that all the above decision-making methods propose strip-layout designated as (c) as the first right choice. The decision makers can choose a method for evaluation of metal stamping layouts.

References

Adamowicz M, Albano A (1976) Nesting two-dimensional shapes in rectangular modules. Computer-aided Design 8:27–33

Albano A, Saruppo G (1980) Optimal allocation of two-dimensional irregular shapes using heuristic search methods. IEEE Transactions on Systems, Man and Cybernetics SMC-10:242–248

Bortfeldt A (2006) A genetic algorithm for the two-dimensional strip packing problem with rectangular pieces. European Journal of Operational Research 172:814–837

Cheok BT, Foong KY, Nee AYC (1996) An intelligent planning aid for the design of progressive dies. Journal of Engineering Manufacture 210:25–35

Cheok BT, Nee AYC (1998) Configuration of progressive dies. Artificial Intelligence for Engineering Design and Manufacturing 12:405–418

Choi JC, Kim BM, Cho HY, Kim C (1998) A compact and practical CAD system for blanking or piercing of irregular-shaped metal products and stator and rotor parts. International Journal of Machine Tools & Manufacture 38:931–963

Chow WW (1979) Nesting of a single shape on a strip. International Journal of Production Research 17:305–322

Ciurana J, Ferrer I, Gao JX (2006) Activity model and computer-aided system for defining sheet metal process planning. Journal of Materials Processing Technology 173:213–222

Donaldson C, Lecain GH, Goold VC (1976) Tool design. McGraw Hill, New York

Edwards W, Newman JR, Snapper K, Seaver D (1982) Multiattribute Evaluation. SAGE Publications, Newbury Park, California

Fogg B, Jamieson GA (1975) Influencing attributes in optimization press tool die layouts and a solution using computer aids. CIRP Annals 24:429–434

Gomes AM, Oliveira JF (2006) Solving irregular strip packing problems by hybridising simulated annealing and linear programming. European Journal of Operational Research 171:811–829

Huang K, Ismail HS, Hon KKB (1996) Automated design of progressive dies. Journal of Engineering Manufacture 210:367–376

Industry Canada (1998) Industry Overview Reports-SIC-E3253. Motor Vehicle Stampings Industry, Ottawa, Canada

Ismail HS, Hon KKB (1992) New approaches for the nesting of two-dimensional shapes for press tool design. International Journal of Production Research 30:825–837

Ismail HS, Chen ST, Hon KKB (1996) Feature-based design of progressive press tools. International Journal of Machine Tools & Manufacture 36:367–378

Jain P, Feynes P, Richter R (1992) Optimal blank nesting using simulated annealing. Journal of Mechanical Design 114:160–165

Joshi S, Sudit M (1994) Procedures for solving single pass strip-layout problems. IIE Transactions 26:27–37

Kamalapurkar RL, Date PP (2006) Minimizing wastage of sheet metal for economical manufacturing. Journal of Materials Processing Technology 177:81–83

Karoupi F, Loftus M (1991) Accommodating diverse shapes within hexagonal pavers. International Journal of Production Research 29:1507–1519

Martin RR, Stephenson PD (1988) Putting objects into boxes. Computer-aided Design 20:506–514

Nee AYC (1984) Computer-aided layout of metal stamping blanks. Journal of Engineering Manufacture. 198:187–194

Nee AYC (1989) PC-based computer aids in sheet metal working. Journal of Mechanical Working Technology 19:11–21

Nee AYC, Venkatesh VC (1984) A heuristic algorithm for optimum layout of metal stamping blanks. CIRP Annals 33:317–320

Nye TJ (2000) Stamping strip-layout for optimal raw material utilization. Journal of Manufacturing Systems 19:239–247

Nye TJ (2001) Optimal nesting of irregular convex blanks in strips via an exact algorithm. International Journal of Machine Tools & Manufacture 41:991–1002

Paquin JR, Crowley RE (1986) Die design fundamentals. Industrial Press, New York

Prasad YKDV, Somasundaram S (1991) CASNS-a heuristic algorithm for the nesting of irregular-shaped sheet metal blanks. Computer-aided Engineering Journal 8:69–73

Prasad YKDV, Somasundaram S (1992) CADDS: an automated die design system for sheet-metal blanking. Computer Control and Engineering Journal 3:185–191

Prasad YKDV, Somasundaram S, Rao KP (1995) A sliding algorithm for optimal nesting of arbitrarily shaped sheet metal blanks. International Journal of Production Research 33:1505–1520
Qu W, Sanders JL (1987) A nesting algorithm for irregular parts and attributes affecting trim losses. International Journal of Production Research 25:381–397
Rao RV (2004) Evaluation of metal stamping layouts using an analytic hierarchy process method. Journal of Materials Processing Technology 152:71–76
Singh R, Sekhon GS (1996) A computerized digraph and matrix approach for evaluation of metal stamping layouts. Journal of Materials Processing Technology 59:285–293
Singh R, Sekhon GS (1998) A low-cost modeler for two-dimensional metal stamping layouts. Journal of Materials Processing Technology 84:79–89
Theodoracates VE, Grimsley JL (1995) The optimal packing of arbitrarily shaped polygons using simulated annealing and polynomial-time cooling schedules. Computational Methods Applications in Mechanical Engineering 125:53–70

21

Selection of Forging Conditions for Forging a Given Component

21.1 Introduction

Forging is the process by which metal is heated and is shaped by plastic deformation by suitably applying compressive force. Usually, the compressive force is in the form of hammer blows using a power hammer or a press. Forging refines the grain structure and improves the physical properties of the metal. With proper design, the grain flow can be oriented in the direction of principal stresses encountered in actual use. Grain flow is the direction of the pattern that the crystals take during plastic deformation. Physical properties (such as strength, ductility and toughness) are much better in a forging than in the base metal, which has crystals oriented randomly.

Forgings are consistent from piece to piece, without porosity, voids, inclusions, and other defects. Thus, finishing operations such as machining do not expose voids, because there are none. Also, coating operations such as plating or painting are straightforward due to a good surface that needs very little preparation. Forgings yield parts that have high strength to weight ratio. Thus, these are often used in, *e.g.*, the design of aircraft frame members. The metal can be forged hot (above recrystallization temperatures) or cold. The common forging processes are open-die forging/hand forging, impression die forging/precision forging, press forging, upset forging, roll forging, swaging and net shape/near-net shape forging.

Since the 1980s, the problem of manufacturing defect-free parts has been tackled with simulation tools. Nowadays, computer simulation is used in metal forming to reduce experimental investigation and tests required in a real trial process. Usually, specialists have some freedom in obtaining the desired forged parts, which is why numerical simulations are applied in most engineering offices and manufacturing industries to evaluate forming difficulties in metal forming processes, including forging processes. Many studies been carried out in this regard. Szyndler and Klimkiewicz (1992) described the method of the design of the open-die elongation process using optimization procedures. Based on numerical analysis and experimental results, the function relations between the technological parameters for various shapes of dies were defined. These relations were the base for the formulation of the objective function for the optimization model.

Jugo and Anza (1994) presented the results of the numerical simulation of some industrial hot forging processes based on the use of commercial finite element codes. Illustrative cases, giving rise to flow defects in industrial conditions, were reproduced and redesigned to eliminate the defects using finite-element method (FEM) as a tool for carrying out the necessary alternatives analysis. Both, two-dimensional axisymmetrical and three-dimensional omplex forging shapes were considered.

Duggirala *et al.* (1994) described a new method for design optimization of process variables in cold forging sequences. An adaptive microgenetic algorithm was implemented to minimize the possibility of the initiation of tensile fracture in the outer race preform of a constant velocity joint manufactured by cold forming operations. The chosen design variables were the preform diameter, the maximum number of forming operations, the number of extrusion and upset operations, the amount of area reduction in each pass, the amount of upset in each upset, and the included angles in the extrusion and upset dies.

Forcellese *et al.* (1996) used a decision-making method based on the analytic hierarchy process (AHP) method for the selection of the best forging condition for manufacturing Al-MMC automotive components. Zhao *et al.* (1997) developed a sensitivity analysis method for preform die shape design in net-shape forging processes using the rigid viscoplastic finite-element method. The preform die shapes were represented by cubic B-spline curves. The control points or coefficients of B-spline were used as the design variables. The optimization problem was to minimize the zone where the realized and desired final forging shapes do not coincide. The sensitivities of the objective function, nodal coordinates, and nodal velocities with respect to the design variables were developed in detail. A procedure for computing the sensitivities of history-dependent functions was presented. The method developed was used to design the preform die shape of H-shaped forging processes, including plane strain and axisymmetric deformations.

Liou and Jang (1997) utilized the finite-element method and robust design methodology to identify the controlling process parameters that could monitor residual stresses in forged products. In the optimizing process of the forging operation, experimental planning was performed by using the orthogonal array and concept of the signal-to-noise ratio. Frictional coefficient, length of die land, reduction percentage, inlet angle, and corner fillet were selected as process parameters. ANOVA showed that the inlet angle, friction coefficient, and length of die land had the most significant effects on the optimum residual stresses.

Roy *et al.* (1997) described a new method for design optimization of process variables in multistage metal forming processes. The selected forming processes were multi-pass cold wire drawing, multipass cold drawing of a tubular profile, and cold forging of an automotive outer race preform. An adaptive microgenetic algorithm (μGA) scheme was implemented for minimizing a wide variety of objective-cost functions relevant to the respective processes. The chosen design variables were die geometry, area reduction ratios, and the total number of forming stages. Significant improvements in the simulated product quality, and reduction in the number of passes were observed to result from the microgenetic algorithms-based optimization process.

Picart *et al.* (1998) presented and applied an optimization technique to metal forming design problems with damage occurrence. Biglari *et al.* (1998) reported a novel shape optimization method for the design of preform die shapes in multistage forging processes using a combination of the backward deformation method and a fuzzy decision-making algorithm. In the backward deformation method, the final component shape was taken as the starting point, and the die was moved in the reverse direction with boundary nodes being released as the die was raised. The optimum die shape was thereby determined by taking the optimum reverse path. A fuzzy decision-making approach was developed to specify new boundary conditions for each backward time increment, based on geometrical features and the plastic deformation of the work piece.

Duan and Sheppard (2002) developed a parameter optimization system in order to increase the reliability and the application range of FEM programs for metal forming by combining the general nonlinear finite-element analysis software with the improved constrained variable metric strategy. Park *et al.* (2001) optimized the powder forging process parameters for an aluminum-alloy piston, namely, the composition, mixing time, sintering time, sintering temperature, shape of the preform, *etc.*, through experiments, and developed a high-strength aluminum-alloy piston with an optimized process.

Sousa *et al.* (2002) presented an approach to optimal shape design in forging. The design problem was formulated as an inverse problem incorporating a finite-element three-dimensional analysis model, and an optimization technique conducted on the basis of design sensitivities. Ou *et al.* (2003) outlined a die shape optimization system for the net-shape forging of aerofoil blades. In forging simulation using finite elements (FEs), a compensation approach was used in order to eliminate the aerofoil thickness errors due to die elasticity. The optimized die shape was obtained by modifying the nominal die shape with a fraction of the die-elastic deflections through an introduced weighting factor.

Dyja *et al.* (2004) discussed the influence of the main parameters of the forging process, and the shape of tools on distribution of the values of local deformations within the volume of forged material. The theoretical analysis of investigations was verified by laboratory tests. The values of the main technological forging parameters were determined, and the application of a group of tools appropriate for the forging process was proposed. The optimization of free hot forging process with respect to relative reduction, relative feed, and tilting of forging was the aim of the investigation.

Ou *et al.* (2004) presented a finite element (FE)-based forging simulation and optimization approach in order to achieve net-shape forging production for aeroengine components. In the hot forging of aerofoil sections, forging errors due to die-elastic deformation, thermal distortion and press elasticity were quantified, revealing distinctive dimensional and shape error patterns. Integrating a general-purpose FE package and an optimization code, a forging simulation and optimization system was developed.

Bariani *et al.* (2004) presented a joint application of finite element-based numerical simulation and real material-based physical simulation techniques for design and optimization of hot forging operations to manufacture high-strength stainless steel turbine blades. Zhao *et al.* (2004) introduced an optimization method for metal forming processes, especially for forging process designs using the finite element-based sensitivity analysis method. An approach for improving the computational efficiency was introduced and demonstrated. After introducing the optimization method of multistage forming processes, the authors presented an optimization method for single stage forming processes.

Castro *et al.* (2004) presented an approach to optimal design in forging. The design problem was formulated as an inverse problem incorporating a finite-element thermal analysis model and an optimization technique, based on an evolutionary strategy. A rigid viscoplastic flow-type formulation was adopted, valid for both hot and cold processes. In order to demonstrate the efficiency of the inverse evolutionary search, specific forging cases were presented, considering the optimization of the process parameters aiming to reduce the difference between the realized and the prescribed final forged shape under minimal energy consumption, and restricting the maximum temperature. Chen *et al.* (2004) presented a novel die-design approach, known as the least squares approach, to minimize the component errors.

Thiagarajan and Grandhi (2005) developed a three-dimensional preform shape optimization method for the forging process using the reduced basis technique. Several critical techniques and new advances that enable the use of the reduced basis technique were presented. The primary objective was to reduce the enormous number of design variables required to define the three-dimensional preform shape. The method was demonstrated on the preform shape optimization of a geometrically complex three-dimensional steering link.

Banaszek and Szota (2005) proposed an optimal shape of profiled anvils for the process of ingot blacksmith forging, and determined appropriate technological parameters that would assure a homogeneous distribution of strain intensities within forgings. The authors presented the results of theoretical studies simulating the operations of stretch forging of large ingots over profiled anvils with specified technological parameters. An assessment of the effectiveness of ingot stretch forming operations was made, and an analysis of energy and force parameters was performed for particular technological parameters. A significant effect of the relative feed and anvil profile angle on the parameters analyzed were found. Conclusions were drawn that proposed the use of optimal anvil shape and technological parameters to enhance the effectiveness of stretch forging operation, and reduce the values of the energy and force parameters of the process.

Vijian and Arunachalam (2006) analyzed the influence of process parameters on surface roughness in squeeze casting of LM6 aluminum alloy using the Taguchi method. Analysis of variance and F-test values were used to determine the most significant process parameters affecting surface roughness. The results indicated that the squeeze pressure and the die preheating temperature were key parameters causing appreciable improvement in the surface finish of the squeeze cast components.

Chastel *et al.* (2006) defined a new state variable to represent grain flow orientation, and implemented in the finite-element code so that forging could be simulated to predict grain flow in a forged part. Khoury *et al.* (2006) developed a finite-element package to solve elasto-plasticity problems with ductile damage in large deformation. An experimental design was presented in order to show the influence of geometric parameters of a turbine engine disk on the total strain energy and the average of the elasto-plastic strain.

Mulyadi *et al.* (2006) examined various methods for parameter optimization in constitutive equations applied to the hot deformation of a popular α–β titanium alloy. The use of direct search and gradient methods were shown to be effective, even with a limited dataset. However, a hybrid approach, whereby genetic algorithms were used to find an initial parameter starting point, and then a direct search (simplex) method was applied to obtain a global minimum, was shown as a more promising approach.

Poursina *et al.* (2006) presented a numerical method for shape optimization of preform dies in two-stage hot forging. The objective of optimization was to eliminate work-piece defects that may arise during the forging process. A two-dimensional finite-element code was developed for the simulation of the mechanical process and prediction of the defects. The optimization method was based on a genetic algorithm supported by an elitist strategy. The scheme developed was used to design optimal preform dies for two axisymmetric examples. The objective function was associated with the quality of the final product.

Choi *et al.* (2006) performed a three-dimensional rigid-plastic finite-element method (FEM) analysis to optimize an open-die forging process in the production of circular shapes. The finite-element method was used to analyze the practice of open-die forging, focusing on the effects of feed rate and rotation angle for optimal forging pass design. The optimal combination of feed rate and rotation angle was determined by quantifying the radius profile in the longitudinal direction and roundness of the product. From an analysis of the results, optimal process conditions were proposed for the production of circular shapes with good dimensional accuracy by open die forging.

It may be said that the optimization of forging process design and forging process plan for various work materials can be based on the maximization of production rate, maximization of product quality, minimization of production cost, minimization of die cost, minimization of forging loads, *etc*. The optimum solution is the one that best satisfies these different requirements. Ideally, it is essential to choose the criteria (or attributes) that are relevant to the particular problem at hand. The actual identification of evaluating attributes may involve discussions with the experts working in the areas of production, die making, tool design, and product design. Furthermore, the relative importance of the one criterion (or attribute) over the other criterion (or attribute) is also required. Very few published studies are available on the aspect of die design problems. In order to solve such a problem, decision-making methods can be used. A number of appropriate alternatives, given by different sets of forging parameters, can be defined. Then, such alternatives can be evaluated by using different criteria (or attributes) that can be properly defined. The alternative with the highest level of satisfaction can be chosen. This is considered in this chapter, using the graph theory and matrix approach (GTMA) and fuzzy MADM methods.

21.2 Example

Forcellese *et al.* (1996) used a decision-making method based on the analytic hierarchy process (AHP) for the selection of the best forging conditions for manufacturing Al-MMC automotive components. Three candidate solutions, in terms of three different sets of initial die and billet temperatures, and die speed, were defined by means of FEM simulations; these were performed within proper strain rate and temperature windows determined by hot formability studies on 6061/Al_2O_3/10p composite. Five criteria (or attributes) for evaluating the different alternatives with respect to the overall goal of selection of the best forging conditions were defined and these were: (i) product quality, (ii) production rate, (iii) die cost, (iv) heating cost, and (v) forging load. The AHP method allowed establishing the alternative characterized by an initial die temperature of 350°C, an initial billet temperature of 400°C, and a die speed of 3 mm/s as the optimum solution. The data of the forging attributes are summarized in Table 21.1. The attributes DC and HC are expressed subjectively in Table 21.1. Hence, appropriately using Table 4.3, the objective values are assigned and shown in Table 21.2.

Table 21.1. Objective and subjective data of the forging conditions selection attributes (from Forcellese *et al.*, 1996; reprinted with permission from Elsevier)

Alternative forging conditions			PQ	PR	DC	HC	FL
IBT(°C)	IDT(°C)	DS (mm/s)	(%)	(pieces/h)			(N)
400	350	3	4.01	73.97	Low	Very low	15,773
500	450	2	2.19	67.92	Very high	Low	9,119
425	350	0.1	1.46	12	Low	Very high	15,110

PQ: product quality PR: production rate DC: die cost
HC: heating cost FL: forging load per unit length
IBT: initial billet temperature IDT: initial die temperature DS: die speed

Table 21.2. Objective data of the forging conditions selection attributes

Alternative forging conditions			PQ	PR	DC	HC	FL
IBT(°C)	IDT(°C)	DS (mm/s)	(%)	(pieces/h)			(N)
400	350	3	4.01	73.97	0.335	0.255	15,773
500	450	2	2.19	67.92	0.745	0.335	9,119
425	350	0.1	1.46	12	0.335	0.745	15,110

21.2.1 Graph Theory and Matrix Approach

Various steps of the methodology, proposed in Section 2.6, are carried out as described below:

Step 1: In the present work, the attributes considered are the same as those used by Forcellese *et al.* (1996) and these are product quality (PQ), production rate (PR), die cast (DC), heating cost (HC), and forging load (FL). The objective values of the attributes, which are given in Table 21.2, are to be normalized. PR is the only beneficial attribute, and the remaining four attributes are considered as non-beneficial. Values of these attributes are normalized, and are given in Table 21.3 in the respective columns.

Table 21.3. Normalized data of the forging conditions selection attributes

Alternative forging conditions IBT(°C)	IDT(°C)	DS (mm/s)	PQ	PR	DC	HC	FL
400	350	3	0.3641	1	1	1	0.5781
500	450	2	0.6667	0.9182	0.45	0.7612	1
425	350	0.1	1	0.1622	1	0.3423	0.6035

Let the decision maker assign equal importance to the attributes as shown below:

	PQ	PR	DC	HC	FL
PQ	-	0.335	0.59	0.865	0.665
PR	0.665	-	0.665	0.955	0.745
DC	0.41	0.335	-	0.745	0.59
HC	0.135	0.045	0.255	-	0.335
FL	0.335	0.255	0.41	0.665	-

Step 2: The forging conditions selection attributes digraph, forging conditions selection attributes matrix of the digraph, and forging conditions selection attributes function for the matrix can be prepared. The value of forging conditions selection index is calculated using the values of A_i and a_{ij} for each alternative forging condition. The forging conditions selection index values of different forging conditions are given below in descending order:

Forging conditions 1 : 4.0834
Forging conditions 2 : 3.8433
Forging conditions 3 : 2.3075

Thus, GTMA suggests forging conditions 1 as the correct choice for the forging conditions selection problem considered and forging conditions 3 as the last choice. This ranking matches well with the results presented by Forcellese *et al.* (1996).

21.2.2 SAW Method

Forcellese *et al.* (1996) used the following weights of importance of the five selection attributes:

$W_{PQ} = 0.236$, $W_{PR} = 0.459$, $W_{DC} = 0.179$, $W_{HC} = 0.037$, and $W_{FL} = 0.089$

Using the same weights of importance of the attributes, SAW leads to the following ranking order:

Forging conditions 1	:	0.8124
Forging conditions 2	:	0.7765
Forging conditions 3	:	0.5558

Thus, the SAW method also suggests forging conditions 1 as the correct choice for the forging conditions selection problem considered. This ranking matches well with the results presented by Forcellese *et al.* (1996).

21.2.3 WPM

For the same weights of importance of the attributes, WPM leads to the following ranking order:

Forging conditions 1	:	0.7503
Forging conditions 2	:	0.7498
Forging conditions 3	:	0.3987

WPM also suggests forging conditions 1 as the correct choice.

21.2.4 AHP Method

Forcellese *et al.* (1996) had already used the AHP method, and the ranking was given as:

Forging conditions 1	:	0.386
Forging conditions 2	:	0.358
Forging conditions 3	:	0.256

AHP also suggests forging conditions 1 as the correct choice.

21.2.5 TOPSIS Method

Following the steps of the methodology given in Section 3.2.6, the TOPSIS method gives the following weighted normalized matrix:

$$\begin{bmatrix} 0.1973 & 0.3357 & 0.0679 & 0.0110 & 0.0593 \\ 0.1078 & 0.3082 & 0.1510 & 0.0145 & 0.0343 \\ 0.0718 & 0.0545 & 0.0679 & 0.0322 & 0.0568 \end{bmatrix}$$

The ideal (best) and negative ideal (worst) solutions are obtained, and these are given as:

$V_{PQ}^{+} = 0.0718$ $\quad$ $V_{PQ}^{-} = 0.1973$

$V_{PR}^{+} = 0.3357$ $\quad$ $V_{PR}^{-} = 0.0545$

$V_{DC}^{+} = 0.0679$ $\quad$ $V_{DC}^{-} = 0.1510$

$V_{HC}^{+} = 0.0110$ $\quad$ $V_{HC}^{-} = 0.0322$

$V_{FL}^{+} = 0.0343$ $\quad$ $V_{FL}^{-} = 0.0593$

The separation measures are:

$S_1^{+} = 0.1279$ $\quad$ $S_1^{-} = 0.2940$

$S_2^{+} = 0.0947$ $\quad$ $S_2^{-} = 0.2709$

$S_3^{+} = 0.2829$ $\quad$ $S_3^{-} = 0.1505$

The relative closeness of a particular alternative to the ideal solution is calculated and named the 'forging conditions selection index (FCSI)' in the present example; these are arranged in descending order as:

Forging conditions 2 : 0.7409
Forging conditions 1 : 0.6968
Forging conditions 3 : 0.3473

TOPSIS method suggests forging conditions 2 as the correct choice for the forging conditions selection problem considered. However, this result seems not to be genuine.

21.2.6 Modified TOPSIS Method

The modified TOPSIS method gives the following results:

The weighted Euclidean distances are:

$D_1^+ = 0.2715$ $D_1^- = 0.4723$
$D_2^+ = 0.2146$ $D_2^- = 0.4357$
$D_3^+ = 0.4361$ $D_3^- = 0.3246$

The FCSI values are calculated, and these are arranged in descending order as:

Forging conditions 2 : 0.6700
Forging conditions 1 : 0.6349
Forging conditions 3 : 0.4267

The results presented by the modified TOPSIS method are exactly the same as those given by the simple TOPSIS method.

It may be noted from the values of the five selection attributes of the three alternative forging conditions, that proposing alternative 1 as the first correct choice seems to be more genuine. Thus, TOPSIS and modified TOPSIS methods are not very suitable for this particular example problem.

References

Banaszek G, Szota P (2005) A comprehensive numerical analysis of the effect of relative feed during the operation of stretch forging of large ingots in profiled anvils. Journal of Materials Processing Technology 169:437–444

Bariani PF, Bruschi S, Negro TD (2004) Integrating physical and numerical simulation techniques to design the hot forging process of stainless steel turbine blades. International Journal of Machine Tools and Manufacture 44:945–951

Biglari FR, O'Dowd NP, Fenner RT (1998) Optimum design of forging dies using fuzzy logic in conjunction with the backward deformation method. International Journal of Machine Tools and Manufacture 38:981–1000

Castro CF, António CAC, Sousa LC (2004) Optimisation of shape and process parameters in metal forging using genetic algorithms. Journal of Materials Processing Technology 146:356–364

Chastel Y, Caillet N, Bouchard PO (2006) Quantitative analysis of the impact of forging operations on fatigue properties of steel components. Journal of Materials Processing Technology 177:202–205

Chen X, Balendra R, Qin Y (2004) A new approach for the optimization of the shrink-fitting of cold-forging dies. Journal of Materials Processing Technology 145:215–223

Choi SK, Chun MS, Tyne CJV, Moon YH (2006) Optimization of open die forging of round shapes using FEM analysis. Journal of Materials Processing Technology 172:88–95

Duan X, Sheppard T (2002) Shape optimization using FEA software: a V-shaped anvil as an example. Journal of Materials Processing Technology 120:426–431

Duggirala R, Shivpuri R, Kini S, Ghosh S, Roy S (1994) Computer aided approach for design and optimization of cold forging sequences for automotive parts. Journal of Materials Processing Technology 46:185–198

Dyja H, Banaszek G, Mróz S, Berski S (2004) Modelling of shape anvils in free hot forging of long products. Journal of Materials Processing Technology 157-158:131–137

Forcellese A, Gabrielli F, Ruffini R (1996) Application of a decision-making method in the forging condition optimization for manufacturing automotive components. Journal of Materials Processing Technology 60:125–132

Jugo EF, Anza J (1994) Industrial applications of numerical simulation to the design and optimization of forging. Journal of Materials Processing Technology 45:81–86

Khoury I, Laurence GM, Lafon P, Labergére C (2006) Towards an optimization of forging processes using geometric parameters. Journal of Materials Processing Technology 177:224–227

Liou JH, Jang DY (1997) Forging parameter optimization considering stress distributions in products through FEM analysis and robust design methodology. International Journal of Machine Tools and Manufacture 37:775–782

Mulyadi M, Rist MA, Edwards L, Brooks JW (2006) Parameter optimization in constitutive equations for hot forging. Journal of Materials Processing Technology 177:311–314

Ou H, Armstrong CG, Price MA (2003) Die shape optimization in forging of aerofoil sections. Journal of Materials Processing Technology 132:21–27

Ou H, Lan J, Armstrong CG, Price MA (2004) An FE simulation and optimization approach for the forging of aeroengine components. Journal of Materials Processing Technology 151:208–216

Park JO, Kim KJ, Kang DY, Lee YS, Kim YH (2001) An experimental study on the optimization of powder forging process parameters for an aluminum-alloy piston. Journal of Materials Processing Technology 113:486–492

Picart P, Ghouati O, Gelin JC (1998) Optimization of metal forming process parameters with damage minimization. Journal of Materials Processing Technology 80-81:597–601

Poursina M, Parvizian J, Antonio CAC (2006) Optimum pre-form dies in two-stage forging. Journal of Materials Processing Technology 174:325–333

Roy S, Ghosh S, Shivpuri R (1997) A new approach to optimal design of multi-stage metal forming processes with micro genetic algorithms. International Journal of Machine Tools and Manufacture 37:29–44

Sousa LC, Castro CF, António CAC, Santos AD (2002) Inverse methods in design of industrial forging processes. Journal of Materials Processing Technology 128:266–273

Szyndler R, Klimkiewicz B (1992) Design of the open-die elongation process using optimization technique. Journal of Materials Processing Technology 34:157–162

Thiyagarajan N, Grandhi RV (2005) Multi-level design process for 3-D preform shape optimization in metal forming. Journal of Materials Processing Technology 170:421–429

Vijian P, Arunachalam VP (2006) Optimization of squeeze cast parameters of LM6 aluminium alloy for surface roughness using Taguchi method. Journal of Materials Processing Technology 180:161–166

Zhao G, Wcorrect E, Grandhi RV (1997) Sensitivity analysis based preform die shape design for net-shape forging. International Journal of Machine Tools and Manufacture 37:1251–1271

Zhao G, Ma X, Zhao X, Grandhi RV (2004) Studies on optimization of metal forming processes using sensitivity analysis methods. Journal of Materials Processing Technology 147:217–228

22

Evaluation of Environmentally Conscious Manufacturing Programs

22.1 Introduction

Environmentally conscious manufacturing (ECM) is concerned with developing methods for designing and manufacturing of new products from conceptual design to final delivery, and ultimately to the end-of-life disposal such that all the environmental standards and requirements are satisfied. In recent years, environmental awareness and recycling regulations have been putting pressure on many manufacturers and consumers to produce, and dispose of products in an environmentally responsible manner. Almost every function within organizations has been influenced by external and internal pressures to become environmentally sound. Issues such as green consumerism and green product development have impacted marketing. Finance, information systems and technology, human resources and training, engineering and research, and development are all organizational functions that have been influenced by these environmental pressures. One of the functions that has been profoundly influenced by environmental pressures is the organizational operations and manufacturing function. The traditional reactive responses to these pressures are now being supplemented and replaced by more proactive, strategic, competitive responses. Many businesses have begun to realize that there is some profitability to environmentally conscious business practices (Sarkis, 1995a; Klassen and McLaughlin, 1996; Cordeiro and Sarkis, 1997; Gungor and Gupta, 1999).

The research topics on ECM programs have focused on managerial practices, business processes, and technology. ECM programs include proactive measures such as life-cycle analysis of products, design for the environment, design for disassembly, total quality environmental management, remanufacturing, ISO14000 certification, and green supply chains. Each of these programs crosses inter- and intra-organizational boundaries. These programs work hand in hand with other environmental alternatives such as development of environmental management systems, and green purchasing (Sarkis and Lin, 1994; Sarkis, 1995b, 1999; Sarkis and Rasheed, 1995).

As mentioned above, many organizations have begun to consider ECM programs from a strategic and competitive advantage perspective. Yet, to determine whether these programs fit within the strategic direction and scope of the organization, and whether they can improve performance, requires some analysis and justification. To help decipher the benefits of these programs, a number of attributes, both environmental and strategic, should be evaluated. For organizations to accept the results, any methodology to help evaluate these systems should be able to handle traditional financial and non-financial measures of performance. In this case, the non-financial measures should include specific attributes that will help determine how well these programs, when implemented, will perform with respect to the natural environment (Sarkis, 1999).

Many precision-based methods for ECM program selection have been developed. Wilhelm *et al.* (1993) described the selection of waste management technologies to implement manufacturing pollution prevention strategies. Presley and Sarkis (1994) used an activity-based strategic justification methodology for ECM technology. Sarkis (1995a) linked the manufacturing strategy with the environmental consciousness. Further, Sarkis (1995b) linked supply chain management aspects with environmentally conscious design and manufacturing. Munoz and Sheng (1995) presented an analytical approach for determining the environmental impact of machining processes. Klassen and McLaughlin (1996) discussed the impact of environmental management on firm performance. Cordeiro and Sarkis (1997) presented the aspects of environmental proactivism and firm performance as evidenced from industry analyst forecasts. Gungor and Gupta (1999) surveyed issues in ECM and product recovery. Sarkis (1998) evaluated environmentally conscious business practices. Further, Sarkis (1999) presented an illustrative problem for evaluating ECM programs for an industrial application. The role of environmentally friendly cutting fluids, which is an important factor in ECM programs, was described in Chapter 8.

Sarkis and Weinrach (2001) evaluated environmentally conscious waste treatment technologies using the DEA method. Khan *et al.* (2002) proposed a holistic and integrated methodology, GreenPro-I, for process/product design by combining the traditional LCA approach with multiple criteria decision-making methods. The methodology was simple and applicable at the early design stage, and was more robust against uncertainty in the data. Madu *et al.* (2002) presented a hierarchic metric approach for integration of green issues in manufacturing. A paper recycling application was considered for demonstration. Rao (2004) used digraph and matrix methods for the evaluation of ECM programs. Kuo *et al.* (2006) presented an innovative method, namely, green fuzzy design analysis (GFDA), which involves simple and efficient procedures to evaluate product design alternatives based on environmental consideration using fuzzy logic. The hierarchical structure of environmentally conscious design indices was constructed using the analytical hierarchy process (AHP), which included five aspects: (1) energy, (2) recycling, (3) toxicity, (4) cost, and (5) material. After weighting factors for the environmental attributes were determined, the most desirable design alternative could be selected based on the fuzzy multiple attribute decision-making technique. Morrow *et al.* (2006) presented environmental aspects involved in laser-based and conventional tool and die manufacturing.

The objective of an ECM program selection procedure is to identify the ECM program selection attributes, and obtain the most appropriate combination of the attributes in conjunction with the real requirements of the industrial application. Efforts need to be extended to determine attributes that influence ECM program selection for a given application, using a logical approach, to eliminate unsuitable ECM programs and for selection of a proper ECM program to strengthen the existing ECM program selection procedure. This is considered in this chapter, using graph theory and fuzzy MADM methods.

22.2 Example

Now, an example is considered to demonstrate and validate the proposed procedures. Sarkis (1999) presented an illustrative problem for evaluating ECM programs for an industrial application. Sarkis (1999) assumed that the management had determined its missions, priorities, and objectives in place. It was also assumed that a set of fifteen alternatives had been determined, and that all could be evaluated on each of the six attributes identified for the given industrial application. The problem considering six attributes and fifteen alternative ECM programs is shown in Table 22.1.

Table 22.1. Quantitative data of ECM program selection attributes (from Sarkis 1999; reprinted with permission from Elsevier)

Alternative	C	Q	R	PWR	PGR	RC
1	706,967	2	29	17	0	51
2	181,278	3	5	14	7	45
3	543,399	4	5	3	7	71
4	932,027	7	15	10	17	57
5	651,411	4	19	7	0	21
6	714,917	5	15	6	19	5
7	409,744	1	8	17	1	35
8	310,013	6	23	15	18	32
9	846,595	2	28	16	19	24
10	625,402	3	21	16	7	34
11	285,869	2	1	13	12	54
12	730,637	3	3	4	1	12
13	794,656	5	27	14	14	65
14	528,001	1	6	5	9	41
15	804,090	2	26	6	5	70

C: Costs ($) Q: Quality (% defects) R: Recyclability (% recyclable material) PWR: Process waste reduction (%) PGR: Packaging waste reduction (%) RC: Regulatory compliance (% reduction in violations)

22.2.1 Graph Theory and Matrix Approach

In the present work, the attributes considered are the same those of Sarkis (1999), and these are: costs involved (C), quality (Q, expressed in % of defects), recyclability (R), process waste reduction (PWR), packaging waste reduction (PGR), and regulatory compliance (RC). The quantitative values of the ECM program selection attributes, which are given in Table 22.1, are to be normalized. R, PWR, PGR, and RC are beneficial attributes, and higher values are desirable. Values of these attributes are normalized, and given in Table 22.2 in the respective columns. C and Q are non-beneficial attributes and lower values are desirable. The values of these attributes for different ECM programs are normalized, and given in Table 22.2 in the respective columns.

Let the decision maker make the following assignments of relative importance:

$$
\begin{array}{l} \\ C \\ Q \\ R \\ PWR \\ PGR \\ RC \end{array}
\begin{array}{cccccc} C & Q & R & PWR & PGR & RC \\ \left[\begin{array}{cccccc} - & 0.335 & 0.41 & 0.255 & 0.41 & 0.59 \\ 0.665 & - & 0.59 & 0.335 & 0.665 & 0.745 \\ 0.59 & 0.41 & - & 0.255 & 0.59 & 0.665 \\ 0.745 & 0.665 & 0.745 & - & 0.665 & 0.865 \\ 0.59 & 0.335 & 0.41 & 0.335 & - & 0.665 \\ 0.41 & 0.255 & 0.335 & 0.135 & 0.335 & - \end{array}\right] \end{array}
$$

The ECM selection attributes digraph, ECM selection attributes matrix of the digraph, and ECM selection function for the matrix can be prepared. The value of the ECM selection index is calculated using the values of A_i and a_{ij} for each ECM program.

Table 22.2. Normalized data of the alternative ECM programs

Alternative	C	Q	R	PWR	PGR	RC
1	0.2564	0.5	1	1	0	0.7183
2	1	0.3333	0.1724	0.8235	0.3684	0.6338
3	0.3336	0.25	0.1724	0.1765	0.3684	1
4	0.1945	0.1429	0.5172	0.5882	0.8947	0.8028
5	0.2783	0.25	0.6552	0.4118	0	0.2958
6	0.2536	0.2	0.5172	0.3529	1	0.0704
7	0.4424	1	0.2759	1	0.0526	0.493
8	0.5847	0.1667	0.7931	0.8823	0.9474	0.4507
9	0.2141	0.5	0.9655	0.9412	0.3684	0.4789
10	0.2899	0.3333	0.7241	0.9412	0.3684	0.4789
11	0.6341	0.5	0.0345	0.7647	0.6316	0.7606
12	0.2481	0.3333	0.1034	0.2353	0.0526	0.169
13	0.2281	0.2	0.931	0.8235	0.7368	0.9155
14	0.3433	1	0.2069	0.2941	0.4737	0.5775
15	0.2254	0.5	0.8965	0.353	0.2631	0.9859

The ECM program selection index values of different ECM programs are given below in descending order:

Alternative ECM program 9	:	11.0883
13	:	10.2018
8	:	10.063
1	:	8.9617
11	:	8.6541
2	:	8.6065
7	:	8.38
15	:	8.1909
10	:	8.1322
4	:	7.9691
14	:	7.2478
3	:	5.9276
6	:	5.8365
5	:	5.1647
12	:	3.9799

From the values of the ECM program selection index, it is understood that the ECM program designated as 9 is the best choice among the ECM programs considered for the given industrial application. The next choice is 13, and the last choice is 12. However, Sarkis (1999) had used the ANP and DEA methods together, and had suggested ECM program 11 as the first choice, 8 as the second choice and 12 as the last choice. A closer look at the values of the attributes of these ECM programs reveals that ECM program 9 is better than ECM program 11 for three attributes (*i.e.*, R, PWR, and PGR) and equals in performance for Q, whereas ECM program 11 is better than ECM program 9 only for two attributes (*i.e.*, C and RC). Thus, proposing ECM program 9 as the first choice in the present work appears to be more genuine and the ranking presented by the proposed method is more reliable. Also, the proposed approach ranks the alternatives in a single stage, unlike the two-stage approach used by Sarkis (1999). It may be mentioned that the ranking depends upon the judgements made by the user.

22.2.2 SAW Method

Let the attributes be ranked in order of importance, and 10 points be assigned to the least important attribute RC. Attribute C is given 15 points to reflect its relative importance. PGR, R, Q, and PWR are given 20, 25, 30, and 40 points respectively. Thus, the weights of PWR, Q, R, PGR, C, and RC are calculated as 0.375, 0.1875, 0.125, and 0.3125, respectively. Using these weights, and the normalized data of the attributes for different ECM programs, the ECM program selection index values are calculated, and are arranged in descending order of the index.

Alternative ECM program 9	:	0.7484
8	:	0.6595
1	:	0.6501
13	:	0.6394
7	:	0.6393
10	:	0.5875

11	:	0.5441
2	:	0.5424
15	:	0.5002
4	:	0.4971
14	:	0.4808
6	:	0.4110
5	:	0.3391
3	:	0.2945
12	:	0.2032

From the above values of the ECM program selection index, it is clear that the alternative ECM program, designated as 9 is the best choice among the ECM programs considered.

22.2.3 AHP and its Versions

If the same weights as those of the SAW method are used in this method, then the ranking of ECM programs obtained by using the relative as well as ideal mode AHP method will be the same.

22.2.4 TOPSIS Method

The quantitative values of the ECM program selection attributes, which are given in Table 22.1, are normalized as explained in Section 3.2.6. Let the relative importance of attributes (a_{ij}) be assigned using the AHP method as explained below:

	C	Q	R	PWR	PGR	RC
C	1	3	2	1	2	3
Q	1/3	1	1/3	1/4	1/3	1
R	1/2	3	1	1/2	1/2	2
PWR	1	4	2	1	2	4
PGR	1/2	3	2	1/2	1	2
RC	1/3	1	1/2	1/4	1/2	1

The assigned values are for demonstration purposes only. The normalized weight of each attribute is calculated following the procedure presented in step 4 of Section 3.2.3 and these are: W_C = 0.2613, W_Q = 0.0659, W_R = 0.1371, W_{PWR} = 0.2876, W_{PGR} = 0.1727, and W_{RC} = 0.0754. The value of λ_{max} is 6.1087 and CR = 0.0174. Thus, there is good consistency in the judgements made.

The ideal (best) and negative ideal (worst) solutions are obtained, and these are given as:

V_C^+	=	0.0191	V_C^-	=	0.098
V_Q^+	=	0.0045	V_Q^-	=	0.0317
V_R^+	=	0.0564	V_R^-	=	0.0019
V_{PWR}^+	=	0.1055	V_{PWR}^-	=	0.0186
V_{PGR}^+	=	0.0747	V_{PGR}^-	=	0
V_{RC}^+	=	0.0303	V_{RC}^-	=	0.0021

The separation measures are:

S_1^+	=	0.0934	S_1^-	=	0.1094
S_2^+	=	0.0704	S_2^-	=	0.1110
S_3^+	=	0.1166	S_3^-	=	0.0589
S_4^+	=	0.0984	S_4^-	=	0.0871
S_5^+	=	0.1135	S_5^-	=	0.0542
S_6^+	=	0.0983	S_6^-	=	0.0852
S_7^+	=	0.0865	S_7^-	=	0.1080
S_8^+	=	0.0357	S_8^-	=	0.1297
S_9^+	=	0.0732	S_9^-	=	0.1245
S_{10}^+	=	0.0708	S_{10}^-	=	0.1015
S_{11}^+	=	0.0673	S_{11}^-	=	0.1079
S_{12}^+	=	0.1346	S_{12}^-	=	0.0292
S_{13}^+	=	0.0724	S_{13}^-	=	0.1058
S_{14}^+	=	0.1029	S_{14}^-	=	0.0654
S_{15}^+	=	0.1097	S_{15}^-	=	0.0675

The relative closeness of a particular alternative to the ideal solution is calculated (which can be named the 'ECM program selection index (ECMP-SI)' in the present work) and these are:

$P_1 = 0.5394$, $P_2 = 0.6120$, $P_3 = 0.3355$, $P_4 = 0.4694$, $P_5 = 0.3233$, $P_6 = 0.4644$, $P_7 = 0.5552$, $P_8 = 0.7839$, $P_9 = 0.6298$, $P_{10} = 0.5889$, $P_{11} = 0.6158$, $P_{12} = 0.1784$, $P_{13} = 0.5937$, $P_{14} = 0.3887$, and $P_{15} = 0.3809$.

The alternative ECM programs are arranged in descending order of their ECMP-SI. This can be arranged as 8-9-11-2-13-10-7-1-4-6-14-15-3-5-12.

22.2.5 Modified TOPSIS Method

Following the procedure of the modified TOPSIS method and using the same weights as those selected for the TOPSIS method, the following ECMP-SI values are obtained:

$P_1 = 0.5423$, $P_2 = 0.5617$, $P_3 = 0.3852$, $P_4 = 0.4924$, $P_5 = 0.3429$, $P_6 = 0.4823$, $P_7 = 0.5163$, $P_8 = 0.7074$, $P_9 = 0.6259$, $P_{10} = 0.5727$, $P_{11} = 0.5828$, $P_{12} = 0.2199$, $P_{13} = 0.6185$, $P_{14} = 0.4366$, and $P_{15} = 0.4699$.

The alternative ECM programs are arranged in descending order of their ECMP-SI. This can be arranged as 8-9-13-11-10-2-1-7-4-6-15-14-3-5-12.

It may be observed from the application of the TOPSIS and modified TOPSIS methods, that both methods suggest alternative ECM program 8 as the first choice and 12 as the last choice. One need not be confused when comparing the rankings obtained by using the TOPSIS and modified TOPSIS methods with those presented by other methods such as GTMA, SAW, AHP, *etc.* As mentioned above, the ranking depends upon the judgements of relative importance of attributes made by the decision maker.

References

Cordeiro J, Sarkis J (1997) Environmental proactivism and firm performance: evidence from industry analyst forecasts. Business Strategy and the Environment 6:1–11

Gungor A, Gupta SM (1999) Issues in environmentally conscious manufacturing and product recovery: a survey. Computers & Industrial Engineering 36:811–853

Khan FI, Sadiq R, Husain T (2002) GreenPro-I: a risk-based life-cycle assessment decision-making methodology for process plant design. Environmental Modelling & Software 17:669–692

Klassen RD, McLaughlin CP (1996) The impact of environmental management on firm performance. Management Science 42:1199–1214

Kuo TC, Chang SH, Huang SH (2006) Environmentally conscious design by using fuzzy multi-attribute decision-making. International Journal of Advanced Manufacturing Technology 29:209–215

Madu CN, Kuei C, Ifeanyi E (2002) A hierarchic metric approach for integration of green issues in manufacturing: a paper recycling application. Journal of Environmental Management 64:261–272

Morrow WR, Kim HQI, Mazumder J, Skerlos SJ (2007) Environmental aspects of laser-based and conventional tool and die manufacturing. Journal of Cleaner Production 15:932–943

Munoz AA, Sheng P (1995) An analytical approach for determining the environmental impact of machining processes. Journal of Materials Processing Technology 53:736–758

Presley A, Sarkis J (1994) An activity based strategic justification methodology for ECM technology. International Journal on Environmentally Conscious Design and Manufacturing 3:5–17

Rao RV (2004) Digraph and matrix methods for evaluating environmentally conscious manufacturing programs. International Journal of Environmentally Conscious Design and Manufacturing 12:23–33

Sarkis J (1995a) Manufacturing strategy and environmental consciousness. Technovation 15:79–97

Sarkis J (1995b) Supply chain management and environmentally conscious design and manufacturing. International Journal of Environmentally Conscious Design and Manufacturing 4:43–52

Sarkis J (1998) Evaluating environmentally conscious business practices. European Journal of Operational Research 107:159–174

Sarkis, J (1999) A methodological framework for evaluating environmentally conscious manufacturing programs. Computers & Industrial Engineering 36:793–810

Sarkis J, Lin LA (1994) A general IDEF0 model for the strategic implementation of CIM systems. International Journal of Computer Integrated Manufacturing 7:100–115

Sarkis J, Rasheed A (1995) Greening the manufacturing function. Business Horizons 38:17–27

Sarkis J, Weinrach J (2001) Using data envelopment analysis to evaluate environmentally conscious waste treatment technology. Journal of Cleaner Production 9:417–427

Wilhelm MR, Parsaei HR, Fleischman M, Kaltenhauser A, Watters J (1993) Selection of waste management technologies to implement manufacturing pollution prevention strategies. International Journal of Environmentally Conscious Design and Manufacturing 2:41–47

23

Environmental Impact Assessment of Manufacturing Processes

23.1 Introduction

Recently, there has been a strong move toward environmentally conscious manufacturing with an emphasis on life-cycle assessment. The intention is that life-cycle assessment be integrated into a holistic or systemic approach to product design. Such an approach allows consideration of the total energy expended, the resources used, and the waste created. Hence, enabling the entire environmental impact to be determined and minimized is very important. Currently, design for environment (DFE) and life-cycle assessment (LCA) are the strategies to incorporate environmental concerns into product design and process design. The design of the product, the material selection and the manufacturing method are the critical factors causing impact on the environment.

Munoz and Sheng (1995) presented a model of the environmental impact of machining processes. The analytical model integrated aspects of process mechanics, wear characteristics and lubricant flows. The quantifiable dimensions in the analysis included energy utilization, process rate, work piece primary mass flow, and secondary flow of process catalysts. The generation of multiple waste streams could be compared by examining factors such as toxicity and flammability. The sensitivity of environmental factors to variations in operating parameters such as depth of cut, speed, feed, and tool rake angle were examined. The prioritizations of environmental factors were evaluated for both high-rate transfer line and flexible job shop environment through utility analysis. This model could serve as a framework for decision-making in environmentally conscious manufacturing, including the design of parts for machining, process planning, and the selection of operating parameters.

Choi *et al.* (1997) established an assessment model for manufacturing processes in terms of environmental impact for quantitative evaluation of product design. An assessment methodology was developed on the basis of the 'material balance' of a process, and the relationships among different processes. As a result, the amount of slid waste generated, the energy consumed, waste water incurred, as well as the level of noise were obtained. A case study of the production of a toy train with 12 scenarios was performed to illustrate and examine the assessment model. The aim of the case study was to give an assessment of the product in terms of securing less environmental damage by changing the following: (a) one component of the product being produced by different manufacturing processes, (b) the recycling concept being introduced into the product, and (c) the design of the product being altered.

Hanssen (1998) discussed environmental impacts of product systems from a life-cycle perspective and surveyed five product types based on life-cycle assessment studies. The author summarized the results of 18 LCA studies of a variety of products from Norway and Sweden. The products were categorized into five groups based on functionality and application characteristics, in order to investigate the feasibility of generalization of LCA results. The five product groups were: products that were transformed chemically in their application; stationary, inert products without intrinsic energy consumption in the use phase; stationary, inert products with energy consumption in the use phase; transport products without intrinsic energy consumption in the use phase; and transport products with energy consumption in the use phase. Statistical tests were carried out to analyze the significance of the characterization into product groups, both within the groups and between the groups. It was concluded that there were large differences between product types and life-cycle stages, and that these differences were probably even larger when looking to average European conditions.

Culaba and Purvis (1999) presented a methodology for the life-cycle and sustainability analysis of manufacturing processes. The authors described a general methodology for the life-cycle analysis of manufacturing processes taking into account the flexibility and decision-making potential of knowledge-based systems. Emphasis was placed on on-site waste minimization and associated sustainability characteristics in relation to environmental impact assessment and process improvement. The software model was applied with some success to an initial study of pulp and paper manufacturing.

Karakoussis *et al.* (2001) presented the environmental impact of manufacturing planar and tubular solid oxide fuel cells. The authors examined the environmental impact of manufacturing two types of solid oxide fuel cell (SOFC) systems. The tubular SOFC was based on a 100 kW Siemens–Westinghouse design, and the planar SOFC was based on a 1 kW Sulzer design. Using different levels of details, the environmental impact of the manufacturing of the PEN and interconnect, the balance of the plant, and the production of precursor materials had been assessed for both systems. The results demonstrated that the production and supply of materials for the manufacture of both the balance of plant and the fuel cell were responsible for a significant share of the overall environmental burden associated with each of the fuel cell systems studied. Nonetheless, the total emissions associated with the manufacturing stage contributed an additional 1% to lifetime CO_2 emissions for both fuel cell types. The relative contribution arising from the manufacturing phase relative to several other regulated pollutants was high, but this reflected the low levels associated with the SOFC in use phase, rather than a significant burden arising from manufacture. It was proposed that end-of-life reuse or recycling could play a key role in further reducing environmental burdens.

Ong *et al.* (1999) described the development of a semi-quantitative pre-LCA tool for assessing the environmental impacts of the production of a printer. This tool provides a quick and easy means of assessing the environmental impacts of a complex product. The tool allows a designer to easily compute a total environmental impact value for each of the various alternative designs. The pre-LCA tool computes an environmental impact value by considering factors such as airborne and waterborne emissions, the recyclability potential, waste disposal, the global warming factor, the energy content of raw materials, the divertible plastic waste potential, *etc.* In another work, Ong *et al.* (2001) used the AHP method to derive a single environmental score based on process emissions for each of the products or alternatives evaluated. Based on the environmental scores, the products could be ranked with respect to their environmental merits. The AHP method was incorporated into the pre-LCA tool to assign accurate environmental scores to products. The AHP model developed was applied to a case study, this being a comparative study on polystyrene and porcelain plates. The results showed that the system developed was able to provide sound evaluation.

El-Fadel *et al.* (2001) presented a critical assessment of the Lebanese industrial sector, namely the current status and classification of industrial establishments based on a comparative synthesis and analysis of nationwide surveys and studies pertaining to industrial waste management. Characterization of solid and liquid industrial wastes generated, including hazardous wastes, was presented together with current and projected waste loads, recycling opportunities, and export/import practices. Institutional capacity and needs pertaining to the enforcement of relevant environmental legislation, staffing and resources, monitoring schemes, and public participation were critically evaluated. Finally, realistic options for industrial-waste management were outlined within the context of country-specific institutional economic and technical limitations.

Leão and Pashby (2004) presented a literature survey on the use of dielectric fluids that provide an alternative to hydrocarbon oil. It has been reported that water-based dielectrics might replace oil-based fluids in die sink applications. Gaseous dielectrics such as oxygen might also be an alternative. Nonetheless, these need further research in order to become commercially viable. Donnelley *et al.* (2006) presented a case study of implementing ecodesign principles by means of product-based environmental management system. Socolof *et al.* (2005) presented final impact results from an industry-wide environmental life-cycle assessment of cathode ray tube (CRT) and liquid crystal display (LCD) computer monitors for 20 environmental impact categories. Considering the entire life-cycle of each monitor, water eutrophication and aquatic ecotoxicity impacts for the baseline analysis were greater for the LCD, while all other impact categories (*e.g.*, resource use, energy, ozone depletion, landfill space use, human health toxicity) were greater for the CRT. Comparing the manufacturing stages of each monitor type in the baseline scenario, the LCD had greater relative environmental burdens in eight categories. Energy, global warming, and human health toxicity impacts were also presented in greater detail, showing contributions from each life-cycle stage.

Zackrisson (2005) presented environmental aspects when manufacturing products mainly out of metals and/or polymers. The author examined environmental data from companies manufacturing products mainly from metals and/or polymers. The data were collected in a uniform way by use of special guidelines. Weighting or valuation methods often used in life-cycle assessment served to quantitatively compare and rank environmental aspects. The study suggested that weighting or valuation methods could aid in determining the significance of environmental impacts and aspects within the context of ISO 14001.

English *et al.* (2006) considered how a cold roll forming company could ensure its status as a sustainable and environmentally conscious manufacturer. Hussey and Eagan (2006) presented some insights from structural equation modeling, which was used to evaluate the development of an environmental performance model for small and medium size enterprises (SMEs). The model was based on the Malcolm Baldrige Criteria. The authors reviewed SEM methodology and shared results from a population of SMEs in the plastics manufacturing sector. Fit statistics confirmed the overall model fit, but not all paths in the model were statistically significant. An assessment of the non-significant paths (from leadership and from the system components of the model to environmental results) led the authors to conclude that an improved definition of environmental results was critical. Education of SMEs on the benefits of improved environmental performance was also warranted.

This chapter presents the results of applications of the proposed GTMA and fuzzy MADM methods for the assessment of environmental impact of manufacturing processes.

23.2 Example

Now, to demonstrate and validate the application of the proposed decision-making methods, the case study presented by Choi *et al.* (1997) is considered.

Choi *et al.* (1997) established an assessment model for manufacturing processes in terms of environmental impact for quantitative evaluation of product design. An assessment methodology was developed on the basis of the 'material balance' of a process and the relationships among different processes. As a result, the amount of solid waste generated, the energy consumed, the waste water incurred as well as the level of noise were obtained. A case study of the production of a toy train with 12 scenarios was performed to illustrate and examine the assessment model. The aim of the case study was to give an assessment of the product in terms of securing less environmental damage by changing the following: (a) one component of the product being produced by different manufacturing processes, (b) the recycling concept being introduced into the product, and (c) the design of the product being altered. The product structure is described below along with the related manufacturing processes:

- Engine funnel (two pieces): The product is a plastic one and hence the injection molding process was employed so that machining operations and the EDM process were required for the production of the tooling and the mold set.
- Engine boiler (sheet metal): This was produced by the drawing process and drilling operations. With regard to the former, the machining operation was employed to produce the mold set.

- Engine cabin (zinc): This was produced by the die-casting process and drilling operations. With regard to the former, machining operations and the EDM operation were employed to produce the mold set.
- Steel pin (three pieces, mild steel): This was produced by turning operations.
- Base (zinc): This was produced by the die-casting process and drilling operations. With regard to the former, machining operations and the EDM operation were employed to produce the mold set.
- Bush (six pieces, mild steel): These were produced by turning and drilling operations.
- Wheel (six pieces, mild steel): This was also produced by turning and drilling operations.
- Screw (6 pieces, mild steel): These were produced by turning operations.
- Cover: As this was a plastic product, the injection molding process was employed, so that the machining operation and EDM process were required for the production of the tooling and the mold set.

On the basis of the above information, the assessment of the toy train was carried out for the different scenarios shown in Table 23.1, and the results are shown in Table 23.2.

Table 23.1. Assessment of the toy train for 12 different scenarios (from Choi *et al.* 1997; reprinted with permission from Elsevier)

Scenario	Description of assessment
1	The original toy train model
2	Based on '1', but the production method of the wheel was changed from machining to die casting
3	Based on '1', but the production method of the wheel was changed from machining to injection molding
4	Based on '1', all bushes and screws were removed and three shafts were added
5	Based on '4'. but the production method of the wheel was changed from machining to die casting
6	Based on '4', but the production method of the wheel was changed from machining to injection molding
7	Based on '1', the concept of recycling was introduced. This meant if any material can be reused or recycled, then no solid waste was involved
8	Based on '7', but the production method of the wheel was changed from machining to die casting
9	Based on 7, but the production method of the wheel was changed from machining to injection molding
10	Based on '4', the concept of recycling was introduced. This meant if any material can be reused or recycled, then no solid waste was involved.
11	Based on '10', but the production method of the wheel was changed from machining to die casting
12	Based on '10', but the production method of the wheel was changed from machining to injection molding

Table 23.2. Assessment results for the toy train for 12 different scenarios (from Choi *et al.*, 1997; reprinted with permission from Elsevier)

Scenario	Solid waste (mm^3)	Energy consumption (kWh)	Waste water (m^3)
1	476743	1.252	0.879
2	458449	0.892	0.524
3	473311	0.892	0.524
4	457146	0.815	0.526
5	438852	0.455	0.172
6	453714	0.455	0.172
7	51023	1.252	0.879
8	30593	0.892	0.525
9	30617	0.892	0.525
10	31426	0.815	0.526
11	10996	0.455	0.172
12	11020	0.455	0.172

23.2.1 Graph Theory and Matrix Approach (GTMA)

Various steps of the methodology are carried out as described below:

In the present work, the attributes considered are the same as those of Choi *et al.* (1997), and these are: solid waste (SR), energy consumption (EC), waste water (WW), and noise (N). However, the values of the noise attribute are equal for all the alternative scenarios. Hence this attribute is not considered in the present method. The quantitative values of the attributes, which are given in Table 23.2, are to be normalized. All three attributes considered are non-beneficial, and lower values are desirable. Values of these attributes are normalized, and are given in Table 23.3 in the respective columns.

Table 23.3. Normalized assessment results for the toy train for 12 different scenarios

Scenario	Solid waste (mm^3)	Energy consumption (kWh)	Waste water (m^3)
1	0.02306	0.3634	0.1957
2	0.024	0.51	0.3282
3	0.0232	0.51	0.3282
4	0.02405	0.5583	0.3270
5	0.02505	1	1
6	0.02423	1	1
7	0.2155	0.3634	0.1957
8	0.3594	0.51	0.3276
9	0.3591	0.51	0.3276
10	0.3499	0.5583	0.3270
11	1	1	1
12	0.9978	1	1

Let the decision maker make the following relative importance assignments:

$$
\begin{array}{c} \\ SW \\ EC \\ WW \end{array}
\begin{array}{c}
\begin{array}{ccc} SW & EC & WW \end{array} \\
\begin{bmatrix} - & 0.59 & 0.665 \\ 0.41 & - & 0.59 \\ 0.335 & 0.41 & - \end{bmatrix}
\end{array}
$$

In this example, SW is given higher importance, and EC is given high importance.

The environmental impact assessment attributes digraph, environmental impact assessment attributes matrix of the digraph, and environmental impact assessment function for the matrix can be prepared. The value of the environmental impact assessment index (EIAI) is calculated using the values of A_i and a_{ij} for each scenario. The environmental impact assessment index values of different scenarios are given below in descending order:

Scenario 11	:	1.9350
Scenario 12	:	1.9322
Scenario 5	:	0.7242
Scenario 6	:	0.7232
Scenario 10	:	0.5804
Scenario 8	:	0.5682
Scenario 9	:	0.5681
Scenario 4	:	0.4421
Scenario 2	:	0.4312
Scenario 3	:	0.4309
Scenario 7	:	0.4241
Scenario 1	:	0.3639

Thus, GTMA suggests scenario 11 as the most environmentally friendly, followed by scenario 12; scenario 1 is least environmentally friendly. Choi *et al.* (1997) also suggested scenario 11 as the first, scenario 12 as the second, and scenario 1 as the last choice from an environmental impact point of view.

23.2.2 AHP Method

Let the decision maker prepare the following relative importance matrix:

$$
\begin{array}{c} \\ SW \\ EC \\ WW \end{array}
\begin{array}{c}
\begin{array}{ccc} SW & EC & WW \end{array} \\
\begin{bmatrix} 1 & 2 & 3 \\ 1/2 & 1 & 2 \\ 1/3 & 1/2 & 1 \end{bmatrix}
\end{array}
$$

In this example, SW is given higher importance, and EC is given high importance. The normalized weights of each attribute are calculated, and these are: $W_{SW} = 0.5396$, $W_{EC} = 0.2969$, and $W_{WW} = 0.1635$. The value of λ_{max} is 3.0092 and CR = 0.00793 which is much less than the allowed CR value of 0.1. Thus, there is good consistency in the judgements made.

The value of EIAI is now calculated using the above weights, and the normalized data of the attributes given in Table 23.3. This leads to the ranking given by the revised AHP or ideal mode of AHP. The alternative scenarios are arranged in descending order of the EIAI:

Scenario 11	:	1.0000
Scenario 12	:	0.9988
Scenario 5	:	0.4735
Scenario 6	:	0.4731
Scenario 10	:	0.4088
Scenario 8	:	0.3995
Scenario 9	:	0.3993
Scenario 7	:	0.2567
Scenario 4	:	0.2328
Scenario 2	:	0.2185
Scenario 3	:	0.2180
Scenario 1	:	0.1582

Thus, the AHP method also suggests scenario 11 as the most environmentally friendly, followed by scenario 12; scenario 1 is least environmentally friendly. It may be noted that the ranking depends upon the judgements of relative importance of attributes made by the decision maker.

23.2.3 TOPSIS Method

Various steps of the TOPSIS methodology using the AHP method for assigning the relative importance of attributes are described below:

Step 1: The objective is to evaluate the environmental friendliness of different manufacturing scenarios. The attributes considered are the same as those of Choi *et al.* (1997), and these are: solid waste (SR), energy consumption (EC), and waste water (WW).

Step 2: The next step is to represent all the information available on attributes in the form of a decision matrix. The data are shown in Table 23.2.

Step 3: The quantitative values of the environmental impact assessment attributes, which are given in Table 23.2, are to be normalized. All three attributes are of non-beneficial type. The values of these attributes for different vendors are normalized but not shown here.

Step 4: Let the decision maker select the following assignments regarding the relative importance of attributes (a_{ij}), using AHP method:

$$\begin{array}{c} \\ SW \\ EC \\ WW \end{array} \begin{array}{c} \begin{array}{ccc} SW & EC & WW \end{array} \\ \begin{bmatrix} 1 & 2 & 3 \\ 1/2 & 1 & 2 \\ 1/3 & 1/2 & 1 \end{bmatrix} \end{array}$$

The normalized weights of each attribute are calculated, and these are: W_{SW} = 0.5396, W_{EC} = 0.2969, and W_{WW} = 0.1635. The value of λ_{max} is 3.0092 and CR = 0.00793, which is much less than the allowed CR value of 0.1.

Step 5: The weighted normalized matrix is calculated, and is shown below:

$$\begin{bmatrix} 0.2279 & 0.1277 & 0.0789 \\ 0.2191 & 0.0909 & 0.0470 \\ 0.2262 & 0.0909 & 0.0470 \\ 0.2185 & 0.0831 & 0.0472 \\ 0.2098 & 0.0464 & 0.0154 \\ 0.2168 & 0.0464 & 0.0154 \\ 0.0244 & 0.1277 & 0.0789 \\ 0.0146 & 0.0909 & 0.0471 \\ 0.0146 & 0.0909 & 0.0471 \\ 0.0150 & 0.0831 & 0.0472 \\ 0.0053 & 0.0464 & 0.0154 \\ 0.0053 & 0.0464 & 0.0154 \end{bmatrix}$$

Step 6: The next step is to obtain the ideal (best) and negative ideal (worst) solutions. These are given as:

$V_1^+ = 0.00526$ $\quad$ $V_1^- = 0.2279$

$V_2^+ = 0.0464$ $\quad$ $V_2^- = 0.1277$

$V_3^+ = 0.0154$ $\quad$ $V_3^- = 0.0789$

Step 7: The next step is to obtain the separation measures, and these are:

$S_1^+ = 0.2453$ $\quad$ $S_1^- = 0$

$S_2^+ = 0.2207$ $\quad$ $S_2^- = 0.0494$

$S_3^+ = 0.2276$ $\quad$ $S_3^- = 0.0486$

$S_4^+ = 0.2187$ $\quad$ $S_4^- = 0.0555$

$S_5^+ = 0.2045$ $\quad$ $S_5^- = 0.1047$

$S_6^+ = 0.2116$ $\quad$ $S_6^- = 0.1037$

$S_7^+ = 0.1049$ $\quad$ $S_7^- = 0.2035$

$S_8^+ = 0.0555$ $\quad$ $S_8^- = 0.2187$

$S_9^+ = 0.0555$ $\quad$ $S_9^- = 0.2187$

$S_{10}^+ = 0.0495$ $\quad$ $S_{10}^- = 0.2198$

$S_{11}^+ = 0$ $\quad$ $S_{11}^- = 0.2453$

$S_{12}^+ = 0.000011$ $\quad$ $S_{12}^- = 0.2453$

Step 8: The relative closeness of a particular alternative to the ideal solution is calculated, and these are:

$P_1 = 0$ $\quad$ $P_2 = 0.18285$ $\quad$ $P_3 = 0.1761$ $\quad$ $P_4 = 0.20235$

$P_5 = 0.33862$ $\quad$ $P_6 = 0.32891$ $\quad$ $P_7 = 0.65987$ $\quad$ $P_8 = 0.79764$

$P_9 = 0.79763$ $\quad$ $P_{10} = 0.8161$ $\quad$ $P_{11} = 1.0000$ $\quad$ $P_{12} = 0.9999$

This relative closeness to ideal solution is named the 'environmental impact assessment index (EIAI)' in the present example.

Step 9: The scenarios are arranged in descending order of their EIAI. This can be arranged as 11-12-10-8-9-7-5-6-4-2-3-1.

From the above values of EIAI, it is understood that scenarios 11, 12, and 10 are more environmentally friendly, and scenario 1 is the least environmentally friendly.

23.2.4 Modified TOPSIS Method

Using the same weights of attributes as those of the AHP and TOPSIS methods, the modified TOPSIS method leads to the following ranking order:

Scenario 11	:	1.0000
Scenario 12	:	0.9999
Scenario 10	:	0.7488
Scenario 8	:	0.7299
Scenario 9	:	0.7299
Scenario 7	:	0.5595
Scenario 5	:	0.4391
Scenario 6	:	0.4297
Scenario 4	:	0.2700
Scenario 2	:	0.2505
Scenario 3	:	0.2439
Scenario 1	:	0.0000

Thus, the modified TOPSIS method suggests the same ranking as that proposed by the simple TOPSIS method.

Once the product has been identified, all of the related manufacturing processes can be determined. The consequences of each process, including the amount of solid waste, energy consumption, waste water generated, *etc.*, can be obtained, and the environmental impact of each manufacturing process can be assessed using the decision-making methods described in this chapter.

References

Choi ACK, Kaebernick H, Lai WH (1997) Manufacturing processes modelling for environmental impact assessment. Journal of Materials Processing Technology 70:231–238

Culaba AB, Purvis MRI (1999) A methodology for the life-cycle and sustainability analysis of manufacturing processes. Journal of Cleaner Production 7:435–445

Donnelly K, Zoe BF, Traeger S, Okrasinski T, Holman S (2006) Eco-design implemented through a product-based environmental management system. Journal of Cleaner Production 14:1357–1367

El-Fadel M, Zeinati M, El-Jisr K, Jamali D (2001) Industrial-waste management in developing countries: The case of Lebanon. Journal of Environmental Management 61:281–300

English M, Castellucci M, Mynors DJ (2006) Eco-efficiency of the cold roll formed product supply chain. Journal of Materials Processing Technology 177:626–629

Hanssen OJ (1998) Environmental impacts of product systems in a life-cycle perspective: a survey of five product types based on life-cycle assessments studies. Journal of Cleaner Production 6:299–311

Hussey DM, Eagan PD (2006) Using structural equation modeling to test environmental performance in small and medium-sized manufacturers: can

SEM help SMEs? Journal of Cleaner Production doi:10.1016/j.jclepro.2005.12.002

Karakoussis V, Brandon NP, Leach M, Vorst RV (2001) The environmental impact of manufacturing planar and tubular solid oxide fuel cells. Journal of Power Sources 101:10–26

Leão FN, Pashby IR (2004) A review on the use of environmentally-friendly dielectric fluids in electrical discharge machining. Journal of Materials Processing Technology 149:341–346

Munoz AA, Sheng P (1995) An analytical approach for determining the environmental impact of machining processes. Journal of Materials Processing Technology 53:736–758

Ong SK, Koh TH, Nee AYC (1999) Development of a semi-quantitative pre-LCA tool. Journal of Materials Processing Technology 89090:574–582

Ong SK, Koh TH, Nee AYC (2001) Assessing the environmental impact of materials processing techniques using an analytical hierarchy process method. Journal of Materials Processing Technology 113:424–431

Socolof ML, Overly JG, Geibig JR (2005) Environmental life-cycle impacts of CRT and LCD desktop computer displays. Journal of Cleaner Production 13:1281–1294

Zackrisson M (2005) Environmental aspects when manufacturing products mainly out of metals and/or polymers. Journal of Cleaner Production 13:43–49

24

Evaluation of Aggregate Risk in Green Manufacturing

24.1 Introduction

There is a growing interest in green manufacturing (also called environmentally conscious manufacturing). The current focus on green manufacturing is different from the traditional focus on pollution control. Here, the emphasis is on life-cycle assessment. Products or processes are seen as interacting with the environment, and could have chain reaction effects on environmental pollution. Thus, rather than looking at any product or process in isolation, the manufacturer needs to adopt a cradle-to-grave approach for the product or process. For example, how much energy is expended in unit product manufacturing, how much resources are used, how much waste is created, and what are the product requirements for transportation and distribution? These are not issues that product designers are accustomed to considering. Their traditional role has been to look at the product on its own, and design products that meet specific guidelines and that may become environmental pollution laws. Today's focus is different. Manufacturers must take a product stewardship approach, and this will predict their survival in today's competitive environment (Madu *et al.*, 2002).

Industrial economies have generated a tremendous amount of waste that is often not reused or properly disposed of. Industrial societies are increasingly faced with the problems of hazardous waste management, locating new landfills, and the depletion of raw materials. Rather than continuing with this cycle of waste and extravagance, industrial economies should find better ways to convert wastes from one industry into input in another industry. This implies interdependence between industries, where one industry's output could become another's input. This cycle of dependence or reuse of material is generally referred to as recycling, and its goal is to eliminate or reduce waste.

The manufacturing industries must seek to minimize environmental impact and resource consumption during the entire product cycle. Industrial risk and the diversification of risk types have both increased with industrial development. At the same time, the risk acceptability threshold of the population has decreased. In response, industry has developed methodologies for risk prevention and protection (Tixer *et al.*, 2002). Green manufacturing was first proposed about 10 years ago, so there are only few examples that can be used to evaluate risks, and many uncertain factors. Because of this incomplete and uncertain knowledge, decision-making methods based on probabilities to represent risk, which need many examples, cannot be used for green manufacturing projects (Hua *et al.*, 2005). In addition, green manufacturing involves a very wide range of topics, such as environmental consciousness, life-cycle thinking, and sustainable development, which all increase the risk. Therefore, risk decision-making in green manufacturing projects must consider multiple indicators. Hua *et al.* (2005) reported that industries are implementing green manufacturing projects for sustainable production for four types of risk categories: technological, organizational, financial and circumstantial. Each category is related to certain risk factors. These risks are described below:

- Technological risk - Since the concept of green manufacturing is relatively new, its theories and technologies are still being developed. Only experience will show whether, or not, each technology can be used in green manufacturing projects to create extended benefits for industry, society, and in ecology. Therefore, there are many technological risk factors, including reliability, maintenance, and applicability.
- Organizational risk - Green manufacturing is a new manufacturing mode with the product cycle extending to the entire product life (raw materials, production, use, recycling, and disposal), so traditional management methods are not suitable. Therefore, the management system must be reformed to successfully implement green manufacturing, which will lead to unpredictable risks. The main organizational risk factors are the integration of the management approach, the knowledge level of the lead group, and the knowledge level of the personnel.
- Financial risk - Green manufacturing projects require a very long investment period due to the length of the entire product cycle, which increases the risk. Corporate income is gained by saving energy and materials, protecting the environment and workers, improving productivity and product quality, reducing costs, and by accurate market timing.
- Circumstantial risk - Green manufacturing projects are constrained not only by internal resources, but also by external resources. Many uncertain circumstantial factors can cause critical risks. Such external factors include laws, regulations, macro-economic changes, and industrial development.

Hua *et al.* (2005) applied a fuzzy multiple attribute decision-making (FMADM) method for evaluating aggregate risk in green manufacturing. The authors presented a case study of a refrigerator company that made a strategic plan to implement ISO14000 series standards to achieve sustainable green manufacturing production. The standards were divided into three phases: ISO14001 certification, life-cycle assessment, and environmental labeling. For the ISO14001 certification phase, decision makers analyzed three scenarios for implementing ISO14001 certification. The first project obtains ISO9001 certification and then ISO14001 certification. The second project obtains ISO9001 certification, and ISO14001 certification at the same time. The third project obtains ISO14001 certification and then ISO9001 certification. Different risk grades were assigned by Hua *et al.* (2005) to various risk items described above. The FMADM method was then used to evaluate the aggregate risk in implementing each of these three scenarios. It was concluded by the authors that the third scenario was associated with the highest risk, and the first scenario had the lowest risk. However, even though the method used by the authors is efficient, it is complicated and requires more computation. To simplify the procedure, different risk grades assigned by Hua *et al.* (2005) to various risk items are rewritten "using" Table 4.3, and are given in Table 24.1. Table 24.2 shows the corresponding objective data obtained by using Table 4.3.

Table 24.1. Subjective risk grades of the three scenarios (from Hua *et al.*, 2005; with permission from Elsevier)

Attributes (risk factor)	Scenario 1	Scenario 2	Scenario 3
Technological risk:			
Reliability	AA	H	L
Development	VL	A	H
Maintenance	H	A	AA
Applicability	L	AA	H
Organizational risk:			
Integration	L	L	L
Knowledge level of lead group	A	AA	A
Knowledge level of personnel	BA	VH	EH
Financial risk:			
Investment period	AA	A	H
Investment capital	L	VL	AA
Corporate income	A	EL	VL
Circumstantial risk:			
Government	EL	XL	XL
Market	H	A	XL

XL: exceptionally low, EL: extremely low, VL: very low, L: low, BA: below average, A: average, AA: above average, H: high, VH: very high, EH: extremely high

Table 24.2. Objective risk grades of the three scenarios

Attributes (risk factor)	Scenario 1	Scenario 2	Scenario 3
Reliability	0.59	0.665	0.335
Development	0.255	0.5	0.665
Maintenance	0.665	0.5	0.59
Applicability	0.335	0.59	0.665
Integration	0.335	0.335	0.335
Knowledge level of lead group	0.5	0.59	0.5
Knowledge level of personnel	0.41	0.745	0.865
Investment period	0.59	0.5	0.665
Investment capital	0.335	0.255	0.59
Corporate income	0.5	0.135	0.255
Government	0.135	0.045	0.045
Market	0.665	0.5	0.135

24.2 Example

Now, to demonstrate and validate the application of the proposed decision-making methods, the case study presented by Hua *et al.* (2005) is considered here. For a start, GTMA is applied, and subsequently a few MADM methods are applied to evaluate the aggregate risk in green manufacturing.

24.2.1 Graph Theory and Matrix Approach (GTMA)

In the present work, the attributes considered are the same as those of Hua *et al.* (2005), and these are: reliability (R), development (D), maintenance (M), applicability (A), integration (I), knowledge level of lead group (K), knowledge level of personnel (P), investment period (V), investment capital (C), corporate income (E), government (G) and market (T). As all these attributes represent different risks, lower values are preferred. The objective values of these attributes are normalized, as given in Table 24.3.

Table 24.3. Normalized data of the three scenarios

Attributes (risk factor)	Scenario 1	Scenario 2	Scenario 3
Reliability	0.5678	0.5037	1
Development	1	0.51	0.3834
Maintenance	0.7519	1	0.8474
Applicability	1	0.5678	0.5038
Integration	1	1	1
Knowledge level of lead group	1	0.8474	1
Knowledge level of personnel	1	0.5503	0.4739
Investment period	0.8474	1	0.7519
Investment capital	0.7612	1	0.4322
Corporate income	0.27	1	0.5294
Government	0.3333	1	1
Market	0.2030	0.27	1

Let the decision maker make the following assignments of relative importance:

	R	D	M	A	I	K	P	V	C	E	G	T
R	-	0.255	0.410	0.335	0.745	0.500	0.410	0.745	0.665	0.500	0.500	0.665
D	0.745	-	0.665	0.590	0.955	0.745	0.665	0.955	0.865	0.745	0.745	0.865
M	0.590	0.335	-	0.410	0.865	0.590	0.500	0.665	0.665	0.590	0.590	0.665
A	0.665	0.410	0.590	-	0.865	0.665	0.590	0.745	0.745	0.665	0.665	0.745
I	0.255	0.045	0.135	0.135	-	0.255	0.135	0.500	0.500	0.255	0.255	0.500
K	0.500	0.255	0.410	0.335	0.745	-	0.335	0.665	0.665	0.500	0.500	0.665
P	0.590	0.335	0.500	0.410	0.865	0.665	-	0.865	0.865	0.665	0.665	0.865
V	0.255	0.045	0.335	0.255	0.500	0.335	0.135	-	0.410	0.255	0.255	0.410
C	0.335	0.135	0.335	0.255	0.500	0.335	0.135	0.590	-	0.255	0.255	0.410
E	0.500	0.255	0.410	0.335	0.745	0.500	0.335	0.745	0.745	-	0.500	0.665
G	0.500	0.255	0.410	0.335	0.745	0.500	0.335	0.745	0.745	0.500	-	0.500
T	0.335	0.135	0.335	0.255	0.500	0.335	0.135	0.590	0.590	0.335	0.335	-

The aggregate risk attributes digraph, aggregate risk attributes matrix of the digraph, and aggregate risk function for the matrix can be prepared. The value of the aggregate risk(less) index is calculated using the values of A_i and a_{ij} for each scenario. The aggregate risk(less) index values of different scenarios are given below in descending order:

Scenario 2 : 64713.2933
Scenario 1 : 61279.2662
Scenario 3 : 60057.4801

From the above values of the aggregate risk(less) index, scenario 3 is understood to have the highest aggregate risk (*i.e.*, lowest aggregate risk(less) index). Scenario 2 is considered to have the lowest aggregate risk for the values of relative importance among the attributes considered here.

24.2.2 AHP Method

Let the decision maker prepare the following pair-wise comparison matrix:

	R	D	M	A	I	K	P	V	C	E	G	T
R	1	1/5	1/3	1/4	3	1	1/3	3	2	1	1	2
D	5	1	3	2	7	5	3	7	6	5	5	6
M	3	1/3	1	1/2	5	3	1	5	4	3	3	4
A	4	1/2	2	1	6	4	2	6	5	4	4	5
I	1/3	1/7	1/5	1/6	1	1/3	1/5	1	1/2	1/3	1/3	1/2
K	1	1/5	1/3	1/4	3	1	1/3	3	2	1	1	2
P	3	1/3	1	1/2	5	3	1	5	4	3	3	4
V	1/3	1/7	1/5	1/6	1	1/3	1/5	1	1/2	1/3	1/3	1/2
C	1/2	1/6	1/4	1/5	2	1/2	1/4	2	1	1/2	1/2	1
E	1	1/5	1/3	1/4	3	1	1/3	3	2	1	1	2
G	1	1/5	1/3	1/4	3	1	1/3	3	2	1	1	2
T	1/2	1/6	1/4	1/5	2	1/2	1/4	2	1	1/2	1/2	1

The normalized weights of each attribute are calculated following the procedure presented in Section 3.2.3, and these are: $W_R = 0.0534$, $W_D = 0.2478$, $W_M = 0.1250$, $W_A = 0.1811$, $W_I = 0.0211$, $W_K = 0.0534$, $W_P = 0.1250$, $W_V = 0.0211$, $W_C = 0.0325$, $W_E = 0.0534$, $W_G = 0.0534$, and $W_T = 0.0325$. The value of λ_{max} is 12.2678 and CR = 0.01645 which is much less than the allowed CR value of 0.1. Thus, there is good consistency in the judgements made.

The value of the aggregate risk(less) index is now calculated using the above weights and the normalized data of the attributes given in Table 24.3. The alternative scenarios are arranged in the descending order of the aggregate riskless index:

Scenario 1: 0.8379
Scenario 2: 0.6819
Scenario 3: 0.6228

From the above values of aggregate risk(less) index, scenario 3 is understood to have the highest aggregate risk (*i.e.*, lowest aggregate risk(less) index). Scenario 1 is considered to have lowest aggregate risk for the values of relative importance among the attributes considered here. These results match well with those obtained by Hua *et al.* (2005) using the FMADM method.

24.2.3 TOPSIS Method

The weights selected for the AHP method are used in this method. The ideal (best) and negative ideal (worst) solutions are obtained, and these are given as:

$V_R^+ = 0.0188$ $V_R^- = 0.0374$
$V_D^+ = 0.0726$ $V_D^- = 0.1894$
$V_M^+ = 0.0613$ $V_M^- = 0.0815$
$V_A^+ = 0.0639$ $V_A^- = 0.1268$
$V_I^+ = 0.0122$ $V_I^- = 0.0122$
$V_K^+ = 0.0290$ $V_K^- = 0.0342$
$V_P^+ = 0.0423$ $V_P^- = 0.0892$
$V_V^+ = 0.0104$ $V_V^- = 0.0138$
$V_C^+ = 0.0114$ $V_C^- = 0.0265$
$V_E^+ = 0.0125$ $V_E^- = 0.0463$
$V_G^+ = 0.0161$ $V_G^- = 0.0483$
$V_T^+ = 0.0052$ $V_T^- = 0.0257$

The separation measures are:

$S_1^+ = 0.0568$ $S_1^- = 0.1413$
$S_2^+ = 0.0948$ $S_2^- = 0.0737$
$S_3^+ = 0.1424$ $S_3^- = 0.0493$

The relative closeness of a particular alternative to the ideal solution is calculated (which can be named as ‘aggregate risk(less) index (ARI)’ in the present work) and these are arranged in descending order as:

Scenario 1: 0.7132
Scenario 2: 0.4373
Scenario 3: 0.2570

As in the case of the AHP method, TOPSIS method also suggests scenario 3 as having the highest aggregate risk, and scenario 1 as having the lowest aggregate risk.

24.2.4 Modified TOPSIS Method

For the same weights as those used in the AHP method, the modified TOPSIS method gives the following ranking order:

Scenario 1: 0.5604
Scenario 2: 0.5177
Scenario 3: 0.4069

The modified TOPSIS method also suggests scenario 3 as having the highest aggregate risk, and scenario 1 as having the lowest aggregate risk.

In this particular example of evaluating aggregate risk in green manufacturing, proposing scenario 1 as the least risk scenario, and scenario 3 as the highest risk scenario seems to be more logical and genuine. However, it may be noted that the weights of importance assigned to the attributes play an important role in the selection process.

References

Hua L, Weiping C, Zhixin K, Tungwai N, Yuanyuan L (2005) Fuzzy multiple attribute decision-making for evaluating aggregate risk in green manufacturing. Tsinghua Science & Technology 10:627–632

Madu CN, Kuei C, Ifeanyi E (2002) A hierarchic metric approach for integration of green issues in manufacturing: a paper recycling application. Journal of Environmental Management 64:261–272

Tixer J, Dusserre G, Salvi O (2002) Review of 62 risk analysis methodologies of industrial plants. Journal of Loss Prevention in the Process Industries 15:291–303

25

Selection of Best Product End-of-Life Scenario

25.1 Introduction

Since the last two decades, dramatic changes are taking place in our vision of the factory of the future. This vision was created as a way of satisfying the market demands for shorter lead times, shorter and precise delivery times, just-in-time production, flexibility in product variants, *etc.*, ensuring a better global competitiveness. A new vision of the manufacturing enterprise is characterized by the following (Sohlenius, 1989; Alting and Zhang, 1991; Alting 1993):

- A factory with human beings as decision makers, supported by computer integration.
- Concurrent engineering approaches, that is, the major functional tasks are carried out simultaneously.
- Company organizations are loosened up to become more dynamic and supporting concurrently.
- Product design is performed as a life-cycle design; that is, all life-cycle phases (design, production, distribution, usage, and disposal/recycling) are considered from the beginning at the conceptual stage to ensure fulfillment of the environmental requirements.

Azapagic (1999) reviewed some of the emerging applications of life-cycle assessment (LCA). A number of case studies indicated that the process selection must be based on considerations of the environment as a whole, including indirect releases, consumption of raw materials, and waste disposal. This approach goes beyond the practice of choosing the best practicable environmental option (BPEO), by which it is possible to reduce the environmental impacts directly from the plant, but also to increase them elsewhere in the life-cycle. These issues were discussed and demonstrated by examples of end-of-pipe abatement techniques for SO_2, NO_x, and VOCs, and processes for the production of liquid CO_2 and O_2. The integration of LCA into the early stages of process design, and optimization were also reviewed and discussed. The approach was outlined and illustrated with real case studies related to the mineral and chemical industries. It was shown that the life-cycle process design (LCPD) tool offers a potential for technological innovation in process concept and structure through the selection of best material and process alternatives over the whole life-cycle.

Zussman *et al.* (1994) applied multiple objective utility theory to support product design for EOL. Marttunen and Hamalainen (1997) described LCA methodology and application and discussed how the integration of decision analysis and LCA could improve LCA as a tool for decision-making. Sarkis and Weinrach (2001) used the data envelopment analysis method to evaluate environmentally conscious waste treatment technologies. Nagel *et al.* (1999) argued that in the near future, the original equipment manufacturers in many countries will be financially and organizationally responsible for the take-back of their products when these products reach the end of their life-cycle. This follows the principle of extended producer responsibility, according to which producers should be responsible for the entire life-cycle of their products, and especially for take-back, the recycling and the final disposal of their products. Hence, choice of the most appropriate scenario for treating products at their end of life should not only be based on economic considerations but should also take into account environmental and social aspects in order to ensure compliance with legislation, and the satisfaction of customers.

Adda *et al.* (2002) pointed out that the problem of selecting a good scenario for treating products at their end of life concerns different types of users, such as authorities, recycling companies, re-manufacturers, and original equipment manufacturers. Each user has his or her own objectives and priorities, and a good scenario for one user is not necessarily good for another. Even for the same set of EOL scenarios and the same family of criteria, the weight of a criterion may differ from one user to another and within a criterion, a given score of an EOL scenario has not necessarily the same importance for all users. Hence, the type of user will play a key role in this respect (Bufardi *et al.*, 2003).

Khan *et al.* (2002) proposed an integrated methodology, 'GreenPro-I, for process/product design by combining the traditional LCA approach with multiple criteria decision-making methods. The methodology was applicable at the early design stage and was robust against uncertainty in the data.

Keoleian and Kar (2003) demonstrated the life-cycle design (LCD) framework for enhancing design analysis and decision-making through a collaborative effort between the University of Michigan, a cross functional team at Ford and the US Environmental Protection Agency. The LCD framework was used to evaluate three air intake manifold designs: a sand cast aluminum, brazed aluminum tubular, and nylon composite. Life-cycle inventory, life-cycle cost and product/process performance analyses highlighted significant tradeoffs among alternative manifolds, with respect to system design requirements. The life-cycle cost analysis estimated Ford manufacturing costs, customer gasoline costs, and end-of-life management costs. A total of 20 performance requirements were used to evaluate each design alternative.

Desai and Mittal (2003) presented a comprehensive methodology to enhance disassemblability of products. Disassemblability of a product was expressed as a function of several parameters, such as exertion of manual force for disassembly, degree of precision required for effective tool placement, weight, size, material and shape of components being disassembled, use of hand tools, *etc.* A systematic methodology to incorporate disassembly considerations into product design, and enable quantitative evaluation of the design was presented.

Arena *et al.* (2003) used life-cycle assessment to investigate the environmental performance of alternative solid waste management options that could be used in an area in the south of Italy suffering from a situation of weighty solid waste emergency. The extreme delicacy of the decision-making process to which the results had to contribute suggested increasing the reliability of the assessment conclusions by using high-quality data, and a deepened analysis of technical processes. An analytical comparison between three selected scenarios was reported with reference to some crucial environmental impact categories. The results quantified the relative advantages and disadvantages of different management schemes, and suggested some possible improvements in design and operating criteria.

The treatment of the end-of-life phase of the life-cycle of a product is raising ever interest from producers, consumers and authorities. The amount of worn-out products generated each year is increasing, and landfills are becoming saturated, while their expansion is not always possible. In addition to the problem of finding landfills to dispose of the huge volume of these products, the problem arises of addressing the hazardous nature of some of their components. Hence, alternative options to landfilling should be taken into account (Bufardi *et al.*, 2003, 2004).

Bufardi *et al.* (2004) argued that the economic, environmental and social impacts of a product during its life-cycle also depend on the way the product is treated. To achieve this goal, the most important EOL alternatives should be considered and compared on the basis of their performances with respect to relevant criteria and the preferences of the user, in order to select the best compromise EOL alternative. Byggeth and Hochschorner (2006) analyzed 15 different ecodesign tools to ascertain whether a valuation was included in the tools, in what way the tools give support in different types of tradeoff situations and whether the tools provide support from a sustainability perspective. Nine of the 15 tools analyzed included a valuation, and were able to provide support in a tradeoff situation, but the support was not sufficient.

Very few studies are available in the literature dealing specifically with selecting the best product end-of-life scenarios. There is a need for a simple, systematic, and logical scientific method, or mathematical tool, to guide user organizations in taking a proper EOL scenario selection decision. The objective of an EOL scenario selection procedure is to identify the EOL scenario selection attributes, and obtain the most appropriate combination of EOL scenario selection attributes in conjunction with the real requirement. Thus, efforts need to be extended to determine attributes that influence EOL scenario selection, using a simple logical approach to eliminate unsuitable EOL scenarios, and for selection of a proper EOL scenario to strengthen the existing EOL scenario selection procedure.

Now, an example problem presented by Bufardi *et al.* (2003) is considered to demonstrate the applicability of GTMA and fuzzy MADM methods to the selection of best product end-of-life scenarios.

25.1 Example

Bufardi *et al.* (2003) presented an illustrative example of selecting a best product end-of-life scenario using ELECTRE-III method. The product considered by the authors was a telephone with various elements including components, functional components and subassemblies (handset and its components, base and its components, mainboard, buzzer speaker, buzzer case, keys, silicon contacts, screws and cables). Possible EOL options associated with these elements were: functional reclamation (FNC), remanufacturing/reuse (REM), recycling (REC), incineration with energy recovery (INC1), incineration without energy recovery (INC2), and disposal to landfill (LND). Different elements had different EOL options. The authors defined five EOL scenarios by combining elements of the telephone and EOL options. The first EOL scenario suggested that the whole product should be disposed to landfill. However, due to legislation restrictions, that scenario was not possible and, hence, was eliminated. The second scenario suggested REC for certain elements, and LND for the remaining elements of the telephone. The third scenario suggested INC1 for most of the elements and LND for a few elements of the telephone. The fourth scenario suggested REC for most of the elements, and INC1 for the remaining elements of the telephone. The fifth scenario suggested REM, REC, LND, and FNC for different elements of the telephone.

The attributes considered for the evaluation of EOL scenarios were categorized into economic, social and environmental categories. The attributes considered under the economic category were: logistics cost, disassembly cost, product value and product cost. The attributes considered under the social category were: number of employees, and exposure to hazardous materials. The attributes considered under the environmental category were: CO_2 emissions, SO_2 emissions, and energy consumption. Table 25.1 presents the EOL scenarios and the attributes data.

Table 25.1. Data of the EOL scenarios on considered attributes (from Bufardi *et al.*, 2003; with permission from Taylor & Francis Ltd., http:www.tandf.co.uk/journals)

Attributes	S2	S3	S4	S5
CO_2 emissions (kg)	0.12	0.15	0.13	0.10
SO_2 emissions (kg)	0.23	0.45	0.32	0.22
Energy consumption (kWh)	0.65	0.96	0.87	0.98
Logistics cost (Euro)	0.34	0.25	0.28	0.30
Disassembly cost (Euro)	0.42	0.44	0.43	0.45
Product value (Euro)	0.21	0.12	0.16	0.26
Product cost (Euro)	0.25	0.19	0.23	0.22
No. of employees to perform the scenario	3	2	3	4
Exposure to hazardous materials (on 1–5 scale; 1 for very high, 5 for very low)	3	2	3	3

S2: Scenario 2 S3: Scenario 3 S4: Scenario 4 S5: Scenario 5

25.2.1 Graph Theory and Matrix Approach

Various steps of the GTMA are carried out as described below:

Step 1: The EOL scenario selection attributes considered are the same as those of Bufardi *et al.* (2003), and these are: CO_2 emissions (CE), SO_2 emissions (SE), energy consumption (EC), logistics cost (LC), disassembly cost (DC), product value (PV), product cost (PC), number of employees to perform the scenario (NE), and exposure to hazardous materials (EH).

Step 2: The quantitative values of the EOL scenario selection attributes, which are given in Table 25.1, are to be normalized. PV and NE are beneficial attributes and higher values are desirable. Values of these attributes are normalized, and given in Table 25.2 in the respective columns. The remaining attributes are non-beneficial, and lower values are desirable except for the attribute EH. The attribute EH is non-beneficial, but based on the nature of the scale adapted for assigning the values, its higher values are desired. The values of these remaining attributes for different EOL scenarios are normalized, and are given in Table 25.2.

Table 25.2. Normalized data of the EOL scenario selection attributes

Attributes	S2	S3	S4	S5
CO_2 emissions (kg)	0.8333	0.6667	0.7692	1
SO_2 emissions (kg)	0.9565	0.4889	0.6875	1
Energy consumption (kWh)	1	0.6771	0.7471	0.6633
Logistics cost (Euro)	0.7353	1	0.8929	0.8333
Disassembly cost (Euro)	1	0.9545	0.9767	0.9333
Product value (Euro)	0.8077	0.4615	0.6154	1
Product cost (Euro)	0.76	1	0.8261	0.8636
No. of employees to perform the scenario	0.75	0.5	0.75	1
Exposure to hazardous materials	1	0.6667	1	1

S2: Scenario 2 S3: Scenario 3 S4: Scenario 4 S5: Scenario 5

Relative importance of attributes (a_{ij}) is also assigned values, as explained in Chapter 4. However, Bufardi *et al.* (2003) assigned equal weights of importance to the attributes. To make a comparison between the results of application of GTMA and the results of the ELECTRE-III method used by Bufardi *et al.* (2003), let the decision maker (*i.e.*, user organization) make the following assignments:

	CE	SE	EC	LC	DC	PV	PC	NE	EH
CE	-	0.5	0.5	0.5	0.5	0.5	0.5	0.5	0.5
SE	0.5	-	0.5	0.5	0.5	0.5	0.5	0.5	0.5
EC	0.5	0.5	-	0.5	0.5	0.5	0.5	0.5	0.5
LC	0.5	0.5	0.5	-	0.5	0.5	0.5	0.5	0.5
DC	0.5	0.5	0.5	0.5	-	0.5	0.5	0.5	0.5
PV	0.5	0.5	0.5	0.5	0.5	-	0.5	0.5	0.5
PC	0.5	0.5	0.5	0.5	0.5	0.5	-	0.5	0.5
NE	0.5	0.5	0.5	0.5	0.5	0.5	0.5	-	0.5
EH	0.5	0.5	0.5	0.5	0.5	0.5	0.5	0.5	-

EOL scenario selection attributes digraph, EOL scenario selection attributes matrix of the digraph and EOL scenario selection function for the matrix can be prepared. The value of EOL scenario selection index is calculated using the values of A_i and a_{ij} for each EOL scenario. The EOL scenario selection index values of different EOL scenarios are given below in descending order:

Scenario 5: 1641.5718
Scenario 2: 1485.2711
Scenario 4: 1305.3771
Scenario 3: 1121.5834

From the above values of the EOL scenario selection index, scenario 5 is understood as the best choice among the EOL scenario alternatives considered for the given product. The ranking of EOL scenarios based on the proposed methodology is: scenario 5 - scenario 2 - scenario 4 - scenario 3; by contrast, the ranking presented by Bufardi *et al.* (2003) was: scenario 5/scenario 2 - scenario 3 - scenario 4. Bufardi *et al.* (2003) suggested both the scenarios 5 and 2 were equally best. However, a closer look at the corresponding values of the attributes of scenarios 5 and 2 clearly indicate the superiority of scenario 5 over scenario 2 for equal weights of relative importance of the attributes. Similarly, proposing scenario 4 as the last choice by Bufardi *et al.* (2003) is not genuine. Again, a close look at the corresponding values of the attributes of scenarios 4 and 3 clearly indicate the superiority of scenario 4 over scenario 3 for equal weights of relative importance of the attributes. Thus, the results obtained by using GTMA seem to be more logical and genuine than those results presented by Bufardi *et al.* (2003) using ELECTRE-III method.

Further, it may be mentioned that the ranking presented may change if the decision maker assigns different relative importance values to the attributes. The same is true with the approach proposed by Bufardi *et al.* (2003).

25.2.2 SAW Method

For equal weights of importance of the attributes, the SAW method leads to the following ranking:

Scenario 5: 0.9215
Scenario 2: 0.8714
Scenario 4: 0.8072
Scenario 3: 0.7128

The SAW method also suggests scenario 5 as the right choice for the given EOL scenario selection problem.

For the same weights of relative importance of attributes, the AHP method also leads to the same ranking.

25.2.3 WPM

For equal weights of importance of the attributes, the EOL scenario selection index for each EOL scenario is calculated and the values are arranged as given below:

Scenario 5: 0.9141
Scenario 2: 0.8645

Scenario 4: 0.7981
Scenario 3: 0.6831

The WPM method also suggests scenario 5 as the right choice for the given EOL scenario selection problem.

25.2.4 TOPSIS Method

The quantitative values of the EOL scenario selection attributes, which are given in Table 25.1, are normalized as explained in Section 3.2.6. Equal weights of relative importance of the attributes are considered, *i.e.*, the weight of each attribute is 0.1111.

The ideal (best) and negative ideal (worst) solutions are obtained, and these are given as:

$V_{CE}^{+} = 0.0440$ $V_{CE}^{-} = 0.0660$
$V_{SE}^{+} = 0.0384$ $V_{SE}^{-} = 0.0784$
$V_{EC}^{+} = 0.0413$ $V_{EC}^{-} = 0.0622$
$V_{LC}^{+} = 0.0472$ $V_{LC}^{-} = 0.0642$
$V_{DC}^{+} = 0.0536$ $V_{DC}^{-} = 0.0575$
$V_{PV}^{+} = 0.0742$ $V_{PV}^{-} = 0.0342$
$V_{PC}^{+} = 0.0472$ $V_{PC}^{-} = 0.0621$
$V_{NE}^{+} = 0.0721$ $V_{NE}^{-} = 0.0360$
$V_{EH}^{+} = 0.0599$ $V_{EH}^{-} = 0.0399$

The separation measures are:

$S_1^{+} = 0.0335$ $S_1^{-} = 0.0590$
$S_2^{+} = 0.0760$ $S_2^{-} = 0.0227$
$S_3^{+} = 0.0441$ $S_3^{-} = 0.0407$
$S_4^{+} = 0.0225$ $S_4^{-} = 0.0741$

The relative closeness of a particular alternative to the ideal solution is calculated (which can be named the 'EOL scenario selection index (EOLS-SI)' in the present work), and these are arranged in descending order as:

Scenario 5: 0.7519
Scenario 2: 0.6381
Scenario 4: 0.4797
Scenario 3: 0.2297

The TOPSIS method also suggests scenario 5 as the right choice for the given EOL scenario selection problem.

25.2.5 Modified TOPSIS Method

Following the procedure of the modified TOPSIS method, and using the same weights as those selected for the TOPSIS method, the values of EOLS-SI are obtained and arranged as:

Scenario 5: 0.7477
Scenario 2: 0.6390
Scenario 4: 0.4791
Scenario 3: 0.2429

It may be observed, from the application of the TOPSIS and modified TOPSIS methods, that both methods suggest alternative scenario 5 as the first choice, and scenario 3 as the last choice.

Further, it may be observed that in the present example, all the proposed decision-making methods, *i.e.*, GTMA, SAW, WPM, AHP, TOPSIS, and modified TOPSIS, give the same ranking of the alternative EOL scenarios.

25.2.6 Compromise Ranking Method (VIKOR)

Step 1: The objective is to evaluate the four scenarios and the attributes are: CO_2 emissions (CE), SO_2 emissions (SE), energy consumption (EC), logistics cost (LC), disassembly cost (DC), product value (PV), product cost (PC), number of employees to perform the scenario (NE) and exposure to hazardous materials (EH). The best, *i.e.*, $(m_{ij})_{max}$, and the worst, *i.e.*, $(m_{ij})_{min}$, values of all attributes are also determined.

Step 2: The values of E_i and F_i are calculated using Equations 3.26 and 3.27. As was done by Bufardi *et al.* (2003), equal weights of importance were assigned to the attributes.

$E_2 = 0.0444+0.00483+0+0.1111+0+0.0397+0.1111+0.0556+0 = 0.3667$

$E_3 = 0.1111+0.1111+0.1044+0+0.0741+0.1111+0+0.1111+0.1111 = 0.734$

$E_4 = 0.0667+0.0483+0.0741+0.037+0.037+0.0794+0.0741+0.0556+0= 0.4721$

$E_5 = 0+0+0.1111+0.0617+0.1111+0+0.0556+0+0 = 0.3395$

$E_{i\text{-}min} = 0.3395$ $E_{i\text{-}max} = 0.734$

$R_2 = 0.1111$ $R_3 = 0.1111$ $R_4 = 0.0794$ $R_5 = 0.1111$

$F_{i\text{-}min} = 0.0794$ $F_{i\text{-}max} = 0.1111$

Step 3: The values of P_i are calculated using Equation 3.28 and for $v = 0.5$.

$P_2 = 0.5345$ $P_3 = 1$ $P_4 = 0.1681$ $P_5 = 0.5$

Step 4: The alternatives are arranged in ascending order, according to the values of P_i. Similarly, the alternatives are arranged according to the values of E_i and F_i separately. Thus, three ranking lists are obtained. The best alternative, ranked by P_i, is the one with the minimum value of P_i.

$P_4 = 0.1681$	$E_5 = 0.3395$	$F_4 = 0.0794$
$P_5 = 0.5$	$E_2 = 0.3667$	$F_2 = F_3 = F_5 = 0.1111$
$P_2 = 0.5345$	$E_4 = 0.472$	
$P_3 = 1$	$E_3 = 0.734$	

Step 5: Scenario 4, which is best ranked by the measure P is suggested as it satisfies both conditions given in Section 3.2.7.However, the ranking given by the VIKOR method in this example seems not to be genuine.

References

Adda S, Dahl T, Gjerstad TB, Lamvik T, Miljeteig G, Molden TH, Schnatmeyer M, Snaddon K (2002) Environmental/social cost/value assessment methodology. WP4 D4.1 (version 2)

Alting L (1993) Designing for a lifetime. Manufacturing Breakthrough. May/ June:29-33

Alting L, Zhang H (1991) Trends in the development of the future integrated manufacturing systems. Manufacturing Review 4:1-9

Arena U, Mastellone ML, Perugini F (2003) The environmental performance of alternative solid waste management options: a life-cycle assessment study. Chemical Engineering Journal 96:207-222

Azapagic A (1999) Life-cycle assessment and its application to process selection design and optimization. Chemical Engineering Journal 73:1-21

Bufardi A, Sakara D, Gjeorghe R, Kiritsis D, Xirouchakis P (2003) Multiple criteria decision aid for selecting the best product end of life scenario. International Journal of Computer Integrated Manufacturing 16:526-534

Bufardi A, Gjeorghe R, Kiritsis D, Xirouchakis P (2004) Multi-criteria decision aid approach for product end-of-life alternative selection. International Journal of Production Research 42:3139-3157

Byggeth S, Hochschorner E (2006) Handling trade-offs in ecodesign tools for sustainable product development and procurement. Journal of Cleaner Production 14:1420-1430

Desai A, Mittal A (2003) Evaluation of disassemblability to enable design for disassembly in mass production. International Journal of Industrial Ergonomics 32:265-281

Keoleian GA, Kar K (2003) Elucidating complex design and management tradeoffs through life-cycle design: air intake manifold demonstration project. Journal of Cleaner Production 11:61-77

Khan FI, Sadiq R, Husain T (2002) GreenPro-I: a risk-based life-cycle assessment decision-making methodology for process plant design. Environmental Modelling & Software 17:669-692

Marttunen M, Hamalainen RP (1997) Decision analysis interviews in environmental impact assessment. European Journal of Operational Research 87:551-563

Nagel C, Nilsson J, Boks C (1999) European end-of-life systems for electrical and electronic equipment. In: Proc. 1st International Symposium Environmentally Conscious Design and Inverse Manufacturing, University of Tokyo, Tokyo, pp 197–202

Sarkis J, Weinrach J (2001) Using data envelopment analysis to evaluate environmentally conscious waste treatment technology. Journal of Cleaner Production 9:417-427

Sohlenius G (1989) Needs and possibilities for sciences of manufacturing technology. In: Proc. 21st CIRP Seminar on Manufacturing Systems, The Royal Institute of Technology, Stockholm, Sweden, pp 1-20

Zussman E, Kriwet A, Seliger G (1994) Disassembly-oriented assessment methodology to support design for recycling. CIRP Annals 43:9-14

26

Integrated Project Evaluation and Selection

26.1 Introduction

The evaluation and selection of industrial projects before investment decision is customarily done using marketing, technical, and financial information. Subsequently, environmental impact assessment and social impact assessment are carried out mainly to satisfy the statutory agencies. Because of stricter environmental regulations in developed and developing countries, quite often impact assessment suggests alternate sites, technologies, designs, and implementation methods as mitigating measures. This causes considerable delay to complete project feasibility analysis and selection, as complete analysis requires to be addressed repeatedly till the statutory regulatory authority approves the project. Moreover, project analysis through the above process often results in sub-optimal projects as financial analysis may eliminate better options. Indeed, more environment friendly alternative will always be cost intensive (Dey, 2006).

There is a large literature dedicated to the project selection problem. It includes several approaches that take into account various aspects of the problem. Mehrez and Sinuany-Stern (1983) described an interactive method for presenting a sequence of feasible sets of indivisible projects to a decision-maker. For each set as a whole, the decision-maker evaluated its utilities with respect to each of several attributes; the utilities were then combined to give a single utility for the set. A sequence of zero-one program was used to ensure that the only sets presented were those that were feasible, and that were not contained in larger feasible sets.

Lockett and Stratford (1987) presented the details of ranking of research projects. Khorramshahgole and Steiner (1988) applied a goal programming method for resource analysis in project evaluation. Danila (1989) reviewed the most important families of R&D project evaluation and selection methodologies and associated techniques. For each family, generally one or two methods were chosen to be analyzed from the point of view of integration with the strategy. The strong and weak points of the most prevalent methods were described as they were considered by the users.

Islei *et al.* (1991) developed a computerized decision support system for R&D project ranking, monitoring, and control in the pharmaceutical industry. Using a series of techniques based on a judgmental modeling approach, an overall system was developed that covers several different aspects of the whole decision making process. Illustrative examples were given to show how the components of the model were used in practice, indicating their place in the overall management process. Dey *et al.* (1994) used the AHP method to analyze and assess project risks which are objective or subjective in nature. Regan and Holtzman (1995) described the architecture of R&D decision advisor, a commercial intelligent decision system for evaluating corporate research, and development projects and portfolios. The system guides the user to choose among general project features, but offers flexibility to capture unique project details.

Coffin and Taylor (1996) used fuzzy logic for R&D project selection and scheduling. Chu *et al.* (1996) developed a decision support system (DSS) to help managers select the most appropriate sequences of plans for product research and development projects that have strict constraints on budget, time, and resources. The primary objective of the DSS was to provide an optimal combination of R&D projects. The DSS consisted of several subsystems, each of which had a specific function. At the core of the DSS were a cost model, which covers time–cost tradeoff analysis, and a strategic selection algorithm, which provides an optimal development plan based on dynamic programming for managing R&D projects.

Korpela and Tuominen (1996) demonstrated how AHP method can be used for supporting a generic logistics benchmarking process. First, the customers of a company were interviewed in order to define the logistic critical success factors, and to determine their importance. The performance levels of the companies to be benchmarked were then evaluated with regard to each success factor. Second, the factors enabling the companies to achieve superior logistics performance were determined and prioritized with respect to each success factor. Third, the strengths, weaknesses, and problems of the company conducting the benchmarking process were analyzed and prioritized with respect to each enabler. Then, the potential developmental actions for achieving superior logistics performance were defined and prioritized.

Dey *et al.* (1996a) used a goal programming (GP) method for a project planning problem. The methodology was applied to plan a petroleum pipeline construction project, and its effectiveness was demonstrated. Dey *et al.* (1996b) proposed a methodology for project control through risk analysis, contingency allocation and hierarchical planning models. Risk analysis was carried out through the AHP method, due to the subjective nature of risks in construction projects. The results of risk analysis were used to determine the logical contingency for project control with the application of probability theory. Ultimate project control was carried out by means of the hierarchical planning model, which enabled decision makers to take vital decisions during the changing environment of the construction period. Goal programming was proposed for model formulation because of its flexibility and priority-based structure.

Ramanathan and Geetha (1998) conducted a socio-economic impact assessment in advance to determine the socio-economic consequences of industrial projects. The focus was on the project-affected people. All possible data were collected from census information and academic institutions. Personal interviews were also conducted with the local people and their administrative heads. The main phases of the project addressed were preconstruction, construction and operation. A decision on the acceptability of the project was taken after assessing the positive and negative socio-economic impacts.

Dey and Gupta (1999) presented a decision support system (DSS) for pipeline route selection with the application of the analytical hierarchy process (AHP). This system was demonstrated through a case study of pipeline route selection, from an Indian perspective. A cost-benefit comparison of the shortest route (conventionally selected) and the optimal route established the effectiveness of the model. Mian and Christine (1999) used the AHP method to resolve decision-making issues in project selection. Ghasemzadeh *et al.* (1999) proposed a zero-one integer linear programming model for selecting and scheduling an optimal project portfolio, based on the organization's objectives, and constraints such as resource limitations and interdependence among projects. The proposed model could not only suggest projects that should be incorporated in the optimal portfolio, but it could also determine the starting period for each project.

Meredith and Mantle (2000) thoroughly discussed the strategic intent of the project, factors for project selection, and various qualitative and quantitative project selection models. Ghasemzadeh and Archer (2000) investigated the implementation of an organized framework for project portfolio selection through a decision support system (DSS), which was called Project Analysis and Selection System (PASS). The authors described the results of laboratory tests undertaken to measure its usability and quality, compared to manual selection processes, in typical portfolio selection problems. Dey and Gupta (2001) used the AHP method for feasibility analysis of cross-country pipeline projects.

Loch and Kavadias (2002) developed a dynamic model of resource allocation, taking into account multiple interacting factors, such as 'independent' or 'correlated' status, uncertain market payoffs that change over time, increasing or decreasing returns from the new product development investment, carry-over of the investment benefit over multiple periods, and interactions across market segments. The optimal policies in closed form were characterized, and qualitative decision rules for managers were derived. Dey (2002) demonstrated how the analytic hierarchy process (AHP) can be used for benchmarking project management practices. The entire methodology was applied to benchmark project management practice of Caribbean public sector organizations, with organizations in the Indian petroleum sector, and organizations in the infrastructure sector of Thailand and the UK. This study demonstrated the effectiveness of the benchmarking model using AHP, and suggested improvement measures for effective project management.

Dey (2004) presented a risk-based decision support system (DSS) that reduces the amount of time spent on inspection of an entire pipeline project. The risk-based DSS used the analytic hierarchy process (AHP) to identify the factors that influence failure on specific segments, and analyzed their effects by determining probability of occurrence of these risk factors. The severity of failure was determined through consequence analysis. Walls (2004) used preference analysis as an approach for measuring and applying a corporate risk-taking policy. Mahdi and Alreshaid (2005) developed a decision support system for selecting the proper project delivery method, using the analytical hierarchy process (AHP). Dey (2006) proposed a decision support system that analyzes projects with respect to market, technicalities, and social and environmental impacts in an integrated framework, using the analytic hierarchy process (AHP).

Labuschagne and Brent (2006) introduced a framework to assess the sustainability performances of projects and technology developments in the process industry. The research verified the completeness of the social dimension of that framework to evaluate operational initiatives. Furthermore, the relevance of the social dimension to the process industry was validated. It was found that, to a limited extent, a smaller set of social assessment criteria might be defined for project and technology Life Cycle Management purposes. Also, it was concluded that quantitative social indicators are not practical in the current process industry. Puthamont and Charoenngam (2006) explored and gained an in-depth understanding of the factors influencing three stages of the project selection process, namely, conceptual stage, design stage, and final approval stage. In order to achieve this, data related to factors influencing construction project selection in the Thailand's Ministry of Defence were collected and analyzed.

Huang (2006) proposed a practical tool of incorporating random fuzzy uncertainty into project selection. Investment outlays and annual net cash flows of available projects were regarded as random fuzzy variables. The net present value method was employed, and two types of zero-one integer chance-constrained models with random fuzzy parameters were provided. A hybrid intelligent algorithm integrating genetic algorithm and random fuzzy simulation was designed and numerical examples were presented to illustrate the modeling concept and to show the effectiveness of the algorithm. Carlsson *et al.* (2006) developed a methodology for valuing options on R&D projects when future cash flows are estimated by trapezoidal fuzzy numbers. The authors presented a fuzzy mixed integer programming model for the R&D optimal portfolio selection problem, and discussed how the methodology could be used to build decision support tools for optimal R&D project selection in a corporate environment.

Rescia *et al.* (2006) proposed a methodology for environmental analysis in the selection of alternative corridors in a long-distance linear project – a pipeline. Chapman *et al.* (2006) presented a framework to select public sector projects. Doerner *et al.* (2006) proposed Pareto ant colony optimization with integer linear programming preprocessing in multiobjective project portfolio selection. Talias (2007) examined issues related to various decision-based analytic approaches to sequential choice of projects, with special motivation from and application in the pharmaceutical industry. In particular, the Pearson index and Gittins index were considered as key strategic decision making tools for the selection of R&D projects. This presented a proof of optimality of the Pearson index based on the Neyman-Pearson lemma. Emphasis was also given to how a project manager may differentiate between the two indices, based on concepts from statistical decision theory.

Medaglia *et al.* (2005) proposed an evolutionary method for project selection problems with partially funded projects, multiple (stochastic) objectives, project interdependencies (in the objectives), and a linear structure for resource constraints. The method was compared with the stochastic parameter space investigation method (PSI) and was illustrated by means of an R&D portfolio problem under uncertainty based on Monte Carlo simulation.

The initial approach in project selection can be the identification of pertinent attributes and potential alternative projects by a team of executives at different levels. The values of the attributes (A_i) with the system specifications and requirements can be obtained, and their relative importance (a_{ij}) can be decided. An objective or subjective value, or its range, may be assigned to each identified attribute as a limiting value, or threshold value, for its acceptance for the considered project selection problem. Alternative project with each of its selection attribute meeting the acceptance value of the attribute for the project, may be short-listed. After short-listing the alternative projects, the main task in choosing the alternative project is to see how it serves the considered attributes.

Now, an example is considered to demonstrate the application of GTMA and other decision making methods for project evaluation and selection.

26.2 Example

Dey (2006) proposed a decision support system that analyzes projects with respect to market, technicalities, and social and environmental impacts in an integrated framework using the analytic hierarchy process (AHP). The entire methodology was applied to a cross-country oil pipeline project in India, and its effectiveness was demonstrated. Cross-country petroleum pipelines are the most energy-efficient, safe, environmentally friendly, and economical means for transporting hydrocarbons (gas, crude oil, and finished product) over long distances within a country and between countries. The economy of a country can be heavily dependent on smooth and uninterrupted operation of these pipelines. While pipelines are one of the safest means for transporting bulk energy, failures do occur and sometimes have catastrophic consequences. To avoid failures, pipeline operators choose optimal pipeline routes (Dey and Gupta, 1999). The project studied by Dey (2006) was a cross-country petroleum pipeline project in western India. Its length was 1,300 km plus a 123-km branch line. The pipeline was designed to carry 5 million metric tons per annum (MMTPA) of throughput. The project included three pump stations, one pumping/delivery station, two scraper stations, four delivery stations, and two terminal stations. The project cost was estimated as US $600 million. Dey (2006) considered the following project selection attributes:

Technical attributes (*e.g.*, length of the pipeline; operability, which is affected by route characteristics, augmentation possibility, and expansion capability; maintainability, which is affected by corrosion, pilferage, and third-party activities; approachability, constructability), environmental impact assessment attributes (*e.g.*, during failure of pipelines; during failure of stations; during normal pipeline operation; during pipeline construction), and socio-economic assessment attributes (*e.g.*, effect during planning, which involves compensation, employment, and rehabilitation; effect during construction, which involves employment; effect during operations, which involves employment, and burden on existing infrastructure).

Thus, it can be said that the total number of attributes is 20 (technical attributes, environmental impact assessment attributes, and socio-economic assessment attributes). Four alternative pipeline routes were considered. Table 26.1 presents the data of the attributes for the four alternative routes presented in appropriate form by Dey (2006). The values given in parentheses indicate the normalized weights of the attributes. Dey (2006) used the AHP method, and ranked the alternative pipeline routes in the following order:

Route 4	:	0.309
Route 2	:	0.241
Route 3	:	0.232
Route 1	:	0.218

Now, the different MADM methods described in this book are applied to the same problem, to compare the results. The SAW method gives the same ranking as that of the given by using AHP method (for the same weights of the attributes).

Table 26.1. Project selection data (from Dey 2006; reprinted with permission from Elsevier)

Attributes	Route 1	Route 2	Route 3	Route 4
(i) Technical:				
Route length (0.14)	0.27	0.10	0.37	0.26
Operability:				
Route characteristics (0.019)	0.20	0.22	0.30	0.28
Augmentation possibility (0.04)	0.25	0.36	0.12	0.27
Expansion capability (0.031)	0.26	0.37	0.08	0.29
Maintainability:				
Corrosion (0.065)	0.23	0.30	0.15	0.32
Pilferage (0.027)	0.21	0.24	0.25	0.30
Third-party activities (0.016)	0.21	0.28	0.25	0.26
Approachability (0.045)	0.23	0.33	0.13	0.31
Constructability (0.068)	0.21	0.28	0.17	0.34
(ii) Environmental impact assessment:				
During failure of pipelines (0.102)	0.23	0.30	0.15	0.32
During failure of stations (0.082)	0.21	0.28	0.25	0.26
During normal pipelines operations (0.018)	0.22	0.32	0.18	0.28
During normal station operations (0.02)	0.18	0.32	0.15	0.35
During pipeline construction (0.027)	0.20	0.28	0.22	0.30
(iii) Socio-economic impact assessment:				
Effect during planning:				
Compensation (0.09)	0.21	0.16	0.33	0.30
Employment & rehabilitation (0.039)	0.14	0.24	0.28	0.34
Effect during construction:				
Employment (0.054)	0.25	0.25	0.15	0.35
Effect of construction activities (0.054)	0.12	0.18	0.27	0.43
Effect during operations:				
Employment (0.013)	0.25	0.25	0.25	0.25
Burden on existing infrastructure (0.05)	0.17	0.18	0.30	0.35

26.2.1 Weighted Product Method (WPM)

Using the same weights of the attributes as those of Dey (2006), the weighted product method (WPM) leads to the following ranking of pipeline routes:

Route 4: 0.3067
Route 2: 0.2252
Route 3: 0.2154
Route 1: 0.2145

Thus, WPM also suggests Route 4 as the right choice.

26.2.2 TOPSIS Method

Using the same weights of the attributes as those of Dey (2006), the ideal (best) and negative ideal (worst) solutions are obtained, and these are given as:

$V_1^+ = 0.0966$ $V_1^- = 0.0261$
$V_2^+ = 0.0112$ $V_2^- = 0.0075$
$V_3^+ = 0.0272$ $V_3^- = 0.0091$
$V_4^+ = 0.0211$ $V_4^- = 0.0046$
$V_5^+ = 0.0402$ $V_5^- = 0.0188$
$V_6^+ = 0.0161$ $V_6^- = 0.0112$
$V_7^+ = 0.0089$ $V_7^- = 0.0067$
$V_8^+ = 0.0283$ $V_8^- = 0.0112$
$V_9^+ = 0.0447$ $V_9^- = 0.0224$
$V_{10}^+ = 0.0631$ $V_{10}^- = 0.0296$
$V_{11}^+ = 0.0457$ $V_{11}^- = 0.0343$
$V_{12}^+ = 0.0113$ $V_{12}^- = 0.0063$
$V_{13}^+ = 0.0132$ $V_{13}^- = 0.0057$
$V_{14}^+ = 0.0160$ $V_{14}^- = 0.0107$
$V_{15}^+ = 0.0573$ $V_{15}^- = 0.0278$
$V_{16}^+ = 0.0255$ $V_{16}^- = 0.0105$
$V_{17}^+ = 0.0364$ $V_{17}^- = 0.0156$
$V_{18}^+ = 0.0421$ $V_{18}^- = 0.0117$
$V_{19}^+ = 0.0065$ $V_{19}^- = 0.0065$
$V_{20}^+ = 0.0334$ $V_{20}^- = 0.0162$

The separation measures are:

$S_{Route\ 1}^+ = 0.0619$ $S_{Route\ 1}^- = 0.0530$
$S_{Route\ 2}^+ = 0.0835$ $S_{Route\ 2}^- = 0.0525$
$S_{Route\ 3}^+ = 0.0619$ $S_{Route\ 3}^- = 0.0799$
$S_{Route\ 4}^+ = 0.0306$ $S_{Route\ 4}^- = 0.0837$

The relative closeness of a particular alternative route to the ideal solution is calculated (which can be named the 'project selection index' in the present work), and these are $P_{Route\ 1} = 0.4612$, $P_{Route\ 2} = 0.3861$, $P_{Route\ 3} = 0.5636$, $P_{Route\ 4} = 0.7322$

The alternative pipeline routes are arranged in descending order of their PSI. This can be arranged as: route 4 - route 3 - route 1 - route 2.

This method also suggests route 4 as the first right choice.

26.2.3 Modified TOPSIS Method

Following the procedure of the modified TOPSIS method, and using the same weights as those of Dey (2006), the ideal (best) and negative ideal (worst) solutions are obtained, and these are given as:

$R_1^+ = 0.6902$ $R_1^- = 0.1865$
$R_2^+ = 0.5920$ $R_2^- = 0.3947$
$R_3^+ = 0.6811$ $R_3^- = 0.2270$
$R_4^+ = 0.6812$ $R_4^- = 0.1473$
$R_5^+ = 0.6184$ $R_5^- = 0.2899$
$R_6^+ = 0.5950$ $R_6^- = 0.4165$
$R_7^+ = 0.5571$ $R_7^- = 0.4178$
$R_8^+ = 0.6295$ $R_8^- = 0.2480$
$R_9^+ = 0.6580$ $R_9^- = 0.3289$
$R_{10}^+ = 0.6184$ $R_{10}^- = 0.2899$
$R_{11}^+ = 0.5571$ $R_{11}^- = 0.4178$
$R_{12}^+ = 0.6257$ $R_{12}^- = 0.3519$
$R_{13}^+ = 0.6617$ $R_{13}^- = 0.2836$
$R_{14}^+ = 0.5920$ $R_{14}^- = 0.3947$
$R_{15}^+ = 0.6367$ $R_{15}^- = 0.3087$
$R_{16}^+ = 0.6529$ $R_{16}^- = 0.2688$
$R_{17}^+ = 0.6736$ $R_{17}^- = 0.2887$
$R_{18}^+ = 0.7791$ $R_{18}^- = 0.2174$
$R_{19}^+ = 0.5000$ $R_{19}^- = 0.5000$
$R_{20}^+ = 0.6689$ $R_{20}^- = 0.3249$

The weighted Euclidean distances are:

$D_{Route\ 1}^+ = 0.2484$ $D_{Route\ 1}^- = 0.1701$
$D_{Route\ 2}^+ = 0.2591$ $D_{Route\ 2}^- = 0.2257$
$D_{Route\ 3}^+ = 0.2608$ $D_{Route\ 3}^- = 0.2380$
$D_{Route\ 4}^+ = 0.0916$ $D_{Route\ 4}^- = 0.3181$

The following PSI values are obtained:

$P_{Route\ 1} = 0.4065$, $P_{Route\ 2} = 0.4656$, $P_{Route\ 3} = 0.4772$, $P_{Route\ 4} = 0.7764$

The alternative pipeline routes are arranged in descending order of their PSI. This can be arranged as route 4 - route 3 - route 2 - route 1. Thus, the modified TOPSIS method also suggests route 4 as the first right choice and the ranking is consistent with that obtained by Dey (2006) using the AHP method.

Application of graph theory and the matrix approach also gives the ranking as route 4 - route 3 - route 2 - route 1. However, the steps are not shown here.

Thus, all decision making methods described above propose pipeline route 4 as the first right choice. It may be mentioned here that the integrated project evaluation and selection procedure emphasizes technical attributes, as well as environmental and social attributes while selecting the best alternative project. Attributes such as capital costs and operating costs can also be considered simultaneously in the evaluation and selection procedure.

References

Carlsson C, Fullér R, Heikkilä M, Majlender P (2006) A fuzzy approach to R&D project portfolio selection. International Journal of Approximate Reasoning doi:10.1016/j.ijar.2006.07.003

Chapman CB, Ward SC, Klein JH (2006) An optimised multiple test framework for project selection in the public sector, with a nuclear waste disposal case-based example. International Journal of Project Management 24:373–384

Chu PYV, Hsu YL, Fehling M (1996) A decision support system for project portfolio selection. Computers in Industry 32:141–149

Coffin MA, Taylor BW (1996) Multiple criteria R&D project selection and scheduling using fuzzy logic. Computers & Operations Research 23:207–221

Danila N (1989) Strategic evaluation and solution of R&D projects. R&D Management 19:47–62

Dey PK (2002) Benchmarking project management practices of Caribbean organizations using analytic hierarchy process. Benchmarking: An International Journal 9:326–356

Dey PK (2004) Decision support system for inspection and maintenance of cross-country petroleum pipeline. IEEE Transactions on Engineering Management 51:47–56

Dey PK (2006) Integrated project evaluation and selection using multiple-attribute decision-making technique. International Journal of Production Economics 103:90–103

Dey PK, Gupta SS (1999) Decision support system for pipeline route selection. Cost Engineering 41:29–35

Dey PK, Gupta SS (2001) Feasibility analysis of cross-country pipeline projects: A quantitative approach. Project Management Journal 32:50–58

Dey PK, Tabucanon MT, Ogunlana SO (1994) Planning for project control through risk analysis: a case of petroleum pipeline laying project. International Journal of Project Management 12:23–33

Dey PK, Tabucanon MT, Ogunlana SO (1996a) Hierarchical approach to project planning. Applied Mathematical Modelling 20:683–698

Dey PK, Tabucanon MT, Ogunlana SO (1996b) Petroleum pipeline construction planning: a conceptual framework. International Journal of Project Management 14:231–240

Doerner KF, Gutjahr WJ, Hartl RF, Strauss C, Stummer C (2006) Pareto ant colony optimization with ILP preprocessing in multiobjective project portfolio selection. European Journal of Operational Research 171: 830–841

Ghasemzadeh F, Archer NP, Iyogun P (1999) A zero-one model for project portfolio selection and scheduling. Journal of Operational Research Society 50:745–755

Ghasemzadeh F, Archer NP (2000) Project portfolio selection through decision support. Decision Support Systems 29:73–88

Huang X (2006) Optimal project selection with random fuzzy parameters. International Journal of Production Economics doi:10.1016/j.ijpe.2006.06.011

Islei G, Lockett G, Cox B, Stratford M (1991) A decision support system using judgmental modeling: a case of R&D in the pharmaceutical industry. IEEE Transaction on Engineering Management 38:202–209

Khorramshahgole R, Steiner HM (1988) Resource analysis in project evaluation: a multicriteria approach. Journal of Operational Research Society 95:795–803

Korpela J, Tuominen M (1996) Benchmarking logistic performance with an application of the analytic hierarchy process. IEEE Transactions on Engineering Management 43:323–333

Labuschagne C, Brent AC (2006) An industry perspective of the completeness and relevance of a social assessment framework for project and technology management in the manufacturing sector. Journal of Cleaner Production doi:10.1016/j.jclepro.2006.07.028

Loch CH, Kavadias S (2002) Dynamic portfolio selection of NPD programs using marginal returns. Management Science 48:1227–1241

Lockett G, Stratford M (1987) Ranking of research projects, experiments with two methods. Omega 15:395–400

Mahdi IM, Alreshaid K (2005) Decision support system for selecting the proper project delivery method using analytical hierarchy process (AHP). International Journal of Project Management 23:564–572

Medaglia AL, Graves SB, Ringuest JL (2005) A multiobjective evolutionary approach for linearly constrained project selection under uncertainty. European Journal of Operational Research doi:10.1016/j.ejor.2005.03.068

Mehrez A, Sinuany-Stern Z (1983) An interactive approach to project selection. Journal of Operational Research Society 34:621–626

Meredith JR, Mantle SJ (2000) Project management: a managerial approach. Wiley, New York

Mian SA, Christine ND (1999) Decision-making over the project life cycle: an analytical hierarchy approach. Project Management Journal 30:40–52

Puthamont S, Charoenngam C (2006) Strategic project selection in public sector: construction projects of the Ministry of Defence in Thailand. International Journal of Project Management doi:10.1016/j.ijproman.2006.05.001

Ramanathan R, Geetha S (1998) Socio-economic impact assessment of industrial projects in India. Impact Assessment and Project Appraisal 16:27–32

Regan P, Holtzman S (1995) R&D decision advisor: an interactive approach to normative decision system model construction. European Journal of Operational Research 84:116–133

Rescia AJ, Astrada EN, Bono J, Blasco CA, Meli P, Adámoli JM (2006) Environmental analysis in the selection of alternative corridors in a long-distance linear project: A methodological proposal. Journal of Environmental Management 80:266–278

Talias MA (2007) Optimal decision indices for R&D project evaluation in the pharmaceutical industry: Pearson index versus Gittins index. European Journal of Operational Research 177:1105–1112

Walls MR (2004) Combining decision analysis and portfolio management to improve project selection in the exploration and production firm. Petroleum Science and Engineering 44:55–65

27

Facility Location Selection

27.1 Introduction

Facility location selection is an integral part of organizational strategies. Facility location decision involves organizations seeking to locate, relocate or expand their operations. The facility location decision process encompasses the identification, analysis, and evaluation of, and selection among alternatives. Factories, warehouses, retail outlets, terminals, and storage yards are the typical facilities to be located. Facility location selection commonly starts with the recognition of a need for additional capacity. A decision is then made to start the search for the correct location. Facility location problems have attracted researchers with diverse backgrounds such as economics, industrial engineering, and geography (Ghosh and Harche, 1993). It has been well recognized that facility location selection has important strategic implications for the operations to be located, because a location decision normally involves long-term commitment of resources and is irreversible in nature. Specifically, the location choice for a manufacturing facility may have a significant impact on the firm's strategic competitive position in terms of operating cost, delivery speed performance, and the firm's flexibility to compete in the market. For example, selecting a production facility location that will allow the company to achieve proximity to suppliers has today become a critical strategic advantage, since proximity to suppliers is important to Just-in-Time (JIT)) production systems, and flexible distribution systems for reduced inventories and improved delivery performance.

The suitability of a specific location for proposed facility operations depends largely on what location factors are selected and evaluated, as well as their potential impact on corporate objectives and operations. There are a large number of location factors that have an influence on location decisions. Facility location attribute is defined as a factor that influences the selection of facility location for a given industry. In the case of plant location selection, these attributes include: cost of land, cost of energy, availability of transportation, cost of transportation, proximity to raw material sources, cost of raw materials, availability of local labor and worker attitude, cost of labor, availability of managerial and technical personnel, nearness to the market, government policies and incentives, tax rates, nearby industries, community environment, availability of water, electricity, environmental conditions, *etc*. Few published studies are available on the actual identification of facility location evaluating attributes and their relative importance (Verter and Dincer, 1992; Chen, 2001; Drezner and Weselowsky, 2001; Verter and Dasci, 2002; Thomas *et al*., 2002; Bhatnagar and Sohal, 2005; Bhattacharya *et al*., 2003; Ghosh, 2003; Klose and Drexl, 2005; ReVelle and Eiselt, 2005; Yong, 2005; Averbakh *et al*., 2006; Rao, 2006; Farahani and Asgari, 2007; Boffey *et al*., 2007).

Some researchers have concentrated on solving capacitated facility selection problems in which the capacity of the facility is taken into consideration when the firm decides to open new facilities and wishes to determine which facilities to open for production within the facility's capacities. Some of the related research works are those of Ertogral and Wu (2000), Hinojosa *et al.* (2000), Diaz and Fernandez (2002), Ghiani *et al.* (2002), Jaramillo *et al.* (2002), Cao and Chen (2006), Arostegui *et al.* (2006), and Barreto *et al.* (2006).

Facility location selection is a typical multiple criteria decision-making problem in which managerial preference among performance attributes plays a key role in the final decision. Now, two examples are considered to demonstrate and validate the proposed GTMA and fuzzy MADM methods for facility location selection.

27.2 Examples

Two facility location selection problems are considered.

27.2.1 Example 1

The example problem formulated by Bhattacharya *et al.* (2003), involving five attributes and six alternative facility locations, is shown in Table 27.1.

27.2.1.1 Graph Theory and Matrix Approach (GTMA)

In the present work, the attributes considered are: cost of land (CL), cost of energy (CE), cost of raw material (CRM), cost of transportation (CT) and cost of labor (CLR). The quantitative values of the facility location selection attribute, given in Table 27.1, are to be normalized. In this example, all five attributes considered are non-beneficial, and hence lower values are desirable. The quantitative values of these attributes are normalized, and are given in Table 27.2 in the respective columns.

Table 27.1. Quantitative data of the facility location attributes of example 27.2.1

Facility location	CL	CE	CRM	CT	CLR
P1	3,300,000	2.5	142	6	214
P2	2,500,000	3.1	179	5.8	175
P3	5,200,000	3.6	138	7.8	325
P4	2,500,000	2.8	195	8.4	252
P5	2,000,000	3.2	167	6.3	155
P6	5,700,000	3.7	181	5.5	160

CL: Cost of land (Rs) CE: Cost of energy (Rs/B.O.T. unit)
CRM: Cost of raw material (Rs/kg)
CT: Cost of transportation (Rs/item) CLR: Cost of labor (Rs/worker)

Table 27.2. Normalized data of the facility location selection attributes of example 27.2.1

Facility location	CL	CE	CRM	CT	CLR
P1	0.6061	1	0.972	0.9167	0.7243
P2	0.8	0.8065	0.7710	0.9483	0.886
P3	0.3846	0.6944	1	0.705	0.477
P4	0.8	0.893	0.7077	0.655	0.615
P5	1	0.7813	0.8263	0.873	1
P6	0.351	0.6757	0.7624	1	0.97

Let the decision maker select the following assignments of relative importance:

	CL	CE	CRM	CT	CLR
CL	-	0.745	0.5	0.665	0.665
CE	0.255	-	0.255	0.335	0.335
CRM	0.5	0.745	-	0.665	0.665
CT	0.335	0.665	0.335	-	0.5
CLR	0.335	0.665	0.335	0.5	-

Facility location selection attributes digraph, facility location selection attributes matrix of the digraph, and facility location selection function for the matrix can be prepared. The value of the facility location selection index is calculated using the values of A_i and a_{ij} for each facility location. The facility location selection index values of different facility locations are given below in descending order:

P5: 6.7966
P2: 6.0824
P1: 6.0695
P4: 4.8592

P6: 4.8429
P3: 3.9897

From the values of the facility location selection index, it is understood that facility location, designated as 5 is the best choice among the six facility locations considered for the given industrial application. The next choice is 2, and the last choice is 3. However, the ranking obtained using GTMA differs from that of Bhattacharya *et al.* (2003), according to who the first choice was P3. It is observed that Bhattacharya *et al.* (2003) simply mentioned the use of 'scale of relative importance' suggested by Saaty (1980), but not really used this in making their judgements. These were incorrectly made, and there was no consistency check in the example given by these authors. Further, even a cursory look at the values of CL, CE, CT, and CLR for P3 clearly indicates that these values are comparatively high. Thus, except for the value of CRM, in no case is P3 better than the other alternative facility locations. Thus, the interpretation P3 as best choice by Bhattacharya *et al.* (2003) is incorrect and the GTMA results presented here are more logical. Further, GTMA makes provision for dealing with fuzziness involved in decision-making.

27.2.1.2 SAW Method

The procedure suggested by Edwards and Newman (1982) to assess weights for each attribute to reflect its relative importance to the facility location selection decision is followed. For a start, the attributes are ranked in order of importance and 10 points are assigned to the least important attribute, CE. The attributes CT and CLR are considered as equally important in the present example, and given 20 points each to reflect their relative importance. CL and CRM are considered as equally important, and given 40 points each. The final weights are obtained by normalizing the sum of the points to one. Thus, the weights of CL, CRM, CT, CLR, and CE are calculated as 0.3076, 0.3076, 0.1538, 0.1538, and 0.0769, respectively. Using these weights, and the normalized data of the attributes for different facility locations, the facility location selection index values are calculated, and are arranged in descending order of the index.

P5: 0.9099
P2: 0.8274
P1: 0.8147
P4: 0.7278
P6: 0.6974
P3: 0.6611

Thus, the SAW method also suggests P5 as the right choice among the facility locations considered here.

27.2.1.3 WPM

Using the same weights of the attributes as those selected for the SAW method, the following ranking of facility locations is obtained:

P5: 0.9061
P2: 0.8253
P1: 0.7979
P4: 0.7236

P6:	0.6438
P3:	0.6129

The ranking is the same as that obtained by using the SAW method in this example.

27.2.1.4 AHP and its Versions

If the same weights as those selected for SAW method are used in this method, then the ranking of facility locations obtained by using the relative as well as ideal mode AHP method will be the same. The multiplicative AHP method yields the same ranking as that given by WPM.

However, let the decision maker prepare the following matrix:

	CL	CE	CRM	CT	CLR
CL	1	5	1	3	3
CE	1/5	1	1/5	1/3	1/3
CRM	1	5	1	3	3
CT	1/3	3	1/3	1	1
CLR	1/3	3	1/3	1	1

CL and CRM are considered to have equal importance in the above matrix. Similarly, CT and CLR are considered to have equal importance. The normalized weights for each attribute are: W_{CL} = 0.3439, W_{CE} = 0.0544, W_{CRM} = 0.3439, W_{CT} = 0.1289, and W_{CLR} = 0.1289. The value of λ_{max} is 5.0555 and CR = 0.0124, which is much less than the allowed CR value of 0.1. Thus, there is good consistency in the judgements made.

The value of the facility location selection index is now calculated using the above weights and the normalized data of the attributes given in Table 27.2. This leads to the ranking given by the revised AHP or ideal mode of AHP method. The alternative facility locations are arranged in the descending order of the facility location selection index.

P5:	0.9120
P2:	0.8206
P1:	0.8086
P4:	0.6964
P6:	0.6736
P3:	0.6663

From the above values of the facility location selection index, it is clear that the facility location designated as 5 is the best choice among the facility locations considered here.

For the above weights of importance of attributes, the multiplicative AHP method also leads to the same ranking order.

27.2.1.5 TOPSIS Method

Following the steps of the methodology given in Section 3.2.6, the TOPSIS method gives a ranking of P5-P2-P1-P4-P3-P6. This method also suggests P5 as the right choice. However, P6 is proposed as the last choice, and P3 as the fifth choice.

27.2.1.6 Modified TOPSIS Method

This method leads to the following ranking, which is the same as that proposed by the other methods, excepting the TOPSIS method:

P5:	0.8432
P2:	0.7686
P1:	0.6804
P4:	0.6245
P6:	0.3440
P3:	0.2698

It can be observed that all the above decision-making methods propose facility location designated as P5 as the first right choice. The decision makers can choose a method for evaluation of facility locations.

27.2.2 Example 2

Now, another example is considered to further demonstrate the application of the GTMA and fuzzy MADM methods. This example problem considers eight facility location selection attributes, and three alternative facility locations. The objective and subjective information of the attributes is given in Table 27.3. All the attributes are expressed subjectively in linguistic terms, except for cost of labor, and these attributes are assigned objective values with the help of Table 4.3. The objective data of the attributes are given in Table 27.4. It may be mentioned here that the fuzzy judgements (average, above average, high and very high) made in Table 27.3 can be understood in appropriate equivalent terms such as good, very good, *etc.* with respect to the attributes. Except for CLR, all seven attributes are beneficial, and higher values are desirable.

Table 27.3. Data of the facility location attributes of example 27.2.2

L*	CM	CR	LT	AT	CLR	AL	E	BC
P1	H	VH	H	AA	250	H	AA	VH
P2	VH	H	H	VH	265	AA	H	VH
P3	A	H	VH	AA	255	AA	VH	H

L*: Location CM: Closeness to market CR: Closeness to raw material
LT: Land transportation AT: Air transportation
CLR: Cost of labor (Rs/worker) AL: Availability of labor
E: Community education BC: Business climate
A: Average AA: Above average H: High VH: Very high

Table 27.4. Objective data of the facility location attributes of example 27.2.2

L*	CM	CR	LT	AT	CLR	AL	E	BC
P1	0.665	0.745	0.665	0.59	250	0.665	0.59	0.745
P2	0.745	0.665	0.665	0.745	265	0.59	0.665	0.745
P3	0.5	0.665	0.745	0.59	255	0.59	0.745	0.665

L*: Location

27.2.2.1 Graph Theory and Matrix Approach

The quantitative values of the eight attributes are normalized, and are given in Table 27.5 in the respective columns.

Table 27.5. Normalized data of the facility location selection attributes of example 27.2.2

L*	CM	CR	LT	AT	CLR	AL	E	BC
P1	0.8926	1	0.8926	0.7919	1	1	0.7919	1
P2	1	0.8926	0.8926	1	0.9434	0.8872	0.8926	1
P3	0.6711	0.8926	1	0.7919	0.9804	0.8872	1	0.8926

L*: Location

Let the decision maker select the following assignments of relative importance:

	CM	CR	LT	AT	CLR	AL	E	BC
CM	-	0.5	0.59	0.665	0.59	0.665	0.335	0.255
CR	0.5	-	0.59	0.665	0.59	0.665	0.335	0.255
LT	0.41	0.41	-	0.59	0.5	0.59	0.41	0.335
AT	0.335	0.335	0.41	-	0.41	0.5	0.335	0.255
CLR	0.41	0.41	0.5	0.59	-	0.5	0.5	0.41
AL	0.335	0.335	0.41	0.5	0.5	-	0.5	0.41
E	0.665	0.665	0.59	0.665	0.5	0.5	-	0.41
BC	0.745	0.745	0.665	0.745	0.59	0.59	0.59	-

The value of the facility location selection index is calculated using the values of A_i and a_{ij} for each facility location. The facility location selection index values of different facility locations are given below in descending order:

P2: 324.8389
P1: 311.9973
P3: 291.2581

From the values of the facility location selection index, it is understood that the facility location, designated as 2 is the best choice among the six facility locations considered for the given industrial application. The next choice is 1, and the last choice is 3.

27.2.2.2 AHP and its Versions

Let the decision maker prepare the following relative importance matrix:

	CM	CR	LT	AT	CLR	AL	E	BC
CM	1	1	2	3	2	3	1/3	1/4
CR	1	1	2	3	2	3	1/3	1/4
LT	1/2	1/2	1	2	1	2	1/2	1/3
AT	1/3	1/3	1/2	1	1/2	1	1/3	1/4
CLR	1/2	1/2	1	2	1	1	1	1/2
AL	1/3	1/3	1/2	1	1	1	1	1/2
E	3	3	2	3	1	1	1	1/2
BC	4	4	3	4	2	2	2	1

The normalized weights of each attribute are: $W_{CM} = 0.1267$, $W_{CR} = 0.1267$, $W_{LT} = 0.0883$, $W_{AT} = 0.0517$, $W_{CLR} = 0.0929$, $W_{AL} = 0.0706$, $W_E = 0.1668$, and $W_{BC} = 0.2764$. The value of λ_{max} is 8.7086 and CR = 0.0723, which is much less than the allowed CR value of 0.1.. Thus, there is good consistency in the judgements made.

The value of the facility location selection index is calculated, and the alternative facility locations are arranged in descending order of the facility location selection index.

P2: 0.9458
P1: 0.9315
P3: 0.8946

Thus, AHP method also suggests P2 as the right choice among the facility locations considered here.

For the above weights of importance of attributes, the multiplicative AHP method also leads to the same ranking order.

It may be observed that the above ranking is for the given preferences of the decision maker. The ranking depends upon the judgements of relative importance of the attributes made by the decision maker.

References

Arostegui MA, Kadipasaoglu SN, Khumawala BM (2006) An empirical comparison of Tabu search, simulated annealing, and genetic algorithms for facilities location problems. International Journal of Production Economics 103:742–754

Averbakh I, Berman O, Drezner Z, Wesolowsky GO (2006) The uncapacitated facility location problem with demand-dependent setup and service costs and customer-choice allocation. European Journal of Operational Research doi:10.1016/j.ejor.2005.11.041

Barreto S, Ferreira C, Paixão J, Santos BS (2006) Using clustering analysis in a capacitated location-routing problem. European Journal of Operational Research doi:10.1016/j.ejor.2005.06.074

Bhatnagar R, Sohal AS (2005) Supply chain competitiveness: measuring the impact of location factors, uncertainty and manufacturing practices. Technovation 25:443–456

Bhattacharya A, Sarkar B, Mukherjee SK (2003) Selection of plant location through an MCDM methodology. Industrial Engineering Journal (India) 32:15–20

Boffey B, Galvão R, Espejo L (2007) A review of congestion models in the location of facilities with immobile servers. European Journal of Operational Research 178:643–662

Cao D, Chen M (2006) Capacitated plant selection in a decentralized manufacturing environment: a bilevel optimization approach. European Journal of Operational Research 169:97–110

Chen CT (2001) A fuzzy approach to select the location of the distribution center. Fuzzy Sets and Systems 118:65–73

Diaz J, Fernandez E (2002) A branch-and-price algorithm for single source capacitated plant location problem. Journal of the Operational Research Society 53:728–739

Drezner Z, Weselowsky GO (2001) On the collection depots location problem. European Journal of Operational Research 130:510–518

Edwards W, Newman JR, Snapper K, Seaver D (1982) Multiattribute Evaluation. SAGE Publications, Newbury Park, California

Ertogral K, Wu S (2000) Auction-theoretic coordination of production planning in the supply chain. IIE Transactions 32:931–940

Farahani RZ, Asgari N (2007) Combination of MCDM and covering techniques in a hierarchical model forfacility location: a case study. European Journal of Operational Research 176:1839–1858

Ghiani G, Guerriero F, Musmanno R (2002) The capacitated plant location problem with multiple facilities in the same site. Computers & Operations Research 29:1903–1912

Ghosh D (2003) Neighborhood search heuristics for the uncapacitated facility location. European Journal of Operational Research 150:150–162

Hinojosa Y, Puerto J, Fernández FR (2000) A multiperiod two-echelon multicommodity capacitated plant location problem. European Journal of Operational Research 123:271–291

Jaramillo J, Bhadury J, Batta R (2002) On the use of genetic algorithms to solve location problems. Computers & Operations Research 29:761–769

Klose A, Drexl A (2005) Facility location models for distribution system design. European Journal of Operational Research 162:4–29

Rao RV (2006) Plant location selection using fuzzy digraph and matrix methods. International Journal of Industrial Engineering 13:166–176

Revelle CS, Eiselt HA (2005) Location analysis: a synthesis and survey. European Journal of Operational Research 165:1–19

Saaty TL (1980) The analytic hierarchy process. McGraw Hill, New York

Thomas P, Chan Y, Lehmkuhl L, Nixon W (2002) Obnoxious-facility location and data-envelopment analysis: a combined distance-based formulation. European Journal of Operational Research 141:495–514

Verter V, Dasci A (2002) The plant location and flexible technology acquisition problem. European Journal of Operational Research 136:366–382

Verter V, Dincer MC (1992) An integrated evaluation of facility location, capacity acquisition and technology selection for designing global manufacturing strategies. European Journal of Operational Research 60:1–18

Yong D (2005) Plant location selection based on fuzzy TOPSIS. International Journal of Advanced Manufacturing Technology 28:839–844

28

Operational Performance Evaluation of Competing Companies

28.1 Introduction

The performance of a company during a stated period of time is usually reflected by various financial ratios summarized from its financial statements, such as the balance sheet, the income statement and the trading account. These ratios provide useful information to the stakeholders of the company, and reflect the company's performance from various perspectives (Barnes, 1987). For a specific company, these ratios do not always evolve in the same direction, and very often an improvement in one ratio can be achieved only at the expense of deterioration in another. The overall performance of competing companies cannot be meaningfully evaluated or ranked without simultaneous consideration of all these conflicting ratios (van der Wijst, 1990; Davis and Kay, 1990).

Van der Wijst (1990) described a method of inter-firm comparison in small business that was not based on ratio analysis, but rather on the use of less restricted models. The method was aimed at removing a number of disadvantages of the common, ratio-based methods of inter-firm comparison. Models were specified for all major items of the income statement and balance sheet. The specification was partly based on financial and other theory, but also on practical experience in small business. Together, these models enabled a fairly detailed and complete assessment of small-business performance. The use of the models for inter-firm comparisons was illustrated with an example, and their applicability and relation with bankruptcy prediction models were discussed.

Multivariate techniques have been widely used for the explanation and prediction of a firm's behavior. However, in practice, the comparative evaluation and ranking of companies is usually based on the consideration of a single measure of corporate success. Nevertheless, the definition of the most appropriate measure has given rise to considerable debate. Diakoulaki *et al.* (1992) utilized the results of a multicriteria analysis, applied to a large sample of Greek pharmaceutical industries, in order to indicate how suitable some common financial ratios were as indices of the firm's overall performance. The results showed that profitability constitutes the most representative measure for the differentiation and ranking of companies. Labor productivity and market share were the best indicators of the business' success, while business' failure was more closely related to ratios indicating long- and short-term solvency. This means that a sound capital structure is necessary but insufficient to ensure the profitable and effective operation of the firm.

Smith (1990) extended the traditional ratio analysis to permit the incorporation of any number of dimensions of performance, using data envelopment analysis. The method produced measures of corporate efficiency, together with a wealth of supporting information. The strengths and weaknesses of the method applied to financial statements were appraised.

Deng *et al.* (2000) reported that simultaneous consideration of multiple financial ratios is required to adequately evaluate and rank the relative performance of competing companies. The authors formulated the inter-company comparison process as a multicriteria analysis model, and presented an effective approach by modifying TOPSIS for solving the problem. An empirical study of a real case in China was conducted to illustrate how the approach can be used for the inter-company comparison problem. To ensure that the evaluation result is not affected by the interdependence of financial ratios, objective weights were used. As a result, the comparison process was conducted on a commonly accepted basis, and was independent of subjective preferences of various stakeholders.

28.2 Example

Now, to demonstrate and validate the application of the proposed decision-making methods, the case study presented by Deng *et al.* (2000) is considered here. Seven companies in the textile industry in Wuhan, China were compared. Four financial ratios, profitability, productivity, market position, and debt ratio, were identified as the pertinent attributes. The first three attributes are beneficial, and higher values are desirable. The performance ratings are given in Table 28.1. The ratings of debt ratio were adjusted by taking the reversal of the original values, so that this attribute could be treated as being beneficial.

Table 28.1. Performance ratings of companies (from Deng *et al.*, 2000; reprinted with permission from Elsevier)

Company	Profitability	Productivity	Market position	Debt ratio
A_1	0.12	49469	0.15	1.21
A_2	0.08	34251	0.14	1.23
A_3	0.04	32739	0.09	1.12
A_4	0.16	44631	0.11	1.56
A_5	0.09	33151	0.13	1.09
A_6	0.15	31408	0.07	1.39
A_7	0.13	30654	0.17	1.16

28.2.1 Application of Graph Theory and Matrix Approach (GTMA)

Various steps of the methodology proposed in Section 2.6 are carried out as described below.

Step 1: In the present work, the attributes considered are the same as those of Deng *et al.* (2000) and these are: profitability (PR), productivity (PD), market position (MP), and debt ratio (DR). The quantitative values of the attributes, which are given in Table 28.1, are to be normalized. All four attributes are considered as beneficial, and higher values are desirable. Values of these attributes are normalized, and are given in Table 28.2 in the respective columns.

Table 28.2. Normalized performance ratings of companies

Company	Profitability	Productivity	Market position	Debt ratio
A_1	0.75	1.00	0.8823	0.7756
A_2	0.50	0.6924	0.8235	0.7885
A_3	0.25	0.6618	0.5294	0.7179
A_4	1.00	0.9022	0.6470	1.00
A_5	0.5625	0.6701	0.7647	0.6987
A_6	0.9375	0.6349	0.4118	0.8910
A_7	0.8125	0.6197	1.00	0.7436

Let the evaluator select the following relative importance assignments:

$$\begin{array}{c} \\ PR \\ PD \\ MP \\ DR \end{array}\begin{array}{c}\begin{array}{cccc} PR & PD & MP & DR \end{array}\\ \left[\begin{array}{cccc} - & 0.745 & 0.665 & 0.865 \\ 0.255 & - & 0.335 & 0.665 \\ 0.335 & 0.665 & - & 0.745 \\ 0.135 & 0.335 & 0.255 & - \end{array}\right]\end{array}$$

In this case, PR is considered more important than the other three attributes.

Step 2:

The inter-company comparison attributes digraph, inter-company comparison attributes matrix of the digraph, and inter-company comparison attributes function for the matrix can be prepared. The value of the inter-company comparison index is calculated, using the values of A_i and a_{ij} for each company. The inter-company comparison index values are given below in descending order:

A_4	2.4560
A_1	2.2342
A_7	1.9666
A_6	1.6673
A_2	1.5792
A_5	1.5028
A_3	1.0716

Thus, GTMA suggests a ranking order of A_4 - A_1 - A_7 - A_6 - A_2 - A_5 - A_3. However, this ranking doesn't fully correspond to that presented by Deng *et al.*

(2000) using entropy measure to determine the objective weights (*i.e.*, A_4 - A_7 - A_1 - A_6 - A_5 - A_2 - A_3). This difference in ranking is due to the different relative importance weights used for these methods.

28.2.2 SAW Method

Deng *et al.* (2000) used the entropy measure method to determine the objective weights of the attributes. The weights obtained were: W_{PR} = 0.54, W_{PD} = 0.13, W_{MP} = 0.28, and W_{DR} = 0.06. Using the same objective weights as used by Deng *et al.* (2000), the SAW method leads to the following values of the inter-company comparison index:

A_4	0.8984
A_7	0.8439
A_1	0.8286
A_6	0.7575
A_5	0.6469
A_2	0.6379
A_3	0.4123

The SAW method suggests a ranking order of A_4 - A_7 - A_1 - A_6 - A_5 - A_2 - A_3, which matches exactly with that presented by Deng *et al.* (2000) using entropy measure to determine the objective weights.

28.2.3 WPM

Using the same weights of the attributes as those selected for the SAW method, the inter-company comparison index value for each company is calculated, and the values are arranged as given below:

A_4	0.8734
A_7	0.8252
A_1	0.8141
A_6	0.7052
A_5	0.6317
A_2	0.6087
A_3	0.3678

The WPM method also suggests a ranking order of A_4 - A_7 - A_1 - A_6 - A_5 - A_2 - A_3, which matches exactly with that presented above for the AHP method, and Deng *et al.* (2000) using the entropy measure to determine the objective weights.

28.2.4 AHP and its Versions

The AHP method may use the same weights as those of the SAW method. In that case, the ranking of the companies will be the same.

For the above weights of importance of attributes, multiplicative AHP leads to the same ranking order.

28.2.5 TOPSIS Method

Following the steps of the methodology given in Section 3.2.6, the TOPSIS method gives the following ranking order:

A_4	0.8143
A_7	0.7513
A_6	0.6919
A_1	0.6840
A_5	0.4371
A_2	0.3853
A_3	0.0688

The TOPSIS method suggests a ranking order of A_4 - A_7 - A_6 - A_1 - A_5 - A_2 - A_3. The positions of companies 1 and 6 are mutually exchanged in this ranking, as compared to those given by the other methods.

28.2.6 Modified TOPSIS Method

This methods leads to the following ranking order:

A_4	0.7596
A_7	0.7186
A_1	0.6968
A_6	0.6066
A_5	0.4423
A_2	0.4143
A_3	0.0911

The modified TOPSIS method also suggests a ranking order of A_4 - A_7 - A_1 - A_6 - A_5 - A_2 - A_3, which matches exactly with that presented by the other methods, except for the simple TOPSIS method.

It may be noted here that the modified TOPSIS method proposed in this book is different from the 'modified' TOPSIS method used by Deng *et al.* (2000). The normalization procedure used by Deng *et al.* (2000) was different.

Thus, the methods proposed in this chapter enable simultaneous consideration of multiple financial ratios to adequately evaluate and rank the relative performance of competing companies.

References

Barnes A (1987) The analysis and use of financial ratios: a review article. Journal of Business and Finance Accounting 14:449-461

Davis E, Kay J (1990) Assessing corporate performance. Business Strategy Review, Summer:1-16

Deng H, Yeh CH, Willis RJ (2000) Inter-company comparison using modified TOPSIS with objective weights. Computers & Operations Research 27: 963-973

Diakoulaki D, Mavrotas G, Papayannakis L (1992) A multicriteria approach for evaluating the performance of industrial firms. Omega 20:467-474

Smith P (1990) Data envelopment analysis applied to financial statements. Omega 18:131-138

van der Wijst D (1990) Modelling inter-firm comparisons in small business. Omega 18:123-129

29

Vendor Selection in a Supply Chain Environment

29.1 Introduction

In today's highly competitive and interrelated manufacturing environment, the performance of the vendor becomes a key element in a company's success, or failure. Vendor selection decisions are an important component of production and logistics management for many companies. These decisions are typically complicated, for several reasons. First, potential options may need to be evaluated on more than one criterion. A second complication is the fact that individual vendors may have different performance characteristics for different criteria. A third complication arises from internal policy constraints, and externally imposed system constraints placed on the buying process. The nature of vendor selection decision usually is complex, unstructured, and inherently a multiple criteria problem.

Weber and Ellram (1993) explored the use of a multi-objective programming approach as a method for supplier selection in a just-in-time (JIT) setting. Weber (1996) used the data envelopment analysis (DEA) approach for measuring vendor performance. Roodhooft and Konings (1996) used an activity-based costing approach for vendor evaluation. Weber *et al.* (1998, 2000) described however, in certain situations, two multicriteria analysis tools, *i.e.*, multi-objective programming and DEA, could be used together for the supplier selection and negotiation process. Verma and Pullman (1998) examined the difference between managers' rating of the perceived importance of different supplier attributes, and their actual choice of suppliers in an experimental setting.

Ghodsypour and O'Brien (1998) proposed an integration of the analytic hierarchy process (AHP) method and linear programming to consider both tangible and intangible factors in choosing the best suppliers, and placing the optimum order quantities among these such that the total value of purchasing is maximized. In another work, Ghodsypour and O'Brien (2001) presented a mixed integer nonlinear programming model to solve a multiple sourcing problem, which took into account the total cost of logistics, including net price, and storage, transportation and ordering costs. Buyer limitations on budget, quality, service, *etc.* could also be considered in the model.

Boer *et al.* (1998) studied the application of outranking methods in support of vendor selection. Motwani *et al.* (1999) attempted to fill a void in supplier selection research by developing a model for sourcing and purchasing in an international setting, particularly in developing countries. Yahya and Kingsman (1999) presented vendor rating for an entrepreneur development program using the AHP method. Liu *et al.* (2000) compared suppliers using data envelopment analysis (DEA). Bragilia and Petroni (2000) described a multi-attribute utility theory based on the use of DEA, aiming at helping purchasing managers to formulate viable sourcing strategies in a changing market. Boer *et al.* (2001) presented a review of decision methods reported in the academic literature for supporting the vendor selection process.

Akarte *et al.* (2001) developed an approach to evaluate casting suppliers, using the analytical hierarchy process (AHP) method. The approach was implemented in a prototype web-based system. Handfield *et al.* (2002) illustrated the use of AHP as a decision support model to help managers understand the tradeoffs between environmental dimensions. Gunasekaran *et al.* (2001) established a framework consisting of three-level indices: strategic performance, tactical performance, and operational performance. Feng *et al.* (2001) presented a stochastic integer programming approach for simultaneous selection of tolerances and suppliers based on the quality loss function and process capability indices. Oliveria and Lourenco (2002) developed a multicriteria model for assigning new orders to service vendors. Kwang *et al.* (2002) combined a scoring method and fuzzy expert systems for vendor assessment, and presented a case study. Cebi and Bayraktar (2003) structured vendor selection problem in terms of an integrated lexicographic goal programming (LGP) and AHP model, including both quantitative and qualitative conflicting factors.

Cengiz *et al.* (2003) applied the fuzzy AHP method for solving the vendor selection problem. Ibrahim and Ugur (2003) used activity-based costing and fuzzy present-worth techniques for vendor selection. Kumar *et al.* (2004) presented a fuzzy goal programming approach for the vendor selection problem in a supply chain. Ge *et al.* (2004) developed an integrated AHP and preemptive goal programming (PGP)-based multicriteria decision-making methodology to account both qualitative and quantitative factors in supplier selection. Pi and Low (2005) presented a supplier evaluation and selection approach using Taguchi's loss function and AHP.

Degraeve *et al.* (2005) used total cost of ownership information for evaluatinga firm's strategic procurement options. The approach was used to develop a decision support system at a European multinational steel company. Shyur and Shih (2006) proposed a hybrid MCDM model using ANP and TOPSIS methods for strategic vendor selection. Sucky (2006) proposed a dynamic decision-making approach based on the principle of hierarchical planning for strategic vendor selection. Cao and Wang (2006) discussed the aspects of optimizing vendor selection in a two-stage outsourcing process. Wadhwa and Ravindran (2006) presented multi-objective optimization methods including goal programming and compromise programming. Amid *et al.* (2006) proposed a multi-objective linear model for supplier selection in a supply chain. Rao (2007) proposed a combined AHP and genetic algorithm (GA) method for the vendor selection problem.

As described above, many precision-based methods have been proposed in the past for vendor selection. However, there is need for a systematic, simple and convenient procedure acceptable to companies for efficient and effective evaluation of vendors. The aim of the vendor selection procedure is to identify the vendor selection attributes, and obtain the most appropriate combination of these in conjunction with the real requirements of the company. Efforts need to be extended to determine those attributes that influence vendor selection for supplying a given product, using a systematic approach, to eliminate unsuitable vendors and for selection of an appropriate vendor to strengthen the existing vendor selection procedure. This is considered in this chapter using GTMA and other fuzzy MADM methods.

A vendor selection attribute is defined as a factor that influences the selection of a vendor for supplying a given product in a supply chain environment. The attributes include the vendor's technical capability, financial position, procedural compliance, reputation and position in industry, attitude, flexibility, packaging ability, labor relation record, communication systems, management and organization, geographical location, production facilities and capacity, warranties and claim policies, repair service, environmentally friendly features of the product, *etc*. The vendor selection attributes can be identified and the vendors can be short-listed on the basis of the attributes identified satisfying the requirements. A quantitative or qualitative value, or its range, can be assigned to each identified attribute as a limiting value, or threshold value, for its acceptance. A vendor, meeting each of these limiting or threshold values of the attributes identified, can be short-listed.

29.2 Example 1

Now, an example is considered to demonstrate and validate the proposed decision-making methods for solving the vendor selection problem in a supply chain environment.

Liu *et al.* (2000) presented a case study to demonstrate the vendor performance evaluation using the data envelopment analysis (DEA) method for a firm that manufactures agricultural and construction equipment. The company purchases a significant number of parts used on the assembly line from domestic and foreign vendors, in addition to materials purchased for manufacturing major frame components in-house. The company had divided all purchased parts into 18 commodity groups, including hydraulic valves, fasteners, electrical, weldments, stampings, machined parts, and fabrication groups. Within each commodity group, the vendors were viewed as similar vendors, and interchangeable. The interchangeability was expressed by a capability matrix for various vendors to supply various parts in each commodity group.

To collect data in each commodity group, the company first listed all parts supplied by each vendor to obtain the supply variety. If a vendor supplies more than one commodity group, then the supply variety of this vendor in each group is the sum of the number of parts in all the groups, as this represents the comprehensive supplying ability of this vendor. Table 29.1 presents the details of alternative vendors, and the information on the vendor selection attributes for a hydraulic valve group.

The aggregate quality of a vendor was represented by the weighted percentage of non-defective parts supplied by the vendor with regard to the commodity group, where the weights of various parts were based on their annual quantities. For the price comparison, the company had given each part in a commodity group a price evaluation relative to the average market price. To prevent massive data collection effort, one of three estimated levels of prices, 120, 100, and 80, relative to market average prices were assigned to each part by the materials department of the company. The delivery performance was represented by the percentage of purchase orders delivered within the delivery window according to the purchase orders.

Table 29.1. Quantitative data of vendor selection attributes (from Liu *et al.* 2000; with permission from Emerald Insight)

Vendor	P	Q	DP	D	SV
1	100	100	90	249	2
2	100	99.79	80	643	13
3	100	100	90	714	3
4	100	100	90	1809	3
5	100	99.83	90	238	24
6	100	96.59	90	241	28
7	100	100	85	1404	1
8	100	100	97	984	24
9	100	99.91	90	641	11
10	100	97.54	100	588	53
11	100	99.95	95	241	10
12	100	99.85	98	567	7
13	100	99.97	90	567	19
14	100	91.89	90	967	12
15	80	99.99	95	635	33
16	100	100	95	795	2
17	80	99.99	95	689	34
18	100	99.36	85	913	9

P: Price ($) Q: Quality (%) D: Delivery performance (%)
D: Distance (miles) SV: Supply variety

29.2.1 Graph Theory and Matrix Approach

In the present work, the attributes considered are the same as those of Liu *et al.* (2000), and these are: price (P), quality (Q), delivery performance (DP), distance (D), and supply variety (SV).

The quantitative values of the vendor selection attributes, which are given in Table 29.1, are to be normalized. Q, DP, and SV are beneficial attributes, and higher values are desirable. P and D are non-beneficial attributes, and lower values are desirable. The values of these attributes for different vendors are normalized, and shown in Table 29.2.

Table 29.2. Normalized data of the vendors

Vendor	P	Q	DP	D	SV
1	0.8	1	0.9	0.9558	0.0377
2	0.8	0.9979	0.8	0.37	0.2453
3	0.8	1	0.9	0.3333	0.0566
4	0.8	1	0.9	0.1316	0.0566
5	0.8	0.9983	0.9	1	0.4528
6	0.8	0.9659	0.9	0.9876	0.5283
7	0.8	1	0.85	0.1695	0.0188
8	0.8	1	0.97	0.2419	0.4528
9	0.8	0.9991	0.9	0.3713	0.2075
10	0.8	0.9754	1	0.4048	1
11	0.8	0.9995	0.95	0.9876	0.1887
12	0.8	0.9985	0.98	0.4198	0.1321
13	0.8	0.9997	0.9	0.4198	0.3585
14	0.8	0.9189	0.9	0.2461	0.2264
15	1	0.9999	0.95	0.3748	0.6226
16	0.8	1	0.95	0.2994	0.0377
17	1	0.9999	0.95	0.3454	0.6415
18	0.8	0.9936	0.85	0.2607	0.1698

Let the decision maker make the following assignments of relative importance:

	P	Q	DP	D	SV
P	-	0.255	0.335	0.745	0.745
Q	0.745	-	0.665	0.745	0.745
DP	0.665	0.335	-	0.745	0.745
D	0.255	0.255	0.255	-	0.335
SV	0.335	0.255	0.255	0.665	-

The vendor selection attributes digraph, vendor selection attributes matrix of the digraph, and vendor selection function for the matrix can be prepared. The value of the vendor selection index is calculated using the values of A_i and a_{ij} for each vendor. The vendor selection index values of different vendors are given below in descending order:

6	5.3569
5	5.2403
10	5.2356
15	4.7139
17	4.6759
11	4.5581
1	3.9978
13	3.8509
8	3.7665
12	3.5024
9	3.4677
2	3.4020

14	3.1906
18	3.1467
3	3.1239
16	3.0950
4	2.8238
7	2.7684

From the values of the vendor selection index, it is understood that the vendor, designated as 6 is the best choice among the vendors considered for the given industrial application. The next choice is 5, and the last choice is 7.

29.2.2 TOPSIS Method

Various steps of the TOPSIS methodology using the AHP method for assigning the relative importance of attributes are described below:

Step 1: The objective is to evaluate the vendor performance and to select a vendor. The attributes considered are the same as those of Liu *et al.* (2000), and these are: price (P), quality (Q), delivery performance (DP), distance (D), and supply variety (SV).

Step 2: The next step is to represent all the information available of attributes in the form of a decision matrix. These data are shown in Table 29.1.

Step 3: The quantitative values of the vendor selection attributes, which are given in Table 29.1, are to be normalized. Q, DP, and SV are beneficial attributes and P and D are non-beneficial attributes. The values of these attributes for different vendors are normalized as shown below:

$$
\begin{bmatrix}
0.2405 & 0.2377 & 0.2318 & 0.0718 & 0.0224 \\
0.2405 & 0.2372 & 0.2060 & 0.1854 & 0.1453 \\
0.2405 & 0.2377 & 0.2318 & 0.2059 & 0.0335 \\
0.2405 & 0.2377 & 0.2318 & 0.5217 & 0.0335 \\
0.2405 & 0.2373 & 0.2318 & 0.0686 & 0.2682 \\
0.2405 & 0.2296 & 0.2318 & 0.0695 & 0.3129 \\
0.2405 & 0.2377 & 0.2189 & 0.4049 & 0.0112 \\
0.2405 & 0.2377 & 0.2498 & 0.2838 & 0.2682 \\
0.2405 & 0.2375 & 0.2318 & 0.1849 & 0.1229 \\
0.2405 & 0.2318 & 0.2575 & 0.1696 & 0.5922 \\
0.2405 & 0.2375 & 0.2446 & 0.0695 & 0.1117 \\
0.2405 & 0.2373 & 0.2523 & 0.1635 & 0.0782 \\
0.2405 & 0.2377 & 0.2317 & 0.1635 & 0.2123 \\
0.2405 & 0.2184 & 0.2317 & 0.2789 & 0.1341 \\
0.1924 & 0.2376 & 0.2446 & 0.1831 & 0.3687 \\
0.2405 & 0.2377 & 0.2446 & 0.2293 & 0.0224 \\
0.1924 & 0.2376 & 0.2446 & 0.1987 & 0.3799 \\
0.2405 & 0.2361 & 0.2189 & 0.2633 & 0.1006
\end{bmatrix}
$$

Step 4: Let the decision maker select the following assignments regarding the relative importance of attributes (a_{ij}), using the AHP method:

	P	Q	DP	D	SV
P	1	1/5	1/3	4	3
Q	5	1	3	6	5
DP	3	1/3	1	5	4
D	1/4	1/6	1/5	1	1/3
SV	1/3	1/5	1/4	3	1

The assigned values in this chapter are for demonstration purposes only. The normalized weights for each attribute are: $W_P = 0.1361$, $W_Q = 0.4829$, $W_{DP} = 0.2591$, $W_D = 0.0438$, and $W_{SV} = 0.0782$. The value of λ_{max} is 5.3388 and CR = 0.0756, which is much less than the allowed CR value of 0.1. Thus, there is good consistency in the judgements made.

Step 5: The weighted normalized matrix $V1_{ij}$ is calculated.

Step 6: The next step is to obtain the ideal (best) and negative ideal (worst) solutions. These are given as:

$V_1^+ = 0.0262$ $V_1^- = 0.0327$
$V_2^+ = 0.1148$ $V_2^- = 0.1055$
$V_3^+ = 0.0667$ $V_3^- = 0.0534$
$V_4^+ = 0.0030$ $V_4^- = 0.0228$
$V_5^+ = 0.0462$ $V_5^- = 0.0009$

Step 7: The next step is to obtain the separation measures using Equations 3.12 and 3.13, and these are:

$S_1^+ = 0.0454$ $S_1^- = 0.0228$
$S_2^+ = 0.0382$ $S_2^- = 0.0202$
$S_3^+ = 0.0449$ $S_3^- = 0.0180$
$S_4^+ = 0.0488$ $S_4^- = 0.0116$
$S_5^+ = 0.0269$ $S_5^- = 0.0304$
$S_6^+ = 0.0240$ $S_6^- = 0.0319$
$S_7^+ = 0.0491$ $S_7^- = 0.0111$
$S_8^+ = 0.0278$ $S_8^- = 0.0269$
$S_9^+ = 0.0381$ $S_9^- = 0.0206$
$S_{10}^+ = 0.0084$ $S_{10}^- = 0.0502$
$S_{11}^+ = 0.0382$ $S_{11}^- = 0.0253$
$S_{12}^+ = 0.0409$ $S_{12}^- = 0.0224$
$S_{13}^+ = 0.0313$ $S_{13}^- = 0.0249$
$S_{14}^+ = 0.0392$ $S_{14}^- = 0.0158$
$S_{15}^+ = 0.0184$ $S_{15}^- = 0.0350$
$S_{16}^+ = 0.0456$ $S_{16}^- = 0.0188$
$S_{17}^+ = 0.0178$ $S_{17}^- = 0.0354$
$S_{18}^+ = 0.0411$ $S_{18}^- = 0.0162$

Step 8: The relative closeness of a particular alternative to the ideal solution is calculated, and these are:

$P_1 = 0.3343$ $P_2 = 0.3458$ $P_3 = 0.2863$ $P_4 = 0.1919$
$P_5 = 0.5299$ $P_6 = 0.5706$ $P_7 = 0.1847$ $P_8 = 0.4919$
$P_9 = 0.3504$ $P_{10} = 0.8566$ $P_{11} = 0.3984$ $P_{12} = 0.3539$
$P_{13} = 0.4432$ $P_{14} = 0.2873$ $P_{15} = 0.6551$ $P_{16} = 0.2914$
$P_{17} = 0.6654$ $P_{18} = 0.2824$

This relative closeness to ideal solution is named the 'vendor selection index (VSI)' in the present example.

Step 9: The vendors are arranged in the descending order of their VSI. This can be arranged as 10-17-15-6-5-8-13-11-12-9-2-1-16-14-3-18-4-7.

From the above values of VSI, it is understood that vendor 10 is the first choice for the supply of items in the hydraulic valve group considered under the given conditions. The second choice is vendor 17, and the third choice is vendor 15 and these results match with those of Liu *et al.* (2000). Transferring of parts can be made from the vendors with low ranking to the peer vendors with higher ranking, keeping in mind the capacities and limitations of the high-ranking vendors. However, Liu *et al.* (2000) had suggested that only five vendors, *i.e.*, 1, 10, 12, 15, and 17 were efficient, the remaining vendors were inefficient, and vendors 2 and 14 were most inefficient. In the present work, vendor 7 is proposed as the last choice. A closer look at the values of the attributes for vendors 7 and 14 reveals that vendor 7 is inferior to 14 with respect to three attributes, DP, D, and SV; better than 14 with respect to Q; and equal to 14 with respect to P. Thus, proposing vendor 7 as the last choice seems to be justified. However, as mentioned above, this all depends on the values of relative importance judiciously decided by the decision maker. If different values of relative importance are assigned to the attributes, then the ranking will change. The same is true in the case of the DEA approach used by Liu *et al.* (2000). In their work, Liu *et al.* (2000) assigned different weights of relative importance to the attributes considered and these weights were different from those of the present work and hence the ranking obtained was somewhat different from those presented in this chapter. Further, the authors had not indicated the values of weights of relative importance assigned to the attributes, and the basic concept that summation of all relative importance weights should be equal to 1.0 was not followed. Also, DEA requires more computation, can not handle the qualitative attributes adequately, and does not offer any provision for checking the consistency in judgements of relative importance.

One need not confused about the differences in rankings of vendors presented by the GTMA and TOPSIS methods. Indeed, the differences are due to assignment of different values of relative importance to the attributes in these methods.

As mentioned in Section 29.1, in a supply chain environment, the decision-making is related with the selection of vendors and the quantities to be purchased from those vendors. A number of vendor selection attributes are to be considered and the objectives are to be formulated in this regard. The attributes identified may be classified into two categories. The first category of attributes include those factors that the company will formulate as its objectives, such as net purchasing costs, net rejections, net late deliveries, *etc.* The second category of attributes includes those factors that will help the company in short-listing the vendors, such as the vendor's technical capability, financial position, procedural compliance, reputation and position in industry, attitude, flexibility, packaging ability, labor relation record, communication systems, management and organization, geographical location, production facilities and capacity, warranties and claim policies, repair service, environmentally friendly features of the product, *etc.* These attributes have already mentioned in Section 29.1. The vendor selection attributes can be identified, and the objectives can be formulated based on the attributes identified of first category. The vendors can be short-listed based on the identified attributes of second category. A quantitative or qualitative value, or its range, can be assigned to each identified attribute of second category as a limiting value, or threshold value, for its acceptance. A vendor, meeting each of these limiting or threshold values of the identified attributes in the second category, can be short-listed.

The objectives of a company in a supply chain environment correspond to the vendor selection attributes of the first category. The objective may be minimization of net purchasing cost or minimization of net late deliveries, or minimization of net rejections, or any such criterion of interest. A combined objective function can be formulated considering all the objectives of interest, and the decision maker may assign equal, or different, weights to the objectives depending on the policies of the company regarding relative importance of the objectives. This is to be done preferably using a logical and systematic approach and AHP is one such method. However, AHP method is not suitable to solve constrained multi-objective problems. Hence, Rao (2007) used AHP only to assign the weights of importance to the objectives and the genetic algorithm (GA) method to constrained multi-objective vendor selection problem. The next section briefly describes the basics of the GA method.

29.3 Genetic Algorithms

Over the last decade, genetic algorithms (GAs) have been extensively used as search and optimization tools in various problem domains, including the sciences, commerce, and engineering. The primary reasons for their success are their broad applicability, ease of use, robustness and global perspective (Goldberg 1989; Mitchell, 1996; Gen and Cheng, 1997; Vose, 1999; Deb, 2002). The genetic algorithms are inspired by Darwin's theory evolution. The algorithm is started with a set of solution (represented by chromosomes) called a population. Solutions from one population are used to form a new population. This is motivated by that the new population will be better than the old one. Solutions to forming new solutions (offsprings) are selected according to their fitness. The more suitable they are, the more chances they have of reproducing. The iteration is stopped after the completion of maximal number of iterations (generations) or on the attainment of the best result.

The decision variables of multiple objective, multiple variable, constrained or unconstrained optimization problems solved by GAs may be represented by either binary coding or real coding. GAs employ three important genetic operators for solving optimization problems, and these operators are briefly described below.

Reproduction or selection operator: GA begins with a set of solutions called population (represented by chromosomes or strings). The primary objective of the reproduction operator is to make duplicates of good solutions, and eliminate bad solutions in a population, while keeping the population size constant. This is achieved by identifying good solutions in a population, making multiple copies of good solutions, and eliminating bad solutions from the population so that multiple copies of good solutions can be placed in the population.

Crossover operator: This operator is applied to the strings of the mating pool after the reproduction operator has been applied. The latter cannot create any new solutions in the population, and it only makes more copies of good solutions at the expense of not-so-good solutions. The creation of new solutions is performed by the crossover operator. In crossover operation, two strings are randomly selected from the mating pool, and some portions of the strings are exchanged between strings to create new strings.

Mutation operator: The crossover operator is mainly responsible for the search aspect of genetic algorithms, even though the mutation operator is also used for this purpose. Mutation is intended to prevent all solutions in the population being concentrated into a local optimum of the solved problem. The bitwise mutation operator changes a 1 into 0, and vice versa, with a small mutation probability. The need for mutation is to maintain diversity in the population.

The three GA operators reproduction or selection, crossover, and mutation, are simple and straight-forward. The reproduction operator selects good strings, while the crossover operator recombines good substrings from two good strings to hopefully form a better spring. The mutation operator alters a string locally to hopefully create a better string. The basic genetic algorithm is outlined below:

1. [Start] Choose a coding to represent problem decision variables, a reproduction or selection operator, a crossover operator, and a mutation operator. Choose population size n, crossover probability p_c, and mutation probability p_m. Initialize a random population of strings of size ‘s’. Choose a maximum allowable generation (*i.e.*, iteration) number t_{max}. Set t=0
2. [Fitness] Evaluate the fitness function of each string in the population
3. [New population] Create a new population by repeating the following steps until the new population is complete

[Reproduction or selection] Select two parent strings from a population according to their fitness (the better fitness, the bigger the chance of being selected)

[Crossover] Crossover the parents to form new offspring (children). If no crossover is performed, then the offspring are the exact copy of parents.

[Mutation] Mutate the new offspring at each locus (position in string).

[Accepting] Place the new offspring in the new population

4. [Replace] Use the newly generated population for a further run of the algorithm
5. [Test] If $t > t_{max}$, or other termination criteria, are satisfied, then terminate and return the best solution in current population
6. [Loop] Go to step 2

The above procedure is repeated until an optimum solution is reached. More details on the genetic algorithms, and their applications can be found in literature (Goldberg 1989, Mitchell 1996, Gen and Cheng 1997, Vose 1999, Deb 2002).

29.4 Proposed Methodology

A methodology using, AHP and GA methods together is proposed in this section, for solving the vendor selection problem in a supply chain environment (from

Rao, 2007; with permission from Inderscience Publishers). AHP is used for logical assignment of weights of relative importance to the objectives, and GA is used to solve the constrained multi-objective vendor selection problem by performing a global search for the optimum values of the decision variables (*i.e.*, vendor order quantities). The novelty of the proposed methodology is that it uses GA, which has not been investigated by earlier researchers to solve the vendor selection problem in a supply chain. Further, the use of AHP for assigning the weights of relative importance to the objectives is relatively new. The main steps of the methodology are given below:

Step 1: Identify the vendor selection attributes of both categories, as explained in Section 29.2, for the given vendor selection problem. The choice of the attributes depends upon the company's requirements.

Step 2: Obtain the data of vendors.

Step 3: Short-list the vendors, based on the attributes identified in the second category. A quantitative or qualitative value, or its range, can be assigned to each identified criterion of second category as a limiting value, or threshold value, for its acceptance. A vendor, meeting each of these limiting or threshold values of the attributes identified in the second category, can be short-listed.

Step 4: Consider the vendor selection attributes of the first category as the objectives of the given vendor selection problem.

Step 5: Formulate the objective functions and constraints set in terms of the decision variables. The decision variables may be the quantities to be purchased from vendors, *i.e.*, vendor order quantities.

Step 6: Apply GA method, and obtain the aspired level of each objective. The aspiration value of an objective indicates the best possible solution of that objective ignoring other objectives. This can be done using a simple heuristic: treat the individual function of that particular objective as the single objective function of the proposed GA method within the given constraint set.

Step 7: Assign the values of relative importance (a_{ij}) to the objectives and find out the relative weights (w_i) of the objectives using the AHP method.

Step 8: Formulate a combined objective function considering the objective functions and their relative weights of importance. Now, treat this combined objective function as the single objective function of the genetic algorithm (GA) with the given constraint set.

Step 9: Apply GA method and obtain the optimum values of the combined objective function and the decision variables.

Step 10: Based on the GA results, decide the number of vendors to employ, and the quantities to be purchased from them.

Step 11: Document the results for future reference.

Now, to demonstrate and validate the proposed methodology for the selection of appropriate vendors and their quota allocations, an example is considered.

29.5 Example 2

Kumar *et al.* (2004) presented a fuzzy goal programming approach for the vendor selection problem in a supply chain. The vendor source data are given in Table

29.3 and data relate to a realistic situation of a manufacturing company dealing with automobile parts. The approach was developed from requirements stipulated during the initial stages of implementing a formal program for better management of supply chain, which subsequently undertook a vendor certification plan for its purchased items. Those vendors who successfully passed the screening processes were eligible for procurement. Four established vendors for a projected demand of the part had been screened for supplying this purchased item. Table 29.3 represents the dataset for the price quoted (p_i), the percentage rejections (q_i), the percentage late deliveries (d_i), vendor capacities (U_i), vendors' quota flexibility (f_i) on a scale of 0–1, vendor rating (r_i) on a scale of 0–1, and budget allocations for the vendors (B_i). The lowest value of flexibility in vendor quota and the least total purchase value of supplied items were policy decisions and dictated by the demand. The lowest value of flexibility in vendor quota was given as $F = fD$ and the least total purchase value of supplied items was given as $P = rD$. Overall flexibility (f) and the overall vendor rating (r) were considered as 0.03 and 0.92, respectively on a scale of 0–1, aggregate demand (D) as 20,000, and the least value of flexibility in vendor quota (F) as well as the lowest total purchase value of supplied items (P) were considered as 600 and 18,400, respectively.

Table 29.3. Vendor source data (from Kumar *et al.*, 2004; reprinted with permission from Elsevier)

V	p_i ($)	q_i (%)	d_i (%)	U_i (units)	f_i	r_i	B_i($)
1	5	0.05	0.04	5,000	0.02	0.88	25,000
2	7	0.03	0.02	15,000	0.01	0.91	10,000
3	6	0.00	0.08	6,000	0.06	0.97	35,000
4	2	0.02	0.01	3,000	0.04	0.85	5,500

V: Vendor

The mixed integer programming vendor selection problem (MIP_VSP) formulation for the three objectives, and the set of system and policy constraints formulated by Kumar *et al.* (2004) are given below:

$$\text{Minimize } Z_1 = \sum_{i=1}^{n} p_i x_i \quad (29.1)$$

$$\text{Minimize } Z_2 = \sum_{i=1}^{n} q_i x_i \quad (29.2)$$

$$\text{Minimize } Z_3 = \sum_{i=1}^{n} d_i x_i \quad (29.3)$$

Subject to:

$$\sum_{i=1}^{n} x_i = D \quad (29.4)$$

$$x_i \le U_i \text{ for all } i = 1,2,\ldots, n \qquad (29.5)$$

$$\sum_{i=1}^{n} f_i x_i \ge F \qquad (29.6)$$

$$\sum_{i=1}^{n} r_i x_i \ge P \qquad (29.7)$$

$$p_i x_i \le B_i \text{ for all } i = 1,2,\ldots, n \qquad (29.8)$$

$$x_i \ge 0 \text{ and integer} \qquad (29.9)$$

x_i = quantity purchased from vendor i
D = aggregate demand for the item during a fixed planning period
n = number of vendors competing for selection
p_i = price of a unit item of the ordered quantity x_i from the vendor i
q_i = percentage of rejected units delivered by vendor i
d_i = percentage of the units delivered late by the vendor i
U_i = upper limit of the quantity available from vendor i
f_i = vendor quota flexibility for vendor i
F = lower limit of flexibility in supply quota that a vendor should have
r_i = vendor rating value for vendor i
P = lower limit to total purchasing value that a vendor should have
B_i = budget allocated to each vendor
Objective function 29.1 minimizes the net cost for all the items.
Objective function 29.2 minimizes the net number of rejected items for the vendors.
Objective function 29.3 minimizes the net number of items delivered late from the vendors.
Constraint 29.4 represents restrictions due to the overall demand of items.
Constraint 29.5 represents restrictions due to the maximum capacity of the vendors.
Constraint 29.6 incorporates the flexibility needed with the vendors' quota.
Constraint 29.7 incorporates the total purchase value constraint for all the ordered quantities.
Constraint 29.8 represents restrictions on the budget amount allocated to the vendors for supplying the items.

The above example is considered to demonstrate and validate the proposed methodology using a combined AHP and GA method for the selection of appropriate vendors and their quota allocations. Different steps of the proposed procedure are carried out as described below.

Steps 1 to 5: These steps had already been carried out by Kumar *et al.* (2004) and hence the objectives and constraints considered to evaluate the vendor performance, and to allocate the quotas are the same as those of Kumar *et al.* (2004). The objectives are: minimization of net purchasing cost (Z_1), minimization of net rejects (Z_2) and minimization of late deliveries (Z_3).

Step 6: The aspired level of an objective is obtained by using a simple heuristic. The individual function of that particular objective is first treated as the single objective function of the proposed genetic algorithm having a constraint set same as that defined by Kumar *et al.* (2004). The decision variables in GA are nothing but the quantities to be purchased from vendors *i.e.*, vendor order quantities. Binary coding is used to represent the decision variables x_1, x_2, x_3 and x_4, and the details of GA employed are given in Table 29.4. The rank selection method is used as a reproduction method and the probabilities of crossover and mutation, 0.9 and 0.01, respectively, are selected based on the various trial runs to obtain better solution. The aspired levels of the three objectives, *viz.* the minimization of net purchasing cost, minimization of net rejects and minimization of net late deliveries, have thereby been obtained as $100,225, 450 units, and 775 units respectively. Each of these values indicates the best possible solution (Z_{min}) of that objective, ignoring other objectives.

Table 29.4. Details of GA used

Variable	Description/value
Reproduction method	: Rank selection method
Crossover type	: Binary GA (single point)
Strategy	: 1 cross site with swapping
Population size	: 100
Total no. of generations	: 100
Crossover probability	: 0.9000
Mutation probability	: 0.0100
String length	: 40
Number of variables, binary	: 4
Sigma-share value	: 0.281
Sharing Strategy	: sharing on Parameter Space
Lower and Upper limits	:
	$0 <= x_1 <= 5000$
	$0 <= x_2 <= 15000$
	$0 <= x_3 <= 6000$
	$0 <= x_4 <= 3000$

Step 7: Relative importance of objectives (a_{ij}) is assigned values, using the AHP method. Let the decision maker select the following assignments:

$$\begin{array}{c} \\ Z_1 \\ Z_2 \\ Z_3 \end{array} \begin{array}{c} \begin{array}{ccc} Z_1 & Z_2 & Z_3 \end{array} \\ \begin{bmatrix} 1 & 2 & 3 \\ 1/2 & 1 & 2 \\ 1/3 & 1/2 & 1 \end{bmatrix} \end{array}$$

The normalized weights of each objective are: $W_{Z1} = 0.54$, $W_{Z2} = 0.3$, and $W_{Z3} = 0.16$. The value of λ_{max} is 3.009203 and CR = 0.00885, which is very much less than the allowed CR value of 0.1. Thus, there is good consistency in the judgements made.

Step 8: A combined objective function, Z, is formulated. The three objectives Z_1, Z_2, and Z_3 are combined, with separate weight factors, W_Z s, for each.

$$Z=W_{Z1}(Z_1/Z_{1min}) + W_{Z2}(Z_2/Z_{2min}) + W_{Z3}(Z_3/Z_{3min}) \tag{29.10}$$

Steps 9 & 10: The combined objective function shown above, having the same constraint set as that defined by Kumar *et al.* (2004), is treated as the single objective function of the genetic algorithm. The GA program is run and the GA results are given in Table 29.5, which also shows the GA results for different values of W_Z s assigned to the three objectives considering four vendors. It can be seen that the values of Z_1, Z_2, and Z_3 depend on the AHP weights assigned. It is evident from Table 3 that in order to decide the weights, the decision maker needs to have a clear idea of the relative importance of objectives. Table 29.5 also shows the results of considering only three vendors (*i.e.*, any of $x_i = 0$).

Table 29.5. Results of application of GA

S. No.	Objective	AHP weights	Value of objective	Vendor order quantity (units) x_1	x_2	x_3	x_4
1	Z_1 ($)	0.3333	115,107	2,869	8,636	5,830	2,665
	Z_2 (units)	0.3333	456				
	Z_3 (units)	0.3333	781				
1.1	Z_1 ($)	0.3333	120,437	0	11,437	5,813	2,750
	Z_2 (units)	0.3333	398				
	Z_3 (units)	0.3333	721				
1.2	Z_1 ($)	0.3333	117,100	4,585	12,669	0	2,746
	Z_2 (units)	0.3333	664				
	Z_3 (units)	0.3333	464				
1.3	Z_1 ($)	0.3333	124,717	4,726	9,443	5,831	0
	Z_2 (units)	0.3333	520				
	Z_3 (units)	0.3333	844				
2	Z_1 ($)	0.54	115,122	2,849	8,651	5,830	2,670
	Z_2 (units)	0.30	456				
	Z_3 (units)	0.16	780				
3	Z_1 ($)	0.4	115,171	2,752	8,719	5,830	2699
	Z_2 (units)	0.3	453				
	Z_3 (units)	0.3	778				

The values of the combined objective function, Z, are calculated using Expression 29.10, considering the levels of aspiration of the three objectives, *viz.*, the minimization of net purchasing cost, minimization of net rejects, and minimization of net late deliveries ($100,225, 450 units, and 775 units, respectively), and the actual values of the objectives obtained by assigning equal or different AHP weights and running the GA program. It is observed that the value of Z is minimum in the case 1.1 of Table 29.5 with a value of 1.00547, and this case offers the optimum solution for the present vendor selection problem in a supply chain environment. It can be seen that for $x_1 = 0$, $Z_1 = 120433$, $Z_2 = 398$, and Z_3=721 and these values are much better than the final results obtained by Kumar *et al.* (2004), in whose case $Z_1 = 124914$, $Z_2 = 420$, and Z_3=700. For this optimal solution in the present work, vendor 2 is allocated 11437 units (76.25% of capacity consuming 80% of his allocated budget), vendor 3 is allocated 5813 units (97% of capacity consuming 99.65% of his allocated budget) and vendor 4 is allocated 2750 units (91.7% of capacity consuming 100% of his allocated budget). In the corresponding solution given by Kumar *et al.* (2004), vendor 2 is allocated 12714 units (85% of capacity consuming 89% of his allocated budget), vendor 3 is allocated 5336 units (89% of capacity consuming 91% of his allocated budget) and vendor 4 is allocated 2750 units (65% of capacity consuming 71% of his allocated budget). Thus the results obtained by the application of GA in the present work are better and more genuine than those by Kumar *et al.* (2004). It may be added here that the percentages of capacity and budget consumption in the case of vendors 2 and 3 were erroneously calculated by Kumar *et al.* (2004).

Further, even though fuzziness was proposed in the work of Kumar *et al.* (2004) for the three objectives, the final values obtained were essentially the same as those for crisp objectives. The values were exactly the same in the case of the second and third objectives, and only a small difference of $86 was present (*i.e.*, $125000 – $124914) in the case of the first objective (Kumar *et al.* 2004). In fact, the fuzzy approach need not be used separately for this problem. A range may be decided for each objective and the problem can be solved using any crisp optimization method. If the final results obtained fall within this range, then this may be considered as a solution to that problem, and there is no need to convert the problem into a fuzzy one. In the present work, AHP can determine the relative importance of objectives, and GA can be used for finding the value of each objective, either ignoring the other objectives or considering all the objectives simultaneously with same, or different, weights of relative importance.

It may be noted from Table 29.5 that the solution is not possible for $x_2 = 0$. This is because that the summation of the remaining x_i s will not be equal to the demand of 20000 units. In general, after deciding the weights of relative importance of the objectives, the management may consider different cases, such as all x_j, any of x_j=0, any two of x_j=0, any three of x_j=0, *etc.* Thus the number of vendors to employ can also be decided. The solution can be easily obtained using the proposed procedure of GA in conjunction with AHP, and the quota allocations can be decided. The management may then evaluate the practical business significance of the values obtained.

29.6 General Remarks

The objective of vendor selection is to identify vendors with the highest potential for meeting a company's needs consistently and at an acceptable cost. Selection is a broad comparison of vendors based on a common set of criteria and measures. However, the level of details used for examining potential vendors may vary depending on a company's needs. The overall goal of selection is to identify high-potential vendors and their quota allocations. An effective and appropriate vendor assessment method is therefore crucial to the competitiveness of companies. Two approaches are suggested in this chapter for solving the vendor selection problem. The first approach is that of a multiple attribute decision-making problem, and GTMA and/or other MADM methods are applied. The second approach is that of a multiple objective decision-making problem. A model using AHP and GA methods together is proposed in this chapter for solving the vendor selection problem in a supply chain environment (treating the problem as a multiple objective decision-making problem). AHP is used for logical assignment of weights of relative importance to the objectives, and GA is used to perform a global search for the optimum values of the decision variables (*i.e.*, vendor order quantities). Thus the solution obtained by GA offers a global optimum, rather than a local optimum. Integration of the proposed approach with the supply chain will lead to reduced costs, improved product quality, improved flexibility to meet the needs of customers, and reduced lead time at different stages of the supply chain. Using the proposed methodology, the number of vendors to be employed can also be decided. The method proposed is a general one, and offers a systematic, more objective, and simple optimization approach that can be used for optimization of any system or process.

References

Akarte M, Surendra N, Ravi B, Rangaraj N (2001) Web based casting supplier evaluation using analytic hierarchy process. Journal of the Operational Research Society 52:511–522

Amid A, Ghodsypour SH, O'Brien C (2006) Fuzzy multiobjective linear model for supplier selection in a supply chain. International Journal of Production Economics 104:394–407

Boer, L, Wegen L, Telgen J (1998) Outranking methods in support of vendor selection. European Journal of Purchasing and Supply Management 4:109–118

Boer L, Labro E, Morlacchi P (2001) A review of methods supporting vendor selection. European Journal of Purchasing and Supply Management 7:75–89.

Braglia M, Petroni A (2000) A quality assurance-oriented methodology for handling trade-offs in supplier selection. International Journal of Physical Distribution & Logistics Management 30:96–112

Cao Q, Wang Q (2006) Optimizing vendor selection in a two-stage outsourcing process. Computers & Operations Research doi:10.1016/j.cor.2006.01.013

Cebi F, Bayraktar D (2003) An integrated approach for vendor selection. Logistics Information Management 16:395–400

Cengiz K, Ufuk C, Ziya U (2003) Multi-criteria vendor selection using fuzzy AHP. Logistics Information Management 16:382–394

Deb K (2002) Multi-objective optimization using evolutionary algorithms. Wiley, Singapore

Degraevel Z, Roodhooft, F, Doveren VB (2005) The use of total cost of ownership for strategic procurement, a company-wide management information system. Journal of the Operational Research Society 56:51–59

Feng CX, Wang J, Wang JS (2001) An optimization model for concurrent selection of tolerances and vendors. Computers & Industrial Engineering 40:15–33

Ge W, Huang SH, Dismukesa JP (2004) Product-driven supply chain selection using integrated multi-criteria decision-making methodology. International Journal of Production Economics 91:1–15

Gen M, Cheng R (1997) Genetic algorithms and engineering design. Wiley, New York

Ghodsypour SH, O'Brien C (2001) The total cost of logistics in vendor selection, under conditions of multiple sourcing, multiple criteria and capacity constraint. International Journal of Production Economics 73:15–27

Ghodsypour SH, O'Brien C (1998) A decision support system for vendor selection using an integrated analytic hierarchy process and linear programming. International Journal of Production Economics 56-57:199–212

Goldberg DE (1989) Genetic algorithms for search, optimization and machine learning. Addison-Wesley, Reading, MA

Gunasekaran A, Patel C, Tirtirouglu E (2001) Performance measures and metrics in a supply chain environment. International Journal of Operations & Production Management 21:71–87

Handfield R, Walton SV, Sroufe R, Melnyk SA (2002) Applying environmental criteria to vendor assessment, a study in the application of the analytical hierarchy process. European Journal of Operational Research 141:70–87

Ibrahim D, Ugur S (2003) Vendor selection using activity-based costing and fuzzy present-worth techniques. Logistics Information Management 16:420–426

Kumar M, Vrat P, Shankar R (2004) A fuzzy goal programming approach for vendor selection problem in a supply chain. Computers & Industrial Engineering 46:69–85

Kwang CK, Ip WH, Chan JWK (2002) Combining scoring method and fuzzy expert systems approach to vendor assessment: a case study. Integrated Manufacturing Systems 13:512–519

Liu F, Ding FY, Lall V (2000) Using data envelopment analysis to compare vendors for vendor selection and performance improvement. Supply Chain Management, An International Journal 5:143–150

Mitchell M (1996) Introduction to genetic algorithms. MIT Press, Ann Arbor, MI

Motwani J, Youssef M, Kathawala Y, Futch E (1999) Vendor selection in developing countries, a model development. Integrated Manufacturing Systems 10:154–162

Oliveria RC, Lourenco JC (2002) A multi-criteria model for assigning new orders to service vendors. European Journal of Operations Research 139:390–399

Pi WN, Low C (2005) Supplier evaluation and selection via Taguchi's loss function and an AHP. International Journal of Advanced Manufacturing Technology doi:10.1007/s00170-004-2227-z

Rao RV (2007) Vendor selection in a supply chain using analytic hierarchy process and genetic algorithm methods. International Journal of Services and Operations Management 3:355–369

Roodhooft F, Konings J (1997) Vendor selection and evaluation an Activity Based Costing approach. European Journal of Operational Research 96:97–102

Shyur HJ, Shih HS (2006) A hybrid MCDM model for strategic vendor selection. Mathematical and Computer Modelling doi:10.1016/j.mcm.2005.04.018

Sucky E (2006) A model for dynamic strategic vendor selection. Computers & Operations Research doi:10.1016/j.cor.2006.01.006

Verma R, Pullman ME (1998) An analysis of the supplier selection process. International Journal of Management Science 26:739–750

Vose MD (1999) Simple genetic algorithm, foundation and theory. MIT Press, Ann Arbor

Wadhwa V, Ravindran AR (2006) Vendor selection in outsourcing. Computers & Operations Research doi:10.1016/j.cor.2006.01.009

Weber CA (1996) A data envelopment analysis approach to measuring vendor performance. Supply Chain Management: An International Journal 1:28–39

Weber CA, Ellram LM (1993) Supplier selection using multi-objective programming: a decision support system approach. International Journal of Physical Distribution and Logistics Management 23:25–38

Weber CA, Current JR, Desai A (1998) Non-cooperative negotiation strategies for vendor selection. European Journal of Operations Research 108:208–223

Weber CA, Current JR, Desai A (2000) An optimization approach to determining the number of vendors to employ. Supply Chain Management: An International Journal 5:90–98

Yahya S, Kingsman B (1999) Vendor rating for an entrepreneur development programme, a case study using the analytic hierarchy process method. Journal of the Operational Research Society 50:916–930

30

Group Decision Making in the Manufacturing Environment

30.1 Introduction

Group decision making (GDM) is the process of making a judgment based upon the opinion of different individuals. Such decision making is a key component to the functioning of an organization, because organizational performance involves more than only one individual's action. Moving from a single decision maker to a multiple decision-maker setting introduces a great deal of complexity into the analysis. Various methods of group decision making are used for a wide set of attributes ranging from the strictly technical, to the psychophysical and social, and finally to the logical and scientifically valid. The group decision making concept can be applied to the graph theory and matrix approach as well as to MADM techniques. There are different ways in which GDM can be carried out (Yu, 1973; Chen and Hwang, 1992; Dyer and Forman, 1992; Csáki *et al.*, 1995; Forman and Penewati, 1998; Chen, 2000; Lai *et al.*, 2002; Jaganathan *et al.*, 2006). In this chapter, the group decision support system presented by Csáki *et al.* (1995) is considered. The method has been described in Section 3.3. However, the same is reproduced below for convenience. In this system, the method of calculating the group utility (group composite performance score) of alternative A_i (for i = 1, 2, ..., N) is as follows:

For each attribute B_j (for j = 1, 2, ..., M), the individual weights of importance of the attributes are aggregated into the group weights w_j (for j = 1, 2, ..., M):

$$w_j = [\sum_{k=1}^{n} l_{g(k)}\, w_j \,/ \sum_{k=1}^{n} l_{g(k)} \qquad j = 1, 2, \ldots, M \qquad (30.1)$$

The group qualification Q_{ij} of the alternative A_i against the attribute B_j is:

$$Q_{ij} = [\sum_{k=1}^{n} l_{g(k)}\, m_{ij} \,/ \sum_{k=1}^{n} l_{g(k)} \qquad j = 1, 2, \ldots, M;\ i = 1, 2, \ldots, N \qquad (30.2)$$

$\sum l_{g(k)}$ need not be equal to 1 in Equations 30.1 and 30.2.

The group utility P_i of alternative A_i is determined as the weighted algebraic mean of the aggregated qualification values with the aggregated weights.

$$P_i = [\sum_{j=1}^{M} w_j Q_{ij} / \sum_{j=1}^{M} w_j \qquad i = 1, 2, \ldots, N \tag{30.3}$$

In addition to the weighted algebraic means used in the above aggregations, weighted geometric means can also be used. The best alternative of group decision is the one associated with the highest value of P_i.

Now, an example of the robot selection problem described in Section 11.2.2 is considered to demonstrate the group decision making approach in the manufacturing environment.

30.2 Example

This example problem considers five robot selection attributes, and three alternative robots. A group consisting of three decision makers is considered. The first decision maker is given an importance weight of 0.5, the second decision maker an importance weight of 0.3, and the third decision maker an importance weight of 0.2. The objective and subjective information of the attributes is given in Table 30.1. The man–machine interface (MI) and programming flexibility (PF) are expressed subjectively in linguistic terms by the group of decision makers, and these attributes are assigned objective values with the help of Table 4.3. The data of the attributes are given in Table 30.2. The objective values of PC, LC, and RE remain intact, as these values are already crisp, and there is no need to have these converted by the decision-making group.

Table 30.1. Robot selection attributes information

Robot	PC ($1,000)	LC (kg)	RE (mm)	MI	PF
Robot 1	73	48	0.15	A, A, AA	H, VH, H
Robot 2	71	46	0.18	AA, AA, BA	VH, H, AA
Robot 3	75	51	0.14	BA, A, AA	H, VH, H

PC: Purchasing cost; LC: Load carrying capacity; R: Repeatability error; MI: Man–machine interface; PF: Programming flexibility
A: Average; AA: Above average; BA: Below average; H: High; VH: Very high

Table 30.2. Data of the robot selection attributes as decided by the group

Robot	PC ($1,000)	LC (kg)	RE (mm)	MI	PF
Robot 1	73	48	0.15	0.5, 0.5, 0.59	0.665, 0.665, 0.745
Robot 2	71	46	0.18	0.59, 0.59, 0.41	0.745, 0.665, 0.59
Robot 3	75	51	0.14	0.41, 0.5, 0.59	0.665, 0.745, 0.665

Using Equation 30.2, the values of MI and PF are aggregated, and the aggregated values are shown in Table 30.3. For example, the value of MI for robot 1 is obtained as 0.5*0.5 + 0.3*0.5 + 0.2*0.59 = 0.518.

Table 30.3. Objective data of the robot selection attributes

Robot	PC ($1,000)	LC (kg)	RE (mm)	MI	PF
Robot 1	73	48	0.15	0.518	0.681
Robot 2	71	46	0.18	0.554	0.69
Robot 3	75	51	0.14	0.473	0.689

30.2.1 Application of Graph Theory and Matrix Approach

In the present work, the attributes considered are PC, LC, R, MI, and PF. The values of the robot selection attributes, which are given in Table 30.3, are to be normalized. LC, MI, and PF are beneficial attributes, and higher values are desirable. Values of these attributes are normalized, as explained in Section 2.4, and are given in Table 30.4 in the respective columns. PC and R are non-beneficial attributes, and lower values are desirable. The values of these attributes for different robots are normalized, and are given in Table 30.4.

Table 30.4. Normalized data of the robot selection attributes

Robot	PC	LC	RE	MI	PF
Robot 1	0.9726	0.9412	0.9333	0.9350	0.9869
Robot 2	1.00	0.9020	0.7777	1.00	1.00
Robot 3	0.9467	1.00	1.00	0.8538	0.9985

Let the three decision makers select the following relative importance assignments:

	PC	LC	RE	MI	PF
PC	-	0.745,0.665,0.745	0.500,0.590,0.410	0.865,0.665,0.745	0.745,0.745,0.665
LC	0.255,0.335,0.255	-	0.255,0.255,0.335	0.590,0.410,0.500	0.500,0.590,0.590
RE	0.500,0.410,0.590	0.745,0.745,0.665	-	0.865,0.745,0.745	0.745,0.665,0.665
MI	0.135,0.335,0.255	0.410,0.590,0.500	0.135,0.255,0.255	-	0.410,0.410,0.500
PF	0.255,0.255,0.335	0.500,0.410,0.410	0.255,0.335,0.335	0.590,0.590,0.500	-

The values of relative importance are aggregated using Equation 30.1, and the relative importance matrix thus obtained is given as:

	PC	LC	RE	MI	PF
PC	-	0.721	0.509	0.781	0.729
LC	0.279	-	0.271	0.518	0.545
RE	0.491	0.729	-	0.805	0.705
MI	0.219	0.482	0.195	-	0.428
PF	0.271	0.455	0.295	0.572	-

For example, the relative importance value of 0.721 for PC over LC is obtained from 0.5*0.745 + 0.3*0.665 + 0.2*0.745.

The robot attributes digraph, robot attributes matrix of the digraph, and robot function for the matrix can be prepared. The value of the robot selection index is calculated, using the values of A_i and a_{ij} for each robot.

The robot selection index values of different robots are given below in descending order:

Robot 3	7.3135
Robot 1	7.2413
Robot 2	6.9304

From the above values of the robot selection index, robot 3 is considered the best choice among the robots considered for the given industrial application under a group decision-making situation. The second choice is robot 1, and the third choice is robot 2.

It may be noted that in example 2 described in Section 11.2.2.1, the ranking given by GTMA was robot 2 - robot 1 - robot 3. However, that was the case when a single decision maker was involved in assigning the values to the qualitative attributes MI and PF, and thereafter assigning the values of relative importance among the attributes.

In the present example, three decision makers are involved in assigning the values to the subjective attributes MI and PF, and thereafter assigning the values of relative importance among the attributes. Thus, the ranking results obtained by group decision making are different in this example.

30.2.2 SAW Method

Let the three decision makers assign the following weights of importance to the attributes:

W_{PC} = 0.40, 0.35, 0.25 W_{LC} = 0.08, 0.10, 0.12 W_R = 0.40, 0.35, 0.40
W_{MI} = 0.05, 0.10, 0.10 W_{PF} = 0.08, 0.10, 0.13

Using Equation 30.1, these weights are aggregated and the aggregated group weights are given below:

$W_{PC} = 0.35$, $W_{LC} = 0.094$, $W_R = 0.385$, $W_{MI} = 0.075$, and $W_{PF} = 0.096$.

Using these weights and the normalized data of the attributes for different robots given in Table 30.4, the robot selection index values are calculated, and are arranged in descending order of the index.

Robot 3	0.9702
Robot 1	0.9531
Robot 2	0.9052

The ranking given by the SAW method is the same as that given by GTMA in Section 30.2.1.

30.2.3 WPM

Using the same weights as those selected for the SAW method, the application of WPM leads to the following ranking:

Robot 3	0.9693
Robot 1	0.9529
Robot 2	0.8990

This method also suggests robot 3 as the first choice.

30.2.4 TOPSIS Method

The quantitative values of the robot selection attributes, which are given in Table 30.4, are normalized. Relative importance of attributes (a_{ij}) is assigned using the SAW method as explained in Section 30.2.2 and these are: W_{PC} = 0.35, W_{LC} = 0.094, W_R = 0.385, W_{MI} = 0.075, and W_{PF} = 0.096.

Ideal (best) and negative ideal (worst) solutions are calculated, and these are given as:

$V_{PC}^+ = 0.1965$ $V_{PC}^- = 0.2076$
$V_{LC}^+ = 0.0572$ $V_{LC}^- = 0.0516$
$V_R^+ = 0.1903$ $V_R^- = 0.2447$
$V_{MI}^+ = 0.0465$ $V_{MI}^- = 0.0397$
$V_{PF}^+ = 0.0557$ $V_{PF}^- = 0.0550$

Separation measures are calculated and these are:

$S_1^+ = 0.0387$ $S_1^- = 0.0178$
$S_2^+ = 0.0124$ $S_2^- = 0.0548$
$S_3^+ = 0.0548$ $S_3^- = 0.0124$

The relative closeness of a particular alternative to the ideal solution is calculated (*i.e.*, robot selection index) and these are:

$P_1 = 0.3153$, $P_2 = 0.8155$, and $P_3 = 0.1847$

The alternative robots are arranged in descending order of their robot selection index. This can be arranged as 2-1-3.

30.2.5 Modified TOPSIS Method

The robot selection index values are calculated and these are given below in descending order:

Robot 2	0.7770
Robot 1	0.3337
Robot 3	0.2236

Thus, both the simple and modified TOPSIS methods suggest robot 2 as the first choice.

30.3 General Remarks

This chapter presents the concept of group decision-making in which a number of decision makers are involved. The example problem considered here is the robot selection for a given industrial application. However, in general, the concept can be applied to any group decision-making situation in the manufacturing environment.

References

Chen CT (2000) Extension of the TOPSIS for group decision making under fuzzy environment. Fuzzy Sets and Systems 114:1–9

Chen SJ, Hwang CL (1992) Fuzzy multiple attribute decision making-methods and applications. Lecture Notes in Economics and Mathematical Systems, Springer, New York

Csáki P, Rapcsák T, Turchányi P, Vermes M (1995) Research and development for group decision aid in Hungary by WINGDSS, a Microsoft Windows based group decision support system. Decision Support Systems 14:205–221

Dyer RF, Forman EH (1992) Group decision support with the analytic hierarchy process. Decision Support Systems 8:99–124

Forman E, Penewati K (1998) Aggregating individual judgements and priorities with the analytic hierarchy process. European Journal of Operational Research 108:165–169

Jaganathan S, Erinjeri JJ, Ker JI (2006) Fuzzy analytic hierarchy process based group decision support system to select and evaluate new manufacturing technologies. International Journal of Advanced Manufacturing Technology doi:10.1007/s00170-006-0446-1

Lai VS, Bo KW, Cheung W (2002) Group decision making in a multiple criteria environment: A case using the AHP in software selection. European Journal of Operational Research 137:134–144

Yu PL (1973) A class of solutions for group decision problems. Management Science 19:936–946

Appendix

Computer Codes

A.1 Computer Code for Calculation of Permanent Function of a Matrix

```
#include<stdio.h>
#include<conio.h>

/*******************************************************************
Variables Used

int      totalColumn         Number of columns and rows in the matrix
double   matrix[50][50]      The matrix input by User
int      z[50]               Variable Array used for calculation of perof2

Functions Used

double   per(int)            Recursive function that calculates the value of permanent
double   perof2();           Calculates the permanent of last 2 elements
*******************************************************************/
int totalColumn;
double matrix[50][50];
int z[50];

double per(int);
double perof2();
/*******************************************************************
Local Variables for main function

int      i                   Temporary integer variable
int      j                   Temporary integer variable
double   k                   Double variable that stores the value returned by per()
*******************************************************************/
void main()
{
        int i,j;
        double k;
```

```
            clrscr();
            printf("Enter the size of the matrix:");
            scanf("%d",&totalColumn);
/*Input from user the elements of Matrix that is n*n, i.e., totalColumn*totalColumn*/
            printf("\nEnter the elements of matrix:\n");
for (i = 0;i < totalColumn-1;i++)
{
                        for(j = 0;j < totalColumn-1;j++)
{
                        printf(" Element [%d][%d]: ", i+1,j+1);
                        scanf("%lf",&matrix[i][j]);
            }
}

Printf("\n\n\n\n\nPress any key to continue....");
Getch();
Clscr();
            /*Show to user the matrix formed or inputted*/
            printf("The matrix is: \n\n");
            printf("\n\t");
            for(i=0;i<=totalColumn-1;i++)
            {
                        for(j=0;j<=totalColumn-1;j++)
                        {
                                    printf("%1f ",matrix[i][j]);

                        }
                        printf("\n\t");

            }
            k=per(totalColumn);
            printf("\n\nThe final Value is %lf",k);
            getch();
}
/*******************************************************************
```

This algorithm works as follows... It calculates the values in a row number, *i.e.*, firstly it will call the value of the permanent for first element in the first row, then adds the permanent of second element of the first row and the process continues...

z keeps the check for per2, what is to be calculated for per2. per will make the values for all z elements '1' except those 2 elements whose per2 has to be calculated.

Variables used in per

int	i	Temporary integer variable
double	res	Double variable for storing the value returned from per()
double	res2	Double variable for storing the value returned from perof2()
double	sum	Variable for calculating the sum

```
*******************************************************************/
double per(int n)
{
            int i;
            double res, res2, sum=0;;
            if((n-2)!=0)
            {
```

```
                int c=0;
                for(i=1;i<=totalColumn;i++)
                {
                        if(z[i]==0)
                        {
                                z[i]=1;
                                z[c]=0;
                                c=i;
                                res=per(n-1);
                                res=(double)(res*matrix[totalColumn-n][i-1]);
                                sum=sum+res;
                        }
                }
                z[c]=0;
        }
        else
        {
                res2=perof2();
                return(res2);
        }
        return(sum);
}
/*********************************************************************
```

This function calculates the value of the matrix 2*2. The 2*2 matrix is defined by the array z[].

Variables used in perof2

int	i,j	Temporary integer variable
int	n,m	Temporary integer variable
int	flag	Contains value 0 or 1, acts as Boolean
double	res	Double variable for storing the value calculated by multiplication

```
*********************************************************************/
double perof2()
{
        int i,j,flag=0,n,m;
double res;
        for(i=1;i<=totalColumn;i++)
        {
                if(z[i]==0)
                {
                        if(flag==0)
                        {
                                n=i;
                                flag=1;
                        }
                        else
                        m=i;
                }
        }
        res=(double)((matrix[totalColumn-2][n-1]*matrix[totalColumn-1][m-
1])+(matrix[totalColumn-2][m-1]*matrix[totalColumn-1][n-1]));
        return res;
}
```

A.2 Computer Code for TOPSIS Method Which Uses AHP Method for Assigning the Values of Relative Importance to the Attributes

```
#include <iostream.h>
#include <process.h>
#include <math.h>
#include <conio.h>
#include <iomanip.h>
#define ERR 7
typedef struct Attribute
{
 int type;
}Attrib;
double **DiMatrix, **RiMatrix, *A2Matrix, *A3Matrix, **V1Matrix;
float **RelMatrix;
int NoAttrib, NoLayout;
long double Lambda, CI, CR;
double RI[] ={0.0, 0.0, 0.52, 0.89, 1.11, 1.25, 1.35, 1.40, 1.45, 1.49, 1.52, 1.54, 1.56, 1.58,
1.59};
Attrib * InfoAttrib;
double *MaxAttrib, *MinAttrib;
double *SeprateMax, *SeprateMin, *Closeness;
void GetAttribLayout()
{
cout<<"Enter the number of attributes to be considered"<<endl;
cin>>NoAttrib;
if(NoAttrib > 15)
{
cout<<"This program is hardcoded for 15 attributes for demonstration purpose and this can
be easily extended to any number of attributes"<<endl;
exit(0);
}
cout<<endl<<"Enter the number of layouts to be considered"<<endl;
cin>>NoLayout;
InfoAttrib = new Attrib[NoAttrib];
char input = '0';
for(int Count =0; Count < NoAttrib; Count++)
{
cout<<"If attribute "<<(Count+1)<<" is NON-BENEFICIAL then enter 1 otherwise press
any key "<<endl;
cin>>input;
if(input == '1')
InfoAttrib[Count].type = 0;
else
InfoAttrib[Count].type = 1;
cin.ignore();
}
}
void ValidateAttribLayout(char &KeyboardResponse)
{
```

```
cout<<endl<<"The entered number of attributes is "<<NoAttrib<<endl;
cout<<"The entered number of layouts is "<<NoLayout<<endl<<endl;
for(int Count =0; Count < NoAttrib; Count++)
{
cout<<"The type of Attribute "<<(Count+1)<<" is "<<(InfoAttrib[Count].type? "Beneficial"
: "Non-Beneficial")<<endl;
}
cout<<"If the information is correct then enter 1 and inorder to correct press 0"<<endl;
cin>>KeyboardResponse;
if(KeyboardResponse == '1' )
;
else
KeyboardResponse = '0';
}
void AllocateMemory()
{
DiMatrix = new double*[NoLayout];
RiMatrix = new double*[NoLayout];
V1Matrix = new double*[NoLayout];
RelMatrix = new float*[NoAttrib];
for(int Count = 0; Count < NoLayout; Count++)
{
DiMatrix[Count] = new double[NoAttrib];
RiMatrix[Count] = new double[NoAttrib];
V1Matrix[Count] = new double[NoAttrib];
}
for(Count = 0; Count < NoAttrib; Count++)
RelMatrix[Count] = new float[NoAttrib];
// if(DiMatrix == NULL || RiMatrix == NULL || V1Matrix == NULL)
//  throw ERR;
}
void GetDiMatrixInput()
{
//file://clrscr();
cout<<"Please enter the row wise values for the Di MATRIX "<<endl;
cout<<"Like shown below for 3 attributes and 3 layouts"<<endl<<endl;
cout<<"6.667 7.778"<<endl;
cout<<"7.89 6.78"<<endl;
cout<<"98.768 78.78 \n"<<endl;
cout<<"You have to enter "<<NoLayout<<" by "<<NoAttrib<<" Matrix"<<endl;
for(int count = 0; count < NoLayout; count++)
{
for(int InnerCount = 0; InnerCount < NoAttrib; InnerCount++)
cin>>DiMatrix[count][InnerCount];
}
}
void ValidateDiMatrix(char &KeyboardResponse)
{
cout<<endl<<"Please confirm the values entered "<<endl;
for(int count = 0; count < NoLayout; count++)
{
for(int InnerCount = 0; InnerCount < NoAttrib; InnerCount++)
```

```
cout<<DiMatrix[count][InnerCount]<<"  ";
cout<<endl;
}
cout<<"If the information is correct then enter 1 and inorder to correct press 0"<<endl;
cin.ignore();
cin>>KeyboardResponse;
if(KeyboardResponse == '1' )
;
else
KeyboardResponse = '0';
}
void CalculateRiMatrix()
{
double *Summation =  new double[NoAttrib];
for(int Count = 0; Count < NoAttrib; Count++)
Summation[Count] = 0.0;
for(Count = 0; Count < NoAttrib; Count++)
{
for(int InnerCount = 0; InnerCount < NoLayout; InnerCount++)
{
Summation[Count] += pow(DiMatrix[InnerCount][Count], 2);
}
Summation[Count] = pow(Summation[Count], 0.5);
}
for(Count = 0; Count < NoLayout; Count++)
{
for(int InnerCount = 0; InnerCount < NoAttrib; InnerCount++)
RiMatrix[Count][InnerCount] = DiMatrix[Count][InnerCount] / Summation[InnerCount];
}
delete [] Summation;
}
void GetRelativeAttribMatrix()
{
// file://clrscr();
float Temp;
cout<<"Please enter the Relative Matrix "<<endl;
cout<<"Like shown below for 3 attributes"<<endl<<endl;
cout<<"6.667 7.778"<<endl;
cout<<"7.89 6.78"<<endl;
cout<<"98.768 78.78 \n"<<endl;
cout<<"U have to enter "<<NoAttrib<<" By "<<NoAttrib<<" MATRIX... Enter fractions
where applicable"<<endl;
for(int Count = 0; Count < NoAttrib; Count++)
{
for(int InnerCount = 0; InnerCount < NoAttrib; InnerCount++)
{
cin>>Temp;
if(Count == InnerCount)
RelMatrix[Count][InnerCount] = 1;
else if(Count > InnerCount)
RelMatrix[Count][InnerCount] = 1 / RelMatrix[InnerCount][Count];
else
```

```
RelMatrix[Count][InnerCount] = Temp;
}
}
}
void ValidateRelativeMatrix(char &KeyboardResponse)
{
cout<<endl<<"Please confirm the values entered "<<endl;
for(int count = 0; count < NoAttrib
; count++)
{
for(int InnerCount = 0; InnerCount < NoAttrib; InnerCount++)
cout<<RelMatrix[count][InnerCount]<<"  ";
cout<<endl;
}
cout<<"If the information is correct then enter 1 and inorder to correct press 0"<<endl;
cin.ignore();
cin>>KeyboardResponse;
if(KeyboardResponse == '1' )
;
else
KeyboardResponse = '0';
}
void CalculateWeightedMatrix() //file://The A2 Matrix
{
A2Matrix = new double[NoAttrib];
double *GM = new double[NoAttrib];
for(int Count = 0; Count < NoAttrib; Count++)
GM[Count] = 1.0;
for(Count = 0; Count < NoAttrib; Count++)
{
for(int InnerCount = 0; InnerCount < NoAttrib; InnerCount++)
{
GM[Count] *= RelMatrix[Count][InnerCount];
}
double Temp1 = 1.0 / NoAttrib, Temp2 = GM[Count];
GM[Count] = pow(Temp2, Temp1);
}
double Sum = 0.0;
for(Count = 0; Count < NoAttrib; Count++)
Sum += GM[Count];
for(Count = 0; Count < NoAttrib; Count++)
{
GM[Count] /= Sum;
A2Matrix[Count] = GM[Count];
}
cout<<"\n\n";
cout<<"The normalised weights of each attribute are ....."<<endl;
for(Count = 0; Count < NoAttrib; Count++)
cout<<A2Matrix[Count]<<endl;
delete [] GM;

}
```

```
void CalculateA3Matrix() //file://Multiplication of A1 and A2
{
A3Matrix = new double[NoAttrib];
for(int Count = 0; Count < NoAttrib; Count++)
A3Matrix[Count] = 0.0;
for(Count = 0; Count < NoAttrib; Count++)
{
for(int InnerCount = 0; InnerCount < NoAttrib; InnerCount++)
{
A3Matrix[Count] += RelMatrix[Count][InnerCount] * A2Matrix[InnerCount];
}
}
cout<<"\n\n";
cout<<"The A3 Matrix is "<<endl;
for(Count = 0; Count < NoAttrib; Count++)
cout<<A3Matrix[Count]<<endl;
}
void CalculateLambdaMax()
{
Lambda = 0.0;
for(int Count = 0; Count < NoAttrib; Count++)
Lambda += (A3Matrix[Count] / A2Matrix[Count]);
Lambda /= NoAttrib;
cout<<"\n\n";
cout<<"The value of lamda is "<<Lambda;
}
void CalculateConsistencyRatio() //file://CR
{
CI = ((Lambda - NoAttrib) / (NoAttrib - 1.0));
CR = CI / RI[NoAttrib - 1];
if( CR > 0.1)
{
cout<<"\n\n";
cout<<"There is/are inconsistencies made in the judgements so please try other values
"<<endl;
getch();
exit(0);
}
else
{
cout<<"\n\n"<<"There is good consistency in the judgements therefore the calculation \nwill
proceed .... and the value of Consistency Ratio is "<<CR<<endl;
}
}
void CalculateNormalisedMatrix()
{
for(int Count = 0; Count < NoLayout; Count++)
{
for(int InnerCount = 0; InnerCount < NoAttrib; InnerCount++)
{
V1Matrix[Count][InnerCount] = RiMatrix[Count][InnerCount] * A2Matrix[InnerCount];
}
```

```
}
cout<<"\n\n";
cout<<"The weighted Normalised Matrix is \n\n";
for(Count = 0; Count < NoLayout; Count++)
{
for(int InnerCount = 0; InnerCount < NoAttrib; InnerCount++)
{
cout<<V1Matrix[Count][InnerCount]<<"    ";
}
cout<<endl;
}
}

void GetMaxMinAttrib()
{
MaxAttrib = new double[NoAttrib];
MinAttrib = new double[NoAttrib];
for(int Count =0; Count < NoAttrib; Count++)
{
MaxAttrib[Count] = V1Matrix[0][Count];
MinAttrib[Count] = V1Matrix[0][Count];
for(int InnerCount = 0; InnerCount < NoLayout; InnerCount++)
{
if(InfoAttrib[Count].type == 1)
{
if(V1Matrix[InnerCount][Count] > MaxAttrib[Count])
MaxAttrib[Count] = V1Matrix[InnerCount][Count];
if(V1Matrix[InnerCount][Count] < MinAttrib[Count])
MinAttrib[Count] = V1Matrix[InnerCount][Count];
}
else if(InfoAttrib[Count].type == 0)
{
if(V1Matrix[InnerCount][Count] > MinAttrib[Count])
MinAttrib[Count] = V1Matrix[InnerCount][Count];
if(V1Matrix[InnerCount][Count] < MaxAttrib[Count])
MaxAttrib[Count] = V1Matrix[InnerCount][Count];
}
}
}
cout<<"\n\n";
cout<<"The ideal(BEST) and Negative-Ideal(WORST) solution are "<<endl;
for(Count = 0; Count < NoAttrib; Count++)
cout<<MaxAttrib[Count]<<"  "<<MinAttrib[Count]<<endl;
}
void CalculateSeprateMeasures()
{
SeprateMax = new double[NoLayout];
SeprateMin = new double[NoLayout];
for(int Count = 0; Count < NoLayout; Count++)
{
SeprateMax[Count] = 0.0;
SeprateMin[Count] = 0.0;
```

```
}
double Temp = 0.0;
for(Count = 0; Count < NoLayout; Count++)
{
for(int InnerCount = 0; InnerCount < NoAttrib; InnerCount++)
{
Temp = V1Matrix[Count][InnerCount] - MaxAttrib[InnerCount];
SeprateMax[Count] += pow(Temp, 2);
Temp = V1Matrix[Count][InnerCount] - MinAttrib[InnerCount];
SeprateMin[Count] += pow(Temp, 2);
}
Temp = SeprateMax[Count];
SeprateMax[Count] = pow(Temp, .5);
Temp = SeprateMin[Count];
SeprateMin[Count] = pow(Temp, .5);
}
cout<<"\n\n";
cout<<"The Separation measures are "<<endl;
for(Count = 0; Count < NoLayout; Count++)
cout<<SeprateMax[Count]<<"  "<<SeprateMin[Count]<<endl;
}
void CalculateRelativeCloseness()
{
Closeness = new double[NoLayout];
double Sum = 0.0;
for(int Count = 0; Count < NoLayout; Count++)
{
Sum = SeprateMax[Count] + SeprateMin[Count];
Closeness[Count] = SeprateMin[Count] / Sum;
}
}
void DisplayResults()
{
double temp;
int *Results = new int[NoLayout], itemp;
for(int j = 0; j < NoLayout; j++)
Results[j] = j + 1;
int i;
for(int Count = 1;  Count < NoLayout; Count++)
{
itemp = Results[Count];
temp = Closeness[Count];
i = Count - 1;
while((i >= 0) && (temp > Closeness[i]))
{
Closeness[i + 1] = Closeness[i];
Results[i + 1] = Results[i];
i--;
}
Closeness[i + 1] = temp;
Results[i + 1] = itemp;
}
```

```
cout<<"\n\n";
cout<<"The final results are "<<endl;
cout<<"StripLayout Index"<<setw(20)<<"No of Layout"<<endl;
for(Count = 0; Count < NoLayout; Count++)
cout<<Closeness[Count]<<setw(20)<<Results[Count]<<endl;
delete [] Results;
}
void CleanUp()
{
for(int Count = 0; Count < NoLayout; Count++)
{
delete [] DiMatrix[Count];
delete [] RiMatrix[Count];
delete [] V1Matrix[Count];
}
for(Count = 0; Count < NoAttrib; Count++)
delete [] RelMatrix[Count];
delete [] DiMatrix;
delete [] RiMatrix;
delete [] V1Matrix;
delete [] RelMatrix;
delete [] MaxAttrib;
delete [] MinAttrib;
delete [] SeprateMin;
delete [] SeprateMax;
delete [] Closeness;
DiMatrix = NULL;
RiMatrix = NULL;
V1Matrix = NULL;
MaxAttrib = NULL;
MinAttrib = NULL;
SeprateMin = NULL;
SeprateMax = NULL;
Closeness = NULL;
}
void main()
{
// file://clrscr();
char KeyboardResponse;
GetAttribLayout();
ValidateAttribLayout(KeyboardResponse);
cin.ignore();
while(KeyboardResponse != '1')
{
//file://clrscr();
GetAttribLayout();
ValidateAttribLayout(KeyboardResponse);
cin.ignore();
}
try
{
AllocateMemory();
```

```
}
catch(...)
{
cout<<"Memory allocation failed"<<endl;
exit(0);
}
GetDiMatrixInput();
ValidateDiMatrix(KeyboardResponse);
cin.ignore();
while(KeyboardResponse != '1')
{
GetDiMatrixInput();
ValidateDiMatrix(KeyboardResponse);
cin.ignore();
}
CalculateRiMatrix();
GetRelativeAttribMatrix();
ValidateRelativeMatrix(KeyboardResponse);
cin.ignore();
while(KeyboardResponse != '1')
{
GetRelativeAttribMatrix();
ValidateRelativeMatrix(KeyboardResponse);
cin.ignore();
}
CalculateWeightedMatrix();
getch();
CalculateA3Matrix();
getch();
CalculateLambdaMax();
getch();
CalculateConsistencyRatio();
getch();
CalculateNormalisedMatrix();
getch();
GetMaxMinAttrib();
getch();
CalculateSeprateMeasures();
getch();
CalculateRelativeCloseness();
DisplayResults();
getch();
CleanUp();
}
```

A.3 Computer Code for TOPSIS Method Which Uses the Weights of Relative Importance inputted by the User (Without Using AHP)

```
#include <iostream.h>
#include <process.h>
#include <math.h>
#include <conio.h>
#include <iomanip.h>
#define ERR 7
typedef struct Attribute
{
int type;
}Attrib;
double **DiMatrix, **RiMatrix, *A2Matrix, *A3Matrix, **V1Matrix,**Wtdmatrix;
double **RelMatrix,*weights;
int NoAttrib, NoLayout;
long double Lambda, CI, CR;
long double RI[] ={0.0, 0.0, 0.52, 0.89, 1.11, 1.25, 1.35, 1.40, 1.45, 1.49, 1.52, 1.54, 1.56,
1.58, 1.59};
Attrib * InfoAttrib;
double *MaxAttrib, *MinAttrib;
double *SeprateMax, *SeprateMin, *Closeness;
void wtdnorm();
void AllocateMemory()
{int s; s=0;
DiMatrix = new double*[NoLayout];
RiMatrix = new double*[NoLayout];
//V1Matrix = new double*[NoLayout];
// Wtdmatrix = new double*[NoLayout];
RelMatrix = new double*[NoAttrib];
weights = new double[NoAttrib];
for(int Count = 0; Count < NoLayout; Count++)
{
DiMatrix[Count] = new double[NoAttrib];
RiMatrix[Count] = new double[NoAttrib];
// V1Matrix[Count] = new double[NoAttrib];
// Wtdmatrix[Count] = new double[NoAttrib];
}
//for(Count = 0; Count < NoAttrib; Count++)
// RelMatrix[Count] = new double[NoAttrib];
// if(DiMatrix == NULL || RiMatrix == NULL || V1Matrix == NULL)
// throw ERR;
}
void GetAttribLayout()
{
cout<<"Enter the number of attributes to be considered"<<endl;
cin>>NoAttrib;
if(NoAttrib > 15)
{
exit(0);
```

```
}
cout<<endl<<"Enter the number of layouts to be considered"<<endl;
cin>>NoLayout;
InfoAttrib = new Attrib[NoAttrib];
char input = '0';
AllocateMemory();
for(int Count =0; Count < NoAttrib; Count++)
{
cout<<"If attribute "<<(Count+1)<<" is NON-BENEFICIAL then enter 1 otherwise press any key "<<endl;
cin>>input;
if(input == '1')
InfoAttrib[Count].type = 0;
else
InfoAttrib[Count].type = 1;
cin.ignore();
}
}
void ValidateAttribLayout(char &KeyboardResponse)
{
cout<<endl<<"The entered number of attributes is "<<NoAttrib<<endl;
cout<<"The entered number of layouts is "<<NoLayout<<endl<<endl;
for(int Count =0; Count < NoAttrib; Count++)
{
cout<<"The type of Attribute "<<(Count+1)<<" is "<<(InfoAttrib[Count].type? "Beneficial" : "Non-Beneficial")<<endl;
}
cout<<"If the information is correct then enter 1 and inorder to correct press 0"<<endl;
cin>>KeyboardResponse;
if(KeyboardResponse == '1' )
;
else
KeyboardResponse = '0';
}
void GetDiMatrixInput()
{
//file://clrscr();
cout<<"Please enter the row wise values for the Di MATRIX "<<endl;
cout<<"Like shown below for 3 attributes and 3 layouts"<<endl<<endl;
cout<<"6.667 7.778"<<endl;
cout<<"7.89 6.78"<<endl;
cout<<"98.768 78.78 \n"<<endl;
cout<<"You have to enter "<<NoLayout<<" by "<<NoAttrib<<" Matrix"<<endl;
for(int count = 0; count < NoLayout; count++)
for(int InnerCount = 0; InnerCount < NoAttrib; InnerCount++)
cin>>DiMatrix[count][InnerCount];
}
void getweights()
{
//file://clrscr();
int flag=0;
while (!flag)
```

```
{
cout << "\n Enter  "<<NoAttrib<<" Weights :";
float s=0.0;
for(int Count =0; Count < NoAttrib; Count++)
{
            cin>>A2Matrix[Count];
            s=s+A2Matrix[Count];
}
            if (s == 1.0)
            {
            for(int Count =0; Count < NoAttrib; Count++)
            cout<< "Weight : "<<Count+1<< "="<<A2Matrix[Count]<<endl;
            getch();
            //wtdnorm();
            flag=1;
            }
    }
}
/*void wtdnorm()
{
for(int InnerCount = 0; InnerCount < NoAttrib; InnerCount++)
for(int count = 0; count < NoLayout; count++)
V1Matrix[count][InnerCount]=A2Matrix[InnerCount]*RiMatrix[count][InnerCount];
            for(int cnt = 0; cnt < NoLayout; cnt++)
            {
            for(int InnerCount = 0; InnerCount < NoAttrib; InnerCount++)
            cout << "   " << V1Matrix[cnt][InnerCount];
            cout << endl;
            }
             getch();
} */
void ValidateDiMatrix(char &KeyboardResponse)
{
cout<<endl<<"Please confirm the values entered "<<endl<<endl;
for(int count = 0; count < NoLayout; count++)
{
for(int InnerCount = 0; InnerCount < NoAttrib; InnerCount++)
}
void getweights()
{
//file://clrscr();
            int flag=0;
            while (!flag)
            {
            cout << "\n Enter  "<<NoAttrib<<" Weights :";
            float s=0.0;
            for(int Count =0; Count < NoAttrib; Count++)
            {

             cin>>A2Matrix[Count];
             s=s+A2Matrix[Count];
            }
```

```
            if (s == 1.0)
            {
            for(int Count =0; Count < NoAttrib; Count++)
            cout<< "Weight : "<<Count+1<< "="<<A2Matrix[Count]<<endl;
            getch();
            //wtdnorm();
            flag=1;
            }
    }
}
/*void wtdnorm()
{
for(int InnerCount = 0; InnerCount < NoAttrib; InnerCount++)
for(int count = 0; count < NoLayout; count++)
V1Matrix[count][InnerCount]=A2Matrix[InnerCount]*RiMatrix[count][InnerCount];
            for(int cnt = 0; cnt < NoLayout; cnt++)
            {
            for(int InnerCount = 0; InnerCount < NoAttrib; InnerCount++)
            cout << "   " << V1Matrix[cnt][InnerCount];
            cout << endl;
            }
             getch();
            } */

void ValidateDiMatrix(char &KeyboardResponse)
{
cout<<endl<<"Please confirm the values entered "<<endl<<endl;
for(int count = 0; count < NoLayout; count++)
{
for(int InnerCount = 0; InnerCount < NoAttrib; InnerCount++)
cout<<DiMatrix[count][InnerCount]<<"  ";
cout<<endl;
}
cout<<"If the information is correct then enter 1 and inorder to correct press 0"<<endl;
cin.ignore();
cin>>KeyboardResponse;
if(KeyboardResponse == '1' )
;
else
KeyboardResponse = '0';
}
/*void CalculateRiMatrix()
{
double *Summation =  new double[NoAttrib];
for(int Count = 0; Count < NoAttrib; Count++)
Summation[Count] = 0.0;
for(Count = 0; Count < NoAttrib; Count++)
{
for(int InnerCount = 0; InnerCount < NoLayout; InnerCount++)
{
Summation[Count] += pow(DiMatrix[InnerCount][Count], 2);
}
```

```
Summation[Count] = pow(Summation[Count], 0.5);
}
for(Count = 0; Count < NoLayout; Count++)
{
for(int InnerCount = 0; InnerCount < NoAttrib; InnerCount++)
{
RiMatrix[Count][InnerCount] = DiMatrix[Count][InnerCount] / Summation[InnerCount];
cout <<RiMatrix[Count][InnerCount] <<"   ";
}
cout << endl;
getch();
}
delete [] Summation;
} */
void CalculateRiMatrix()
{
double *Summation =  new double[NoAttrib];
for(int Count = 0; Count < NoAttrib; Count++)
Summation[Count] = 0.0;
for(Count = 0; Count < NoAttrib; Count++)
{
for(int InnerCount = 0; InnerCount < NoLayout; InnerCount++)
{
Summation[Count] += pow(DiMatrix[InnerCount][Count], 2);
}
Summation[Count] = pow(Summation[Count], 0.5);
}
for(Count = 0; Count < NoAttrib; Count++)
{
for(int InnerCount = 0; InnerCount < NoLayout; InnerCount++)
{
RiMatrix[InnerCount][Count] = DiMatrix[InnerCount][Count] / Summation[Count];
//    cout <<RiMatrix[Count][InnerCount] <<"   ";
}
cout << endl;
}
delete [] Summation;
}
void printRiMatrix()
{
            for(int Count = 0; Count < NoLayout; Count++)
             {
              for(int InnerCount = 0; InnerCount < NoAttrib; InnerCount++)
                       cout <<RiMatrix[Count][InnerCount]<< "  ";
              cout << endl;
             }
}
/*void GetRelativeAttribMatrix()
{
// file://clrscr();
float Temp;
cout<<"Please enter the Relative Matrix "<<endl;
```

```
cout<<"Like shown below for 3 attributes"<<endl<<endl;
cout<<"6.667 7.778"<<endl;
cout<<"7.89 6.78"<<endl;
cout<<"98.768 78.78 \n"<<endl;
cout<<"U have to enter "<<NoAttrib<<" By "<<NoAttrib<<" MATRIX... Enter fractions
where applicable"<<endl;
for(int Count = 0; Count < NoAttrib; Count++)
{
for(int InnerCount = 0; InnerCount < NoAttrib; InnerCount++)
{
cin>>Temp;
if(Count == InnerCount)
RelMatrix[Count][InnerCount] = 1;
else if(Count > InnerCount)
RelMatrix[Count][InnerCount] = 1 / RelMatrix[InnerCount][Count];
else
RelMatrix[Count][InnerCount] = Temp;
}
}
}
*/
/*
void normalizedmatrix()
{
// file://clrscr();
float Temp;
for(int Count = 0; Count < NoAttrib; Count++)
{
for(int InnerCount = 0; InnerCount < NoLayout; InnerCount++)
{
double denom =0.0;
for(int Icnt = 0; Icnt < NoLayout; Icnt++)
denom += DiMatrix[Count][Icnt] * DiMatrix[Count][Icnt] ;
RelMatrix[Count][InnerCount] = DiMatrix[Count][InnerCount] / sqrt(denom);
}
}
} */
/*
void ValidateRelativeMatrix(char &KeyboardResponse)
{
cout<<endl<<"Please confirm the values entered "<<endl;
for(int count = 0; count < NoAttrib
; count++)
{
for(int InnerCount = 0; InnerCount < NoAttrib; InnerCount++)
cout<<RelMatrix[count][InnerCount]<<"  ";
cout<<endl;
}
cout<<"If the information is correct then enter 1 and inorder to correct press 0"<<endl;
cin.ignore();
cin>>KeyboardResponse;
if(KeyboardResponse == '1' )
```

```
;
else
KeyboardResponse = '0';
}
void CalculateWeightedMatrix() //file://The A2 Matrix
{
A2Matrix = new double[NoAttrib];
double *GM = new double[NoAttrib];
for(int Count = 0; Count < NoAttrib; Count++)
GM[Count] = 1.0;
for(Count = 0; Count < NoAttrib; Count++)
{
for(int InnerCount = 0; InnerCount < NoAttrib; InnerCount++)
{
GM[Count] *= RelMatrix[Count][InnerCount];
}
double Temp1 = 1.0 / NoAttrib, Temp2 = GM[Count];
GM[Count] = pow(Temp2, Temp1);
}
double Sum = 0.0;
for(Count = 0; Count < NoAttrib; Count++)
Sum += GM[Count];
for(Count = 0; Count < NoAttrib; Count++)
{
GM[Count] /= Sum;
A2Matrix[Count] = GM[Count];
}
cout<<"\n\n";
cout<<"The normalised weights of each attribute are ....."<<endl;
for(Count = 0; Count < NoAttrib; Count++)
cout<<A2Matrix[Count]<<endl;
delete [] GM;
}
void CalculateA3Matrix()  //file://Multiplication of A1 and A2
{
A3Matrix = new double[NoAttrib];
for(int Count = 0; Count < NoAttrib; Count++)
A3Matrix[Count] = 0.0;
for(Count = 0; Count < NoAttrib; Count++)
{
for(int InnerCount = 0; InnerCount < NoAttrib; InnerCount++)
{
A3Matrix[Count] += RelMatrix[Count][InnerCount] * A2Matrix[InnerCount];
}
}
cout<<"\n\n";
cout<<"The A3 Matrix is "<<endl;
for(Count = 0; Count < NoAttrib; Count++)
cout<<A3Matrix[Count]<<endl;
}
void CalculateLambdaMax()
{
```

```
Lambda = 0.0;
for(int Count = 0; Count < NoAttrib; Count++)
Lambda += (A3Matrix[Count] / A2Matrix[Count]);
Lambda /= NoAttrib;
cout<<"\n\n";
cout<<"The value of lamda is "<<Lambda;
}
void CalculateConsistencyRatio() //file://CR
{
CI = ((Lambda - NoAttrib) / (NoAttrib - 1.0));
cout << "CI :"<< CI; getch();
CR = CI / RI[NoAttrib - 1];
cout << "CR :"<<CR;      getch();
if( CR > 0.1)
{
cout<<"\n\n";
cout<<"There is/are inconsistencies made in the judgements so please try other values
"<<endl;
getch();
exit(0);
}
else
{
cout<<"\n\n"<<"There is good consistency in the judgements therefore the calculation \nwill
proceed .... and the value of Consistency Ratio is "<<CR<<endl;
}
} */
void CalculateNormalisedMatrix()
{
for(int InnerCount = 0; InnerCount < NoAttrib; InnerCount++)
{
for(int Count = 0; Count < NoLayout; Count++)
V1Matrix[Count][InnerCount] = RiMatrix[Count][InnerCount] * A2Matrix[InnerCount];
// cout << V1Matrix[Count][InnerCount]<< "  ";
// cout << endl;
}
cout<<"\n\n";
cout<<"The weighted Normalised Matrix is \n\n";
for(int Count = 0; Count < NoLayout; Count++)
{
for(int InnerCount = 0; InnerCount < NoAttrib; InnerCount++)
{
cout<<V1Matrix[Count][InnerCount]<<"    ";
}
cout<<endl;
}
}
void GetMaxMinAttrib()
{
MaxAttrib = new double[NoAttrib];
MinAttrib = new double[NoAttrib];
for(int Count =0; Count < NoAttrib; Count++)
```

```
{
MaxAttrib[Count] = RiMatrix[0][Count];
MinAttrib[Count] = RiMatrix[0][Count];
for(int InnerCount = 0; InnerCount < NoLayout; InnerCount++)
{
if(InfoAttrib[Count].type == 1)
{
if(RiMatrix[InnerCount][Count] > MaxAttrib[Count])
MaxAttrib[Count] = RiMatrix[InnerCount][Count];
if(RiMatrix[InnerCount][Count] < MinAttrib[Count])
MinAttrib[Count] = RiMatrix[InnerCount][Count];
}
else if(InfoAttrib[Count].type == 0)
{
if(RiMatrix[InnerCount][Count] > MinAttrib[Count])
MinAttrib[Count] = RiMatrix[InnerCount][Count];
if(RiMatrix[InnerCount][Count] < MaxAttrib[Count])
MaxAttrib[Count] = RiMatrix[InnerCount][Count];
}
}
}
cout<<"\n\n";
cout<<"The ideal(BEST) and Negative-Ideal(WORST) solution are "<<endl;
for(Count = 0; Count < NoAttrib; Count++)
cout<<MaxAttrib[Count]<<"  "<<MinAttrib[Count]<<endl;
}
void CalculateSeprateMeasures()
{
SeprateMax = new double[NoLayout];
SeprateMin = new double[NoLayout];
for(int Count = 0; Count < NoLayout; Count++)
{
SeprateMax[Count] = 0.0;
SeprateMin[Count] = 0.0;
}
getweights();
getch();
double Temp = 0.0;
for(Count = 0; Count < NoLayout; Count++)
{
for(int InnerCount = 0; InnerCount < NoAttrib; InnerCount++)
{
Temp = RiMatrix[Count][InnerCount] - MaxAttrib[InnerCount];
SeprateMax[Count] += A2Matrix[InnerCount]*pow(Temp, 2);
Temp = RiMatrix[Count][InnerCount] - MinAttrib[InnerCount];
SeprateMin[Count] += A2Matrix[InnerCount]*pow(Temp, 2);
}
Temp = SeprateMax[Count];
SeprateMax[Count] = pow(Temp, .5);
Temp = SeprateMin[Count];
SeprateMin[Count] = pow(Temp, .5);
}
```

```
cout<<"\n\n";
cout<<"The Separation measures are "<<endl;
for(Count = 0; Count < NoLayout; Count++)
cout<<SeprateMax[Count]<<"  "<<SeprateMin[Count]<<endl;
}
void CalculateRelativeCloseness()
{
Closeness = new double[NoLayout];
double Sum = 0.0;
for(int Count = 0; Count < NoLayout; Count++)
{
Sum = SeprateMax[Count] + SeprateMin[Count];
Closeness[Count] = SeprateMin[Count] / Sum;
}
}
void DisplayResults()
{
double temp;
int *Results = new int[NoLayout], itemp;
for(int j = 0; j < NoLayout; j++)
Results[j] = j + 1;
int i;
for(int Count = 1;  Count < NoLayout; Count++)
{
itemp = Results[Count];
temp = Closeness[Count];
i = Count - 1;
while((i >= 0) && (temp > Closeness[i]))
{
Closeness[i + 1] = Closeness[i];
Results[i + 1] = Results[i];
i--;
}
Closeness[i + 1] = temp;
Results[i + 1] = itemp;
}
cout<<"\n\n";
cout<<"The final results are "<<endl;
cout<<"StripLayout Index"<<setw(20)<<"No of Layout"<<endl;
for(Count = 0; Count < NoLayout; Count++)
cout<<Closeness[Count]<<setw(20)<<Results[Count]<<endl;
delete [] Results;
}
void CleanUp()
{
for(int Count = 0; Count < NoLayout; Count++)
{
delete [] DiMatrix[Count];
delete [] RiMatrix[Count];
//  delete [] V1Matrix[Count];
// delete [] Wtdmatrix[Count];
}
```

```
for(Count = 0; Count < NoAttrib; Count++)
// delete [] RelMatrix[Count];
delete [] DiMatrix;
delete [] RiMatrix;
// delete [] V1Matrix;
// delete [] Wtdmatrix;
delete [] RelMatrix;
delete [] MaxAttrib;
delete [] MinAttrib;
delete [] SeprateMin;
delete [] SeprateMax;
delete [] Closeness;
delete weights;
DiMatrix = NULL;
RiMatrix = NULL;
// V1Matrix = NULL;
MaxAttrib = NULL;
MinAttrib = NULL;
SeprateMin = NULL;
SeprateMax = NULL;
Closeness = NULL;
}
void main()
{
clrscr();
char KeyboardResponse;
GetAttribLayout();
ValidateAttribLayout(KeyboardResponse);
cin.ignore();
while(KeyboardResponse != '1')
{
//file://clrscr();
GetAttribLayout();
ValidateAttribLayout(KeyboardResponse);
cin.ignore();
}
/* try
{
AllocateMemory();
}
catch(...)
{
cout<<"Memory allocation failed"<<endl;
exit(0);
}*/
GetDiMatrixInput();
ValidateDiMatrix(KeyboardResponse);
cin.ignore();
while(KeyboardResponse != '1')
{
GetDiMatrixInput();
ValidateDiMatrix(KeyboardResponse);
```

```
cin.ignore();
}
CalculateRiMatrix();
printRiMatrix();
getch();
// GetRelativeAttribMatrix();
// ValidateRelativeMatrix(KeyboardResponse);
//  normalizedmatrix();
//getweights();
/* cin.ignore();
while(KeyboardResponse != '1')
{
GetRelativeAttribMatrix();
ValidateRelativeMatrix(KeyboardResponse);
cin.ignore();
}*/
// CalculateWeightedMatrix();
// getch();
// CalculateA3Matrix();
// getch();
// CalculateLambdaMax();
// getch();
//CalculateConsistencyRatio();
// getch();
// CalculateNormalisedMatrix();
// getch();
GetMaxMinAttrib();
getch();
CalculateSeprateMeasures();
getch();
CalculateRelativeCloseness();
DisplayResults();
getch();
CleanUp();
}
```

Index

MIX
Papier aus verantwortungsvollen Quellen
Paper from responsible sources
FSC® C105338

If you have any concerns about our products,
you can contact us on
ProductSafety@springernature.com

In case Publisher is established outside the EU,
the EU authorized representative is:
Springer Nature Customer Service Center GmbH
Europaplatz 3, 69115 Heidelberg, Germany

Printed by Libri Plureos GmbH
in Hamburg, Germany